W9-CLZ-403
3 0301 00085519 3

Writing for the Technical Professions

Thomas N. Trzyna
Seattle Pacific University

Margaret W. Batschelet
University of Texas at San Antonio

Wadsworth Publishing Company
Belmont, California
A Division of Wadsworth, Inc.

English Editor: John Strohmeier
Editorial Assistant: Holly Allen
Production Editor: Robin Lockwood, Bookman Productions
Print Buyer: Ruth Cole
Designer: Hal Lockwood
Copy Editor: Loralee Windsor
Technical Illustrator: Carl Brown
Compositor: Thompson Type
Cover: Paula Shuhert
Signing Representative: Karen Buttles

Printed in the United States of America

1 2 3 4 5 6 7 8 9 10—91 90 89 88 87

ISBN 0-534-07884-2

Library of Congress Cataloging-in-Publication Data

Trzyna, Thomas N., 1946–
Writing for the technical professions.

Includes index.
1. Technical writing. I. Batschelet, Margaret.
II. Title.
T11.T748 1987 808'.0666021 86-28946
ISBN 0-534-07884-2

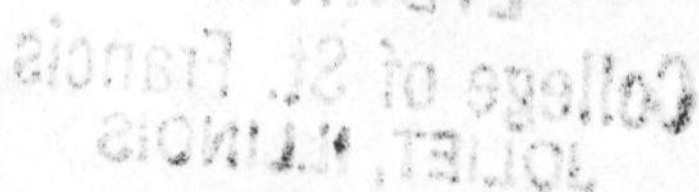

PREFACE

Writing for the Technical Professions is designed for a one-term, upper-division course in technical communication. This text can be used by students who are majoring in a wide range of subjects.

We wrote *Writing for the Technical Professions* to meet two needs we identified in our classes and consulting work. First, we wanted a set of casebook exercises that would provide a full rhetorical context for students' reports. Second, we wanted to cover the analysis of the audience and purposes of reports, and the political or organizational dimensions of writing.

Writing for the Technical Professions has several unique features, including a chapter on clarifying the purposes that can be served by reports, a chapter on organizing and participating in a group writing project, and a chapter on writing environmental impact statements. Our treatment of research outside the library presents a new approach to understanding where information comes from in American business and government, as well as several highly structured methods for locating information that is not normally available in library collections. Our business writing chapter offers extended coverage of the job search and incorporates the practices of professional employment consultants. This chapter takes into consideration the needs of graduates who want to find the right job and the concerns of students whose specialties are in less demand.

Throughout the book we have followed what we feel is a commonsense approach. We have tried to take our own advice by presenting information as simply as we can and by reinforcing important points through emphasis, repetition, and prominently placed chapter summaries. We have also tried to be sensitive to current conditions in the field, including the variety of majors of students who take technical writing and the impact of computers on both writing and information retrieval. Our treatment of organization includes coverage of forecasting techniques, and our chapters on editing and research contain advice on word-processing and data-base technologies. The chapter on audience analysis integrates several popular approaches to this topic, while the chapter on oral presentations includes instruction in the use of scripting techniques. There are many exercises and cases after each chapter; instructors and students will have a wide selection from which to choose. The cases can be used either alone or in conjunction with a traditional term paper or final report.

The *Introduction* that follows presents a chapter-by-chapter overview of the entire text and our rationale for the order in which the parts and chapters appear. We hope that instructors will find enough variety and depth of coverage to meet the needs of the growing—and increasingly diverse—population of students who are finding technical writing an exciting, demanding, and eminently practical subject.

We are indebted to many friends, reviewers, employers, editors, students, and former teachers who directly or indirectly contributed to this text, including our students and colleagues at the Brooks/Cole Publishing Company; California Institute of Public Affairs; City of Seattle; Commercial Design Associates; Datapoint Corporation; Holt, Rinehart & Winston; Interloc Inc.; Keuffel & Esser; Lincoln University; Mariner Kayaks; the Medical College of Wisconsin; The Ohio State University; Panasonic Company; the San Antonio chapter of the Society for Technical Communication; Seattle Pacific University; Texas Instruments Corporation; the University of Texas at San Antonio; the University of Washington; and the University of Washington School of Nursing. We particularly thank Beth Barfield, Peggy Brehm, Matt Broz, Bill Camp, Joyce Q. Erickson, Elizabeth Frame, Tom Howard, Carol Matthews, Dr. George W. Simons, James W. Souther, Dr. T. C. Trzyna, Dr. R. M. Weatherford, and Frederick Williams. We also acknowledge the help of the students who helped develop the casebook exercises, including Amanda Bergman, Catherine Brinkman, Glynn Cooper, Mary Davis, Renate Eubanks, Elizabeth Galliher, Wesley Hopper, Greg Lund, Bryan MacDonald, Rory McGarrity, Michael Reyes, Louis Sorola, Kevin Tillman, Charlotte Waggoner, Jill Kobernick Watson, Sandra

Wells, Warren Westrup, and Ralph Voss. Thanks are also due to our editors and production staff at Wadsworth Publishing, especially John Strohmeier, Karen Buttles, Cedric Crocker, Kevin Howat, and Judith McKibben, and to our reviewers: John C. Thoms, New York Institute of Technology; Jack Selzer, Pennsylvania State University; Erna Kelly, University of Wisconsin-Eau Claire; Robin McNallie, James Madison University; Bruce Appleby, Southern Illinois University; Claude Gibson, Texas A & M University; Marion Smith, Brigham Young University, Main Campus.

CONTENTS

INTRODUCTION

What Is Technical Writing?

Why should you study technical writing? How does it differ from other types of writing? And what will you learn from this text? This brief introductory chapter will answer these questions and, we hope, motivate you to master technical writing, which plays a much larger role in professional life than many students realize.

This chapter presents five perspectives on technical writing:

1. Writing as a part of your career
2. The distinguishing characteristics of technical writing
3. The formats common in technical writing
4. The fields where technical writing is common
5. The contents of this text: what you will learn in this introductory survey of technical communications

Writing in Your Career

Surveys of technical and professional fields show that college-educated professionals spend about 30 percent of their time writing. About 20 percent of their total work hours is spent on informal writing and about 10 percent on formal writing. Often an additional 30 percent of work time is devoted to various other forms of communication, such as telephone conversations, committee and staff meetings, design conferences, and formal oral reports. [Richard M. Davis, "Technical Writing:

Who Needs It?" *Engineering Education*, November 1977, pp. 209–211.]

Writing is so important that in studies performed over the last thirty years, professionals have judged communications and writing skills at or near the top of the skills they need on the job—equal to courses in engineering, math, and basic business skills. [William Kimel and Melford E. Monsees, "Engineering Graduates: How Good Are They?" *Engineering Education*, November 1979, pp. 210–212.] Similarly, one of the most common complaints about unsatisfactory professional employees is that they are unable to communicate what they know. Your analytical abilities are of little use if you can't communicate your findings to others in ways that are appropriate to the audience, clearly focused, well organized, and persuasive.

Keep these general findings in mind. As a beginning professional, you can expect to spend about a quarter of your time communicating orally and in writing. As you are promoted to more responsible positions, you'll find yourself spending up to three quarters of your time writing and speaking. Your ability to communicate effectively is one of your top skills, and to be an effective communicator, you will need to know how professional writing and technical writing differ from the other types of writing you do. This text is designed to help you develop the skills you have and learn the special characteristics, formats, and techniques of technical communication.

CHARACTERISTICS OF TECHNICAL WRITING

Technical writing has several characteristics that distinguish it from much of the personal and college writing you do. Technical writing is highly objective in tone; it grows out of a specific purpose; it is assigned like college work; it is often completed in coordination with other writers; and, of course, its content and vocabulary are generally technical. In addition, technical writing uses formats and conventions that are not common in other contexts, and it is more common in particular fields of work.

Objective Tone

While it is never possible to be completely objective, and it is an error to believe in one's complete objectivity, technical writing does attempt to be more objective than personal writing, creative writing, and some types of informal professional writing such as interoffice memos. This objectivity is apparent in the tone, the choice of vocabulary, the tendency to direct emphasis away from personal opinions, and the clear identification of criteria on which judgments are based. Such formats as descriptions of mechanisms and processes, recommendation reports, and environmental impact statements clearly reveal the attempt to achieve this kind of objectivity.

Specific Function and Purpose

Technical writing always grows out of a specific context and is designed to fulfill a specific purpose. In these respects, technical writing differs from some basic research and from journalistic writing that is designed to inform without necessarily accomplishing any other objective. To put this another way, technical writing tends to promote action and direct decision making. The purposes of technical writing are more instrumental than the purposes of other forms of writing. Chapter 2 of this text presents a careful analysis of the purposes that can be achieved by technical writing.

Assignment

Like most of your college papers, your professional writing is assigned. Sometimes you will be able to decide the topic yourself: when you write a proposal to present your own ideas, for example. Often, however, you will be given a specific task that must be coordinated with the work of many other professionals. You might be asked to write the troubleshooting guide for a manual, or the budget section of a proposal, or one chapter of an impact statement, or the criteria statements for a recommendation report. This characteristic brings us to the next one—technical writing is often a cooperative venture.

Coordination with Other Writers

You will often be asked to write as a member of a group, which means that you must understand how to participate in a group project, how to adjust your style and tone to other writers, and how to edit other writers' work. This characteristic of technical writing is increasingly important as more companies adopt management and problem-solving strategies that encourage group efforts.

Technical Content

Technical writing is also marked by its technical content and the presence of specialized vocabulary and jargon. One of the more important reasons for studying technical writing in a college course is to begin to learn when you should or should not be highly technical. Good technical writers know how to adjust tone, vocabulary, and technical content to different audiences, purposes, and situations. Poor technical writers pour on the jargon no matter what the audience or objective.

Special Writing Formats

A variety of formats have evolved to meet the needs of government, industry, business, and research and development. At one time or another, you will probably be involved in some way with nearly all of these formats, whether as a writer, a receiver, or a commissioner of technical documents. You will learn about most of these formats in this text, but some, such as specifications writing and scientific journalism, are too specialized to warrant treatment in a general text. The formats covered in this text are:

Reports—end of project reports or progress reports, investigation reports, or troubleshooting reports
Proposals—technical, evaluation, cost, and management proposals
Prospectuses, abstracts, and summaries
Handbooks and instructional manuals—technical, training, or other purposes, such as computer systems and software
Environmental impact statements
Short reports, such as incident and trip reports
Recommendations and feasibility studies
Business correspondence—memos, persuasive letters, bad news letters, and letters of application

While you may not need to write all of these documents as a professional, it is important that you recognize these formats, so that you can work with them efficiently when you are asked to participate in a writing group or an evaluation committee.

Common Fields

Technical writing is common in a wide range of fields, not just in high technology and electronics. The examples in this text are drawn from the physical and biological sciences, electronics, computer technology, the health and social sciences, and government. As we define it, technical writing excludes interpretive journalism and creative writing on the one hand and scientific research writing on the other. Creative writing and journalism are not as instrumental as technical writing. Creative writing is not usually intended to direct action or help you make a decision. And academic reports on basic research follow different formats, even though many of the basic principles of audience, purpose, and organization are the same.

How to Use This Text

Before you begin reading this text, you should understand what it contains and why the chapters are presented in this order.

Parts 1 and 2 discuss the most important decisions you must make: defining your audience, determining your purpose, and deciding how and where to find the information you will need to solve your technical and communication tasks. Chapter 1 presents several methods for analyzing audiences. Chapter 2 discusses the purposes that reports can be designed to achieve and how you can write a report that fulfills several purposes. Chapter 3 presents sophisticated methods for gathering data inside libraries. Chapter 4 introduces methods for identifying and contacting experts who can provide you with the information you need.

Part 3 covers the methods you will use to organize your reports. Chapter 5 presents useful techniques for organizing, forecasting what will be presented, and keeping your readers in touch with where your report is going. Chapter 6 reviews style and explains how you can quickly and efficiently edit your own work. Chapter 7 defines the format elements used in formal reports, such as abstracts and letters of transmittal. Chapter 8 reviews the principles of graphics and provides examples of many standard types of graphs, charts, and other visual presentations.

Part 4 presents four important technical formats: proposals, short reports, recommendations and feasibility studies, and descriptions and manuals. Chapter 9 presents guidelines for many common short reports, including the trip report, the incident report, and the investigative report. Chapters 10 and 11 discuss two demanding forms, the recommendation or feasibility study and the formal manual and product description. Chapter 12 covers not only the formats but also the politics and strategy of proposal writing.

Part 5 considers communication on the job, including business writing and oral presentations. Chapter 13 discusses the basic principles of business writing; explains common letter formats; and provides you with a comprehensive manual on the résumé, the letter of application, job search strategies, and correspondence after the interview. Chapter 14 offers advice on the use of audiovisual equipment as well as on the preparation and delivery of oral reports.

Part 6 deals with advanced technical writing formats and skills. Chapter 15 explains how you can be an effective manager or participant in a writing or project group. Chapter 16 introduces the environmental impact statement (EIS) process, summarizes EIS guidelines, and closes with a sample EIS chapter. Chapter 17 supplements the chapter on proposal writing with a discussion of proposal budget terminology and the construction of a proposal budget.

SUMMARY

1. Whatever your field, you can expect to spend about a third of your working hours in formal or informal writing. If you learn the basic principles of technical writing thoroughly, they will help in your career.
2. Technical writing is objective in tone; it grows out of specific contexts and is designed to fulfill a specific purpose; it is done in coordination with other writers; and it is usually technical in its content.
3. Technical writers use formats that are widely known in business and industry.
4. Technical writing is common in business, high technology fields, the social sciences, and the physical and biological sciences.
5. This text introduces basic principles and prewriting strategies first, then covers the elements of your solution to organizational and writing problems, and finally describes specific formats and techniques for communication.

P · A · R · T 1

Defining the Problem

CHAPTER • 1

Audience Analysis

INTRODUCTION

Before you write anything, you need to analyze the composition and needs of your audience: who they are, what they need to know, why they need that information, and when and where they are likely to be when they read your report. Audience analysis is the first of three basic steps for professional and technical writing. Assessing purpose and organizing are the other two steps; they are covered in Chapters 2 and 5. Learn how to perform these three tasks and you will be able to write competently in any format.

This chapter opens with a case study of a technical manual that was written for several different groups of people. The study introduces in detail the issues and problems raised by audiences of technical documents. The remainder of the chapter presents four different yet related approaches to audience analysis: (1) a system for categorizing readers according to their level of knowledge, (2) a way to analyze readers' expectations, (3) a method for charting a report's audiences and for clarifying the relationships among audiences, and (4) suggestions for gathering information about your audiences so that you can write reports that serve your readers. A final section of the chapter emphasizes the importance of remembering that reports are usually written for several audiences, as well as the importance of distinguishing primary from secondary audiences.

A Case Study

The following case study explains how a large public utility used audience analysis to improve the focus, organization, and usefulness of a technical manual. City Light, a large metropolitan electric utility, published a manual that described the legal requirements for obtaining a new electrical service connection. The manual also described construction details that were not covered by electrical codes, as well as industrial specifications that went well beyond what most code books cover. Neither customers nor service technicians, however, were using the manual, which was poorly organized. Over the years, the old manual had grown new sections by a process of amendment and addition. For example, the general introduction, which was intended for all readers, included several highly technical rules for the use of oxide inhibitors on over-sized aluminum conductors. The utility's service manager decided to have the manual reorganized from scratch and gathered a team of engineers, administrators, and writers to do the job.

As the work proceeded, committee members spent much of their time discussing the audience, because they discovered that their decisions about organization and style depended directly on their decisions about the audience and the way they classified members of the audience. When it was used at all, the old manual had been consulted primarily by the technical staff to explain service decisions to customers. Yet the committee recognized that there were at least three possible audiences in addition to the technicians: residential, commercial, and industrial customers. The committee also identified three more audiences that seemed to be secondary in importance: the utility's administrators, the utility's power engineers, and the local government. While none of these secondary audiences would ever need to use or interpret the regulations in the manual, each was required to give formal approval to any changes in the regulations.

The committee first decided to meet the needs of the secondary audiences by writing a short report explaining and identifying the changes that were made, so that these officials could take the necessary legislative action. The problems posed by the primary audiences remained, however. How could the committee write a manual about residential power connections; describe vastly more technical subjects, such as phase control systems for small hydroelectric dams; and at the same time satisfy the expectations of both residential consumers and corporate power engineers? As the committee considered a choice of audience, they also recognized that each audience would use the manual in a different way. A residential consumer would use the manual either to ask questions or to perform simple installations. A corporate power engineer would probably consult the manual with a different set of purposes in mind. The committee's choice of an audience would necessarily require a decision about the purposes the manual was to serve.

The committee's first solution to the audience problem was based on the observation that there were more residential consumers than any other type. Since the utility wanted to foster better relations with its residential customers (rates were rising), it seemed to follow that the manual ought to be written in language that any do-it-yourselfer could understand. On further analysis, however, the committee decided that it would not be possible to use the same simple vocabulary or style for the discussion of more technical topics, such as maximum allowable locked-rotor currents for motors. Besides, no one could recall a time when a residential customer had actually used the manual.

The second solution the committee considered took into account the differences among residential, commercial, and industrial clients. Since these groups seemed to be distinct in terms of the amount of power and types of equipment they used, the committee prepared an outline for a manual that would have three entirely separate sections, one for each class of user. It soon became apparent, though, that this manual would be more than twice the length of the old version, because of the need to repeat dozens of rules that were common to all three types of power installations.

After two further solutions were considered, one of the managers pointed out that in the long run the technicians and installers employed by the utility company would always bear the responsibility of referring to the manual, notifying customers of applicable regulations, and interpreting those regulations. While residential, commercial, and industrial clients were certainly possible audiences, in fact they were not primary at all. The utility's attempts to create good will with customers were admirable, but they were aimed at achieving the wrong purpose, because a technical manual of regulations is not an effective medium for public relations.

The committee finally agreed that the actual audience for the revised manual should be the utility's own field technicians, who had to interpret the regulations. The purpose of the manual was to provide these field technicians with a clearly written and well-organized reference. In the terms presented in Chapter 2 of this text, the purpose of the manual was informational.

The old manual had failed because it was poorly organized and inadequately cross-referenced. In this case, as in most technical writing assignments, the choices of audience, purpose, and organization were interdependent. Until the writers were able to agree on an audience, they were unable to proceed at all.

CASE ANALYSIS

The members of this committee used several methods of audience analysis, although they did not systematically apply any single audience analysis technique. As we review the case, you will see how various approaches to audience analysis were implicit in the questions the committee asked.

First, when they considered the needs of the residential and industrial consumers, the committee members were highly sensitive to these audiences' expectations. Residential customers would expect explanations in simple language, just as industrial power engineers would expect a utility manual to use highly technical vocabulary when such a vocabulary was appropriate and more accurate than any simplifications.

Second, the committee divided the total audience into several groups or categories of audiences, depending on the level of their technical knowledge and the types of decisions they needed to make.

Third, the committee kept returning to this question: who would use the manual most? Which audience was truly primary and which audiences merely seemed to be primary?

Finally, although the committee did not interview any field technicians or corporate power engineers, in a sense the committee members interviewed themselves as they searched their memories for information about how the old manual had been used and misused.

In the next section, you will learn about each of these methods in detail: considering your audience's expectations, dividing the audience on the basis of technical knowledge and decisions made, deciding on your primary audience, and gathering information about that audience. We will start with an examination of four basic types of audiences and their expectations.

AUDIENCE TYPES AND EXPECTATIONS

The four basic types of audiences can be classified according to the level of technical information they understand and the types of decisions they need to make. Audiences expect to see information that they are prepared to understand. For instance, what do you expect when you open a textbook, a popular magazine, a professional journal, or an auto manual? This section analyzes the expectations of the four basic audiences: the general reader, the technician, the technical specialist, and executive or manager. This presentation is based on analyses of reading materials aimed at these audiences; you can check our findings by observing the audience choices that have been made by the authors of the books and magazines you read. When you do, you'll notice an important fact about yourself. You belong to many audiences. In some areas you might be an expert, at other times you are a technician or a general reader.

General Readers

As you saw in the City Light case, the most basic audience is the general reader, in that case the residential consumer of electricity. General readers need technical information to know how to operate equipment designed for their use and perform simple repairs. General readers, however, are not expected to have the knowledge and competence of trained technicians. In fact, general readers often don't make any decisions at all.

Any audience of general readers will include experts from many fields, but a writer is obliged to assume that the general audience is a kind of lowest common denominator. There are many common denominators, of course. Think of the popular magazines you read. Some assume a high school or even some college education; others assume an audience that is barely literate. When you write as a professional for a general audience of your peers, you will probably work at a level somewhere between *Newsweek* and *Scientific American*.

Objectives

The general reader's objectives include satisfying a hunger for general knowledge, searching for ideas that can be transferred to a field in which the reader is an expert, and entertainment. General readers also look for technical information in a highly palatable form. To get an idea of what *palatable* or *easy reading* means in this context, think about the difference between a *Popular Mechanics* article on tuning a car and the same information as it is usually presented in a manufacturer's shop manual.

Background Knowledge

Readers of general presentations expect technical ideas to be explained carefully; they don't want to strain to remember information, and they certainly don't plan to look up background material in reference books. Test your own general background knowledge by quickly defining some of the following terms:

relativity, quark, evolution, ego, brazing, and *soldering*. How well did you do? Could you explain many of these ideas to the satisfaction of your classmates? You probably found out that even when you know about technical and scientific concepts in a general way, you are not necessarily able to provide a clear and detailed account of what you know. In other words, as a general reader you could use some background information.

Language

A general readership book or journal article assumes that readers at least understand their native tongue. The vocabulary, though, will be limited to simpler and shorter words, and technical vocabulary will be carefully explained. *Scientific American* represents the upper limit of "popular" or general reading: It is the most technical magazine you will find in drugstore magazine racks, with the possible exception of some computer magazines. Yet even *Scientific American* is careful to define difficult terms. In one issue, for example, the pressure inside the Cornell particle accelerator was given as 10^{-8} TORR. In parentheses, the authors explained that this value was "10^{11} times less than atmospheric pressure." [Nariman B. Mistry, Ronald Poling, and Edward Thorndike, "Particles with Naked Beauty." *Scientific American*, July 1983, p. 111.] Many magazines, of course, would not even consider using the word *atmospheric*.

Graphics

Graphics for the general audience should be limited to simple graphs and photographs. Pie, bar, and line graphs are common, but line graphs should rarely include more than two lines.

Measurements and Relationships

The reader of general or popular literature also expects to have technical measurements and orders of magnitude either defined or explained by analogies. For this audience, the speed of light is typically defined in terms of meters or miles per second. Other measurements or relationships can be defined by analogies. For example, one writer explained the relationship between the parts of an atom by pointing out that if the nucleus of a hydrogen atom were enlarged to the size of an orange, the electron would orbit at a distance of 30 miles. General readers can also understand a certain amount of symbolic—as opposed to verbal—language, such as mathematics, flow charting, and logic. But even the most advanced general readership publications rarely publish equations; any advanced mathematical ideas are usually presented as geometric diagrams and analogies. In fact, most articles for the general audience present only percentages, proportions, and perhaps simple algebra. The general audience has a limited knowledge of statistics. If you question your friends, you'll find that most will understand the idea of an average and will be able to calculate a simple average. Fewer will understand the concepts of median or mode, and fewer still will be able to explain such slightly advanced concepts as p values or correlation. General audiences will expect to see and understand averages and simple probabilities, but little more.

Organization and Style

The general audience expects organization and sentence structure to meet its particular needs. They expect simple, short sentences and paragraphs limited to four or five sentences each. Large-scale organizational patterns are also tailored to the audience's needs. *Scientific American* frequently begins highly technical articles with background material. Other popular journals, such as *Time* or *Newsweek*, attract readers by emphasizing the personal relevance of a topic. An article on plate tectonics, for example, might open with a few paragraphs about California earthquake predictions or recent American volcanic eruptions. Subtitles, too, will be written to catch the eye; rather than a technical subtitle such as "Plate Movement," a popular article might use "When Will the Next Quake Strike?" The conclusion of a general interest article also takes a form designed to meet readers' needs and expectations. Typically, conclusions make predictions, attempt integrations of familiar and unfamiliar subjects, challenge the reader, make claims on the reader's feelings or sense of moral obligation, or pose large philosophical questions, such as: Why does nature sometimes appear uneconomical and wasteful and at other times utterly simple and economical?

Technicians

Technicians need to know how to operate and maintain complex equipment; they also need to know how to explain technical information to less-specialized audiences. Technicians are usually concerned with present problems, such as installation, assembly, maintenance, and repair, however, rather than with long-term decisions and planning. You become part of the audience of technicians when you set up a new personal computer, connect a video recorder, tune your car, or make Beef Wellington.

Objectives

As an audience, technicians' expectations are governed by two objectives: finding out how to do a task and learning how to explain an operation to others. Writing for technicians is less concerned with the scientific or engineering details—*why* a machine or process functions—than with *how* it functions.

Background Knowledge

As a writer, you can assume that technicians will know their own fields, that welders will know about gas shielding techniques, that electricians will understand the basic codes, and so on. It is more difficult, though, to decide how much a technician might know about closely related fields. An installer for an electric utility needs to know *something* about surveying, property maps, and easements in order to place service lines over or through the right property. An installer also needs to know something about the various types of building construction in order to drill holes in walls and attach masts to structural elements that can stand the tension and weight. But how *much* can you expect a technician to know about these related areas? In the chapter exercises, you'll have a chance to explore this question, to which there is no simple answer.

Language

A simple vocabulary is expected and appreciated. The technical vocabulary, of course, can be large as long as the terms are drawn from the appropriate field. For example, a complex essay on the chemistry of corrosion would be inappropriate in a set of instructions for removing a corroded pipe. You cannot expect a welder to understand *everything* relating to corroded pipes.

You won't find foreign phrases in material for technicians, unless those phrases are common in that area of expertise, as Latin terms and abbreviations are common in medical technology. The mathematics, too, will be limited, because technicians usually don't have time to perform long calculations in the midst of a job. At an electric utility, for example, the calculations for a power system are done by the engineering staff; the technicians then carry out the design by installing the specified components.

Graphics, Measurements, and Relationships

Technicians expect to interpret complex graphics, such as architectural elevations, blueprints, and wiring diagrams—illustrations that are well beyond the understanding of a general audience. Technical measurements and relationships, such as dimensions, voltages, and frequencies, can also be used without definition or explanation, as long as they are common in the field.

Organization and Style

Technicians expect direct, imperative sentences. Sentences and paragraphs will be about the same length as for general audiences, but longer sentences can be broken into tabular form like the last sentence of this paragraph. Informational subtitles should appear frequently, so that the reader can find information quickly, and closely related information should be cross-referenced. In general, the organization should help the reader

solve problems, and
find information quickly.

You can see that the differences between the general reader and the technician include background knowledge, technical vocabulary, the complexity of illustrations, reading objectives, and organizational expectations. The general reader skims to find interesting sections; the technician skims to find necessary information. Expert decision makers have a far different set of expectations.

Technical Specialists

Technical specialists are expert decision makers. They are expected to have highly specialized and extensive knowledge of their subjects, and they need to make decisions about long-term technical, financial, and operational questions. They are characterized by a high level of training, expertise with specialized analytical and decision-making techniques, and familiarity with a specialized technical vocabulary.

Technical specialists need information that can help them make decisions about the technical feasibility of solutions. Articles and reports for specialists are often aimed at small audiences. Highly technical research reports can be intended for a primary audience of five or six other researchers who are trying to answer precisely the same questions. Less technical reports may be designed for the primary use of small groups of managers or executives in a single industry. Because the primary audience of an article for experts is so limited, it follows that those readers may expect the author to spend little if any space on background material or explanations of concepts or vocabulary. For these reasons, articles for experts are demanding reading.

Objectives

Technical specialists usually have two objectives: (1) to get information about a specific topic and (2) to find ideas that can be applied to a related topic. In the second case, the information a specialist might want to apply may be facts, or approaches to organizing research, or methods for data handling, or even useful questions or ways of thinking about a problem. A wheelchair designer, for example, might read about aircraft hydraulic systems to find new ideas to apply to lifting or balancing systems.

Background Knowledge

Technical specialists can be expected to know their own fields well, although as a writer you are on sensitive ground here. One of the most common errors made by beginning writers is to assume that everyone in the same field knows the same things—that all engineers are a single audience, for example. In fact, the field an expert knows may be very limited in scope. Mechanical engineers don't know the same things that electrical engineers do; after the first few quarters of college, their programs are entirely different. Consequently, there is no reason to expect that the two groups should know the same facts and theories. Also, as time passes fields change. The electrical engineering graduate of 1950 learned about vacuum tubes, the graduate of 1960 studied transistors, and the graduates of the seventies and eighties learn about chips. Similarly, an engineer who has spent ten years in management knows a field far different from the one in which he or she was educated. For these reasons, you need to consider what expertise a given audience is likely to have. On the other hand, specialists can usually be expected to know and apply information from related fields. Chemists know some physics, biologists know some chemistry, chemical engineers are bound to know something about industrial engineering, and so on.

Language, Graphics, Measurements, and Relationships

There are few limits on the vocabulary of articles for technical specialists, except the limit of good sense, which requires using plain language whenever possible. In some fields, the audience might expect to read short passages in other languages, such as French, German, or Russian.

More common is the expectation that mathematical and graphic presentations will be highly technical. Advanced journals, such as *The Logistics and Transportation Review, The Journal of Bone and Joint Surgery,* or the *American Journal of Physics*, think nothing of presenting calculations, arguments written in symbolic logic, sections of computer programs, flow charts and minimal path displays, and statistical tables. Complex graphs and visuals are common, and technical systems of measurement are also used freely and without much explanation.

Organization and Style

Expert articles are designed for efficient use. The front matter will include a summary or an abstract, a statement of the problem, or a review of the relevant research. Conclusions or recommendations might appear next, with the body of the presentation placed last for those readers who need to see all the data or follow the entire argument.

Executives and Managers

Like technical specialists executives and managers are expert decision makers. They share some expectations with technical specialists, but they also have specific requirements of their own. It is a mistake to assume that because executives may not have degrees in scientific or technical fields they are not experts. In fact, executives are experts in their particular areas, such as administration or cost accounting. Yet because they may have limited knowledge of some technical materials, they should be treated somewhat differently from technical specialists. Executives need to know information that has a bearing on long-term planning and program development. How should next year's sales campaign be organized? What new technologies should the company investigate?

Objectives

The primary objective for these readers can be summed up in one question: What must I do because of this information? Executives are, above all, decision makers. They read to enable themselves to make informed and intelligent choices. You must include enough information to allow them to make their decisions.

Background Knowledge

Like technical specialists, executives know their own industries well. However, new technologies or unfamiliar material from other fields will require explanation.

Language, Graphics, Measurements, and Relationships

Executives and managers will understand the basic technical vocabularies of their particular industries and fields. More specialized terms should be defined. These definitions can be provided in a glossary if the number of terms is large.

You can include complex mathematics and graphics in the appendix of an executive report, but in the body of the report itself, you should present simplified versions.

Organization and Style

Executives, like technical specialists, want efficient reading. They require a summary at the beginning of a report (sometimes called an executive summary) followed by conclusions and recommendations before the body of the discussion.

A Summary of Reader Expectations

When you begin to write for a new audience, take time to answer these questions.

1. What are my audience's objectives in reading this report?
2. What are their primary fields of expertise? Have they been working in other fields for some time?
3. How much will they know about my field? About subjects related to their main field of expertise?
4. How will they expect to see this report organized? What formats are they accustomed to encounter?
5. What are my audience's expectations for language, vocabulary, and technical terms?
6. How much can I assume that they understand about mathematics?
7. What graphics are they used to seeing in their reports?
8. What measurement systems are familiar to them?

Profiling an Audience

One approach to audience analysis, which helps you answer the questions listed in the preceding section, is audience profiling. Audience profiling consists of two simple steps: (1) analyzing the relationships among the writer and the various readers of a report and (2) making a short list of the important information about each person (or audience) who will be reading the report. This system is applicable to any writing situation, but it is particularly useful when you are writing a report for distribution within your own company. While this analysis might seem time-consuming, it is actually a rapid and effective way to assemble background information on your audience.

Analyzing an Audience

Start by listing all of the people who will be dealing with your report. Begin with yourself and any people with whom you work daily: the people who share your ideas, your hopes, your way of talking about your work. Next list the people in your audience who work closely with your group: managers, supervisors, and members of other departments who are working on projects related to yours. Then list even more distant personnel and administrators who might need to read or evaluate your reports or proposals. This is not a hierarchical listing; people who have high positions in the corporation might be anywhere on the list, depending on how close they are to your work. If business organization is unfamiliar to you, try to remember an occasion when you worked on a lab project or served on a campus committee. In a single committee, you might have worked with other students, graduate students, faculty from several departments, professional staff and administrators, and perhaps a dean or vice president. While these people might have been far apart in terms of the organizational hierarchy of a university, in terms of the work of a particular committee they might be equal partners and members of the same audience. Consider the following example:

Suppose that you are a member of a design team in a pharmaceutical corporation. In order to complete your current assignment more rapidly, you and your team decide to requisition a new chemical analyzer. You need to write a formal requisition because this instrument costs $125,000, and it requires a trained operator. Your first impulse is to write a proposal to the company's management describing in detail the range of reaction products this new device can identify and explaining the technical superiority of the instrument. On second thought, however, you realize that a mass of technical data will not interest anyone else, so you list your possible audiences, as in Figure 1-1, below.

As you see it, your primary audiences—the ones underlined in Figure 1-1—are the people who have the authority to grant your immediate request. These include the technical manager on your own project and the budget manager for the corporation, who has final control of all research and development funds. While you are staring at this fairly short list, however, you realize that there are other people and groups who will need to receive specific information about your request. Some of these audiences, such as the solvent design team in the lab down the corridor, speak your language. Others, like the Product Planning Office, are less interested in technical details.

FIGURE 1-1
List of the audiences for an equipment requisition

1. Close audiences: daily contact

   <u>Technical Manager</u>
   Solvent Team
   <u>Budget Manager</u>

2. Distant audiences: possible interest?

   Labor Relations
   Product Planning
   Legal Affairs
   Personnel

Product Planning, of course, will want to know whether this new machine can be used for any other process in which the company has an interest. Will the solvent development group be able to use this analyzer, too, or is the $125,000 to be used for a one-time, one-product effort? Labor Relations might need to find out if the operators who run these analyzers are affiliated with a union; Personnel will certainly want to know where to advertise for an operator; and Legal Affairs might need to check the purchase with the insurance carriers. By listing your audiences you can pick out two primary audiences (the technical manager and the budget manager) and five secondary audiences (planning, legal, labor, personnel, and the solvent group).

Once you have sketched in the audiences and decided whether each audience is primary or secondary, you are ready for the second step: characterizing each audience in terms of titles, roles, decisions to be made, idiosyncracies, and receptivity.

Profiling Audience Individuals

When you know the individuals to whom you are reporting, writing down their names helps you to remember how they respond to reports and proposals. Job titles help you focus on the reasons they are included in an audience; a job title can also help you decide what audience category an individual (or a group) belongs to, whether executive, technical specialist, technician, or general. Remember that any given individual can change roles. The sales manager who makes technical decisions on one occasion might be part of an executive audience for another report.

You should also try to specify the decision each person needs to make, keeping in mind that different parts of your audience will be making different decisions. In our example, the legal staff will make decisions very different from those faced by the labor relations office or the planning division. Finally, where possible you should consider the personalities and idiosyncrasies of the people who will make the decisions. Some people like long reports, others prefer executive summaries, others favor certain kinds of vocabulary or format.

The physical and emotional condition of the audience is also important, because it affects the receptivity of the audience. One expert failed to convey his message because he didn't consider how jet lag affected his audience. Most of this chapter has focused on *who* will be reading your reports and *why* they will read (for what information). You should also consider *where* a document is likely to be read (in an elevator? a jet? a restaurant?), *when* it will be read (during a sales presentation?

late at night?), and *how* it might be read (in haste? in frustration?).

You might not be able to do all these steps each time, but you should use as detailed an audience analysis as you can. After all, there's no point in writing a sloppy document that doesn't reach its audience at all. Figure 1-2 is a reader profile form that synthesizes much of what you have read in this chapter.

In the example above, you might fill out a profile form for the Executive, Office of Project Planning, as shown in Figure 1-3. This kind of analysis is not hard to develop if you know the people who will read your reports. When you write for audiences you have never met, you will not be able to apply all of these methods for analysis, although by doing a little telephoning and other research, you can often gather enough information about your potential audiences. Whatever method of audience analysis you use, you need a well-organized approach to gathering information that includes skills in interviewing, correspondence, and research in corporate files and libraries. The simplest way to analyze an audience, in fact, is to arrange a personal or telephone conference, so that you can find out what the members of an audience expect. Research techniques are discussed in Chapters 3 and 4. One further aspect of audience analysis is made more explicit in the next section: Most documents are written to be read by several audiences at once.

MULTIPLE AUDIENCES

Every example in this chapter involved reports written for several audiences at once. Even if you think a report has only one immediate audience, it usually has a future audience that needs to be kept in mind even if it consists only of the person who takes the job of the individual for whom you wrote originally. Chapter 2 explores this topic at length, while Chapter 5 explains how you can design reports that will simultaneously serve the needs of several distinct audiences.

FIGURE 1-2
A reader profile form

1. Name
2. Role in organization
3. Audience type (general, technician, specialist)
4. Decision to be made
5. Audience level (primary or secondary?)
6. Idiosyncrasies
7. Receptivity

Figure 1-3
A completed reader profile form

1. Name: Marilyn Walter
2. Role in organization: In charge of long-term planning, sales analysis of new products. Title: Vice President.
3. Audience type: Executive.
4. Decision to be made: She'll support or veto our proposal depending on the chemical analyzer's usefulness to other current research and the firm's long-term plans for chemical product development.
5. Audience level: Primary.
6. Idiosyncrasies: Hates long reports. Better write a cover letter to her.
7. Receptivity: Good mood. She just helped to win a large contract. In and out of town frequently. Rushed. We'd better get straight to the point. Good relations with our working group in the past.

The examples in this chapter also show that you need to distinguish between primary and secondary audiences, as did the writing staff at City Light. Their solution required them to recognize that the technicians were the primary audience and that the residential, commercial, and industrial consumers were all secondary. The various review boards in that case—city engineers, city council, and utility administrators—were actually a tertiary audience that was best served by a separate report and by cover letters that could be detached from the final manual.

Sometimes an audience analysis will help you realize that some of your possible audiences should not be addressed at all. Consider this less technical example. A substitute teacher, Judy Schwartz, planned to write a book about how to be an effective substitute teacher—how to go into a strange class and teach rather than babysit; how to feel at home in an unfamiliar school. Her audience list is reproduced in Figure 1-4.

Ms. Schwartz included all of the audiences in her list because she felt strongly that everyone should know about the plight of the substitute. Students should be kinder; janitors should help open locked doors; teachers should remember to leave class plans, and so on. When she tried to write for all of these audiences at once, however, she found that she could not begin. Her primary audiences, in fact, were other teachers and substitutes, student teachers, and college professors of education. The most probable secondary audience was school administrators, particularly those who return to college to obtain principalship certificates. All of her other audiences could be disregarded. After all, how many students would read a book

FIGURE 1-4
Preliminary audience list for a book on substitute teaching

Students
Janitors
Secretaries
Administrators
Other teachers, full-time
Other substitutes
Parents
College professors of education
Student teachers

about being nice to substitutes? Here audience analysis was used to separate genuine audiences from groups that could not be addressed.

Most of the time you will write for several audiences at once, and you must learn to distinguish the primary from the secondary or even tertiary audiences. Finally, you don't need to write to every audience you can imagine. Sometimes you will need to write several reports, or leave some audiences for another time.

SUMMARY

1. Audience analysis is the first step of a three-part process that includes assessing the purpose of a report and organizing a document that will get the job done.
2. The most common error in audience analysis is to assume that all members of one profession are part of the same audience and know the same things. Engineers, for example, are not a single audience; they have fundamentally different training because of their majors.
3. Most reports are written for several audiences that have distinct needs and different decisions to make. Some audiences are more important than others.
4. There are four basic audience categories: general readers, technicians, technical specialists, and executives or managers. Technical specialists and executives (and managers) are expert decision makers. Most people belong to several of these categories, depending on their reasons for reading specific documents.
5. Your audience will read more carefully if it encounters material that meets its expectations for language, vocabulary,

technical content, graphics, objectives, background provided, and organization. A checklist of specific questions is provided on page 14.

6. Audiences can be listed in terms of their distance from and relationship to the writer.

7. Individual members of an audience, or groups in an audience, can be profiled by listing their names, job titles, audience types, audience priority (primary or secondary), decisions made, idiosyncrasies, and receptivity at the time they will actually be reading a report.

EXERCISES

1. Select an article you can understand from a professional or technical journal in your field. Rewrite the article for a general audience and redesign the illustrations, too. Present a photocopy of the original to the instructor with your rewritten version. Your general version of the technical article should be informative. In other words, if the original article is written to persuade professionals to adopt a new method, you should write your version to inform a general audience about that method and the attempt that is being made to persuade technical specialists to adopt it. Before you begin, review the section of this chapter that describes the expectations of general readers. As you begin to work on the assignment, focus initially on how the change in audience will require changes in both purpose, organization, thoroughness, detail, vocabulary, graphics, and other features.

2. Using the list of questions on page 14, analyze yourself as an audience for this textbook, then answer these questions. What are your idiosyncrasies as a reader? Under what conditions do you normally read assignments and reports? Does this behavior differ when you are working for someone? When are you a technician? An expert? Have you been part of an executive or managerial audience? If so, describe the experience.

3. Interview a technician about his or her knowledge of fields closely related to his primary field of expertise. Write a report on your findings.

4. Make an audience list for the Day-Care Center case in Chapter 14. Try to define the audiences for the two reports. Two major audiences are given in that exercise; consider what other audiences might ultimately be interested in the material. For the purpose of this exercise, assume that the reports are to be written rather than oral.

5. Fill out an audience profile form for the two audiences of the Carlyle Windsor case in Chapter 5.

Audience Case 1

SUPER VANILLA

As a technical writer working for a regional office of the Food and Drug Administration, most of your job involves writing routine reports and lab analyses. But you've been given something new.

It seems that the southwestern region of the United States has been flooded with samples of "super vanilla." Apparently, the vanilla is being sold door to door by salesmen who claim that it has a more intense flavor than the vanilla sold in supermarkets. The new vanilla is also considerably cheaper than pure vanilla extract: as little as $1.50 a quart.

Administration chemists have tested samples of the "super vanilla" and have discovered heavy concentrations of coumarin, or tonka bean extract. They suspect that the coumarin has been brought into the country illegally from Mexico and Central America and either added to small amounts of pure vanilla extract or sold undiluted.

Your boss feels that the public should be alerted both to the fraud involved and to the danger arising from coumarin use. He assigns you to research the subject and write an informational pamphlet to be distributed in the affected areas. This is what you discover in your research.

Coumarin—Dipteryx odorata; Coumarouna odorata; tonka bean; tonco bean; tonquin bean. Prohibited as food or food additive by FDA in 1954 (most common use previously in chocolate products).
Dangers—causes organ damage in test animals, particularly damaging to liver.
Legitimate use—derivative of coumarin, dicumarol, used as active ingredient in some anticoagulants. Also known as warfarin; sometimes used in rat poison. Rats die from internal hemorrhage (no evidence that same effect produced in humans).
Source—Tonka beans come from South American tree—can grow up to 150 feet. Member of pea family. Oval pods about 5 inches long by 3 inches wide. Seeds 3½ inches by 2 inches. Distinct vanilla smell. Only seeds used in flavoring.
Vanilla—Vanilla fragrans. Orchid in vine form growing on trees in southeast Mexico and Central America. Vanilla beans from 7 to 10 inches long and approximately ¼ inch wide. Dark brown to black when cured. Seeds small and round. Entire pod used in flavoring.

FDA Standards—Vanilla extract must contain 13.35 to 15 ounces (depending on moisture content) of vanilla beans per gallon of extracting fluid (35 percent ethyl alcohol; 65 percent water). Vanilla flavoring is more dilute and has different standards. So does artificial vanilla (vanillin). Some additives like sugar are permitted.

The amounts of coumarin in "super vanilla" vary. Purchaser has no way of knowing if toxic amounts are present. Only laboratory tests can detect coumarin.

ASSIGNMENT

Write the informative pamphlet on "super vanilla." Your audience is the general public—anyone who might purchase the coumarin product. Read through the facts. Decide what material would be of most concern to this audience; then order that material in a way that will catch and keep their interest.

Audience Case 2

Mr. Science

As part of your major in science writing, you've landed a summer internship with *Junior Scientist*, a popular magazine aimed at eight- to twelve-year-olds. As your first job you're assigned to assist the contributing editor who writes "Mr. Science," a monthly column that answers questions sent in by readers.

The editor gives you the following letters.

1. Dear Mr. Science:

   Why does my mother's microwave oven cook so fast? Why can't you use regular pans in it? Will it blow up if you do?

   Ted Drivas
   Age 10

2. Dear Mr. Science:

   Last week I saw this movie on TV where there was a giant ant that ate people because it had been close to an atom bomb. Could that ever happen? How could we kill it?

   Lisa Muscovy
   Age 7

3. Dear Mr. Science:

   My brother says thunder is when two clouds bump into each other. Is that true? I think it's when lightning hits the ground.

   Mark Martinez
   Age 9

4. Dear Mr. Science:

   I want to be an astronaut. How long will it be before you can fly to another planet?

   Joanne Leibowitz
   Age 11

ASSIGNMENT 1

The editor tells you to choose two of these letters and write answers to them. He reminds you that the answers must be scientifically correct, but they also have to be understood by the children who read the magazine. He gives you a length limitation of 350 to 400 words per letter.

ASSIGNMENT 2

A week later the editor gives you another letter:

Dear Mr. Science:

Last week my teacher told us the earth was millions of years old, and she showed us some pictures of dinosaurs, and she said that they all died before there were even any people. But in the Bible it says that God created everything in seven days, and it doesn't even talk about dinosaurs. So is my teacher lying? My brother said I should write to you.

Tammy Willis
Age 8

The editor asks you to write an answer. He reminds you that your answer must be scientifically correct (the magazine's policy), but he cautions you to respect the child's religious beliefs. He gives you a length limitation of 350 to 400 words, but he promises to publish your answer if you write something suitable.

CHAPTER • 2

Determining Your Purpose

INTRODUCTION

When you analyze your audience, you must also determine the purpose (or purposes) of your report. Only then can you design and organize a document that will get the job done. Your purpose is the effect you want your writing to have on your readers, the objective you want to achieve. Just as reports, proposals, and other written presentations can have several audiences, a document can be designed to serve many purposes. For example, a proposal that wins a contract first persuades, then serves as a preliminary record of the contract terms. Later still the same proposal might be used as a record of cost-estimating techniques or as a guide for writing other successful proposals.

This chapter presents seven basic purposes that documents can be designed to achieve. Throughout this chapter you will also learn the importance of identifying multiple purposes and deciding which purpose is primary in a given report and which purposes are secondary. Keep one further distinction in mind as you consider the purposes a report might serve. Sometimes there is a difference between your purpose and your employer's purpose, or your purpose and the purposes to which your audience will put your reports. The following case study shows the importance of considering purpose before writing.

Community Health Systems: A Case Study

As part of their course on community health systems analysis, two advanced nursing students were asked to analyze the structure of a small community health service and to report on ways that the service could be improved. The assignment their instructor provided appears in Figure 2-1.

The outline in Figure 2-1 clearly indicates that the report was to be both a recommendation and a source of information about the structure and operation of the health delivery service. The instructor emphasized that it is sometimes more important to spend time redesigning a system that serves hundreds of people than to spend the same number of hours treating the personal health problems of a few individual patients. The instructor did not tell the students whether their audience was to include the directors or staff of the health agency, but the students could safely assume that the agency would be interested in the report. From the order of the outline, it appears that the instructor emphasized good research and analysis. Yet the report was to move toward the formulation of conclusions and recommendations. The students' task, then, was to design a report for two audiences (faculty and agency directors) and two purposes (to inform and to recommend).

Kate and Richard, the two students, chose to analyze the Chinese-American Downtown Lunch Program, a community volunteer agency that provided hot lunches to elderly Chinese-Americans who lived alone. This lunch program asked for help because it hoped both to streamline the meal service and to start a program of medical exams and basic clinical care. Kate and Richard were already registered nurses. When they first visited the lunch room in the basement of the Chinese Baptist church, they were so moved by the poor physical condition of some of the clients that they decided to give each client a full physical examination and make referrals to the public health hospital and other community health facilities. The following is the introduction to the report they wrote for their instructor: a report that consisted of their examination records and a recommendation that the nursing staff of the lunch program be expanded to include an experienced clinician.

As professionals we were shocked at the condition of the clients we saw at the Chinese-American Downtown Lunch Program. Of the fifteen individuals we examined, we referred six (almost 50 percent!) to County Memorial for follow-up, and three more to Judson's outpatient clinic. Obviously, the program, which has several merits, needs the attention of a full-time care giver.

Figure 2-1
Instructions for a term report

Community Health Systems Analysis

Your report should address the following topics in this order:

1. Define the community and the population served by the health delivery system, including demographic data and maps.
2. Describe the health system itself, giving an overview of its parts and listing the professionals, nonprofessionals, and volunteers involved.
3. Describe your methods of research, including interview questions, survey forms, etc.
4. Describe and evaluate your methods for verifying and analyzing your research findings.
5. Define the problems you find. How does this system compare to two of those discussed in class?
6. State your conclusions and develop recommendations. Explain how your recommendations address the problems you identify.
7. Explain how your recommendations might be implemented and explore any problems that might be encountered when they are implemented.
8. Summarize your main points and provide a list of all your sources, including interviews, telephone interviews, correspondence, surveys, and literature searched.

Clearly, these two graduate students failed to achieve the purposes of the report. They did not analyze the health delivery system at all and consequently were unable to answer any of the questions posed in the project outline. Kate and Richard performed a valuable human service, but they missed the point that health service needs to be understood as a system of interacting institutions, organizations, and individuals working together. The best that could be said of their report is that it accomplished part of one of the stated purposes, to recommend changes. Their report failed to inform, to present and interpret the data that supported their recommendations. Note, too, that their opening paragraph is not appropriate for presentation to the directors of the lunch program.

This case also shows that the primary purposes of a document—the effects it is meant to have on readers—may differ from your intermediate objectives while you research and write the report. The physical exams the two nurses gave were a worthy intermediate objective but not the purpose of the assignment and the report. To determine the primary purpose of your report, you should be aware of the various purposes a report can serve.

The Purposes

This section presents seven basic purposes of reports and other documents. The most directive (or assertive) purpose is discussed first, the least directive last. Purposes that offer limited choices to readers are discussed before purposes that allow readers a great range of choice and action.

Directing Action

Directives, work orders, and policy statements start actions. They *order*. Sometimes these documents also explain why an action is to be taken. If starting actions or ordering is your purpose, you should always explain:

- What is to be done
- When and in what order tasks are to be completed
- How the task is to be accomplished
- By what organization or person the work is to be done

You also direct action when you order equipment or supplies and when you teach (as in an instructional manual), because you give your readers the skills necessary to do something. You delegate power to your audience. Notice how the instructions printed below direct, teach, and empower the operator of the instrument—a surveying transit.

Instructions

1. Position the automatic leveling rod on the bench mark and loosen (turn counterclockwise) the band lock sufficiently to disengage the eyelet.
2. Using thumb friction, move the bank until the last two digits of the bench mark are aligned to the telescope crosslines.

[These instructions continue for several steps. Courtesy of Keuffel & Esser Company, Parsippany, New Jersey.]

Note, too, that these instructions, like all purely directive writing, leave the reader with limited choices.

Coordinating

Coordinating documents, which are often memos, provide the information necessary for the successful management and completion of a project. You will use coordinating memoranda to make sure that all the people working on a project have the same information and are working toward the same goals. In other words, these documents coordinate efforts in order to prevent duplication of tasks and to ensure that all parts of the project are covered. Like a directive, a coordinating memorandum can delegate and divide responsibility, restate working agreements for ready reference, provide schedules, describe resources that can be used for analyzing and solving problems, and describe lines of responsibility to be followed routinely or in case of budgetary or other contingencies. Coordinating documents usually are designed as references to be consulted for the life of a project or until they are superseded.

Figure 2-2 is a coordinating document provided to the coauthors of a reference book. Note how the individual responsibilities and authorities are clearly specified. Coordinating documents tell

- Who does what parts of a project
- What is being done
- When the parts of the project are to be finished
- Where resources are located

Coordinating documents do not usually explain the *whys*.

Proposing and Requesting

While directives *tell* a reader what to do, proposals and requests *ask* for approval, funding, or assistance. The reader can choose to accept, refuse, or draft a counterproposal. Proposals can offer a product (a sales proposal), solve a problem, or request help. They are always specific about what is proposed. If your purpose is to propose or request, you should always specify

- Who will do the work
- How much money or assistance is required and how the project will be managed
- What criteria will be used to evaluate the results
- When the project and its parts will be finished
- Why the proposal should be accepted

Requests for assistance or information should also place a clear limit on the amount of help that is needed, and they should also explain how providing help will benefit the donor. These topics are covered in greater detail in Chapters 11 and 13. Here it is enough for you to remember that proposals ask for something specific.

Recommending

A recommendation is a *suggestion*, not an order or a request. Recommendations usually describe and evaluate several alternative actions, so that the audience is given a wide choice. If your purpose is to recommend, you should always

- Specify the action that is recommended
- Name and discuss the alternatives
- Present arguments to support your decision or choice
- Leave the final choice to the readers who commissioned the report

The section of this chapter on informing discusses a report that combines recommendations, information, and matters of record. Recommendations are discussed in detail in Chapter 10.

Providing a Record

Whatever their primary purpose, most reports also provide a record of plans, decisions, and actions. Some report formats, such as the progress report and the final project report, place a heavy emphasis on recording problems that were encountered or anticipated, solutions considered, and alterations in working plans and budgets. Information of this type can simplify later work on a project. Because American workers change jobs roughly every five years, it is common to find that many of the engineers who worked on the original design of a project are gone by the time a new model is needed. Even if it were possible to rely on memory alone, the designers themselves might be working for competitors or for companies in distant parts of the nation or the world. Consequently, clear and readily accessible paper trails are essential. If your purpose is to provide a record, you should always

- Describe important research in detail, covering the who, what, when, and where
- Discuss any problems that occurred
- Explain decisions, solutions to problems, and changes in original plans or procedures
- Indicate where other records are filed

Informing

When you write a purely informational note, you should have no specific expectations of action. Reports often fail to achieve their actual objectives because it is so easy for a writer to think "I'll just let them know what we're doing here," when a little analysis would reveal that a coordinating report or a recommendation is called for. Be sure that an informational report is really what is needed *before* you write, and that you do not expect any action to be taken as a result of your report.

Passages from two different statistics texts are reproduced in Figures 2-3 and 2-4. One of the books (Figure 2-3) is a well-

FIGURE 2-2
Coordinating memorandum

Northwest Handbook Staff: Coordinating Memo
April 30, 1985 (effective date)

Financial Questions. Please refer all financial requests and problems to Dr. Westerby. Limited funding is available for the development of bibliographies and for the purchase of necessary books and computerized information searches.

Editorial Problems. Dr. Redding is the final court of appeal for all editorial matters.

Research Problems. Bring these to Mr. Curtis first, then to Dr. Redding.

Deadline. April 30, 1986 for rough drafts of sections. Earlier is better. The editor reserves the right to refuse late copy and to reassign authorship of the refused sections.

Market. Please keep in mind that the markets will be, in order of priority

Primary: government, university, public, and corporation libraries
Secondary: government offices at all levels, lobbyists, and small corporations.

Purpose. This reference book is designed to offer in one set of covers a range of information that is otherwise not available or available only by consulting many sources that can be found only in special collections or large research libraries.

System of Documentation. APA

Chapter Length Limits. Approximately ten pages per chapter, sixty chapters total. Aim for fifteen to twenty manuscript pages. Cite other references rather than repeat entries at length.

Organization. [At this point the memo goes on for two pages of the chapter titles.]

Staff. [A list of all names, telephone numbers, and assignments, so that team members can consult.]

known guide for working professionals who need to know how to understand and interpret various statistical tests, yet who have no need to calculate any statistics themselves. While this book does teach readers how to interpret, its chief purpose is informational.

Notice how the text passage (Figure 2-4) from a first-year college text differs from the text written for professionals who need only to interpret statistics, not calculate them.

FIGURE 2-3
Informational textbook writing. From Reasoning with Statistics, *Second Edition by Frederick Williams, used by permission of CBS College Publishing.*

Random Sampling

Thus far we have loosely considered a sample as a collection of observations that is somehow representative of a population. Let us now modify the definition of a sample given in Chapter 1.

4.4 Random Sample: a collection of phenomena so selected that each phenomenon in the population had an equal chance of being selected.

Quite apart from any situation involving statistics, you have probably seen random sampling involved in many different types of situations. In almost any type of survey research, various strategies will be employed to try to ensure that every member in the defined population has an equal chance of being selected in the sample. Recall, too, that in the illustration of an experiment involving instructional television (Chapter 1), the two groups of subjects were drawn so that they initially would be from the same population, students in English. This would entail random selection of the subjects from the population for the experiment, then further random assignment to one of the two groups.

In the simplest terms, we can say that if each phenomenon or unit in a population had an equal chance of inclusion in a sample, then that sample will have characteristics that can be used as a basis for estimating population characteristics. In slightly more specific terms, if random sampling is involved, this will enable us to draw upon a body of statistical theory that incorporates the mathematical relations between sample characteristics and population characteristics. Random sampling allows us to employ the logic of statistical inference, that is, estimating parameters from statistics.

The college text clearly directs the reader how to select a random sample from a population. If informing is your purpose, you should be sure to

- Provide the who, what, when, where, why, and how
- Ask yourself if you have another, more important, and more directive purpose to achieve at the same time

Entertaining

Finally, reports don't need to be dull. In fact, you might serve your purpose more effectively by including interesting examples or memorable entertainment. Why? Because an example allows a reader to organize many abstract concepts around a single concrete case. If that case includes some humorous or arresting material, so much the better. Your readers might associate sections of your report with particular stories, cartoons, drawings, or comic remarks.

FIGURE 2-4
Directive textbook writing. Courtesy of Brooks/Cole Publishing Company.

Random Samples

A method called random sampling is commonly used to obtain a sample that is most likely to be representative of the population. *Random* has a technical meaning in statistics and does not mean haphazard or unplanned. A random sample is one in which every potential sample of size N has an equal probability of being selected. To obtain a random sample you must

1. Define the population,
2. Identify every member of the population, and
3. Select a sample in such a way that every member of the population has an equal probability of being chosen to be one member of the sample.

We'll go through these steps with a set of real data—the self-esteem scores of 24 fifth grade children. [The sample continues for two pages.]

For example, one executive wrote at the bottom of an important policy memorandum: "This memo will haunt you." Over a year later the recipients of the memo remembered both the memo itself and the major points of the recommended policy. How much more effective than to have written IMPORTANT or "Please remember and refer to this memo." For maximum effect, devices of this type should be used rarely and with care and judgment. Choose to entertain your readers

- To reinforce important points
- To simplify complex information by organizing it around a memorable example

Most of the examples in this chapter have included more than one purpose. The next section considers how you can attain several purposes at one time.

MULTIPURPOSE REPORTS

While you will sometimes write reports that have a single purpose—to inform or to request action, for example—you will normally write reports that have a hierarchy of purposes. You may want to inform your audience of certain facts and also recommend a course of action based on those facts, or you might wish to coordinate activities in a group and at the same time provide a record of agreements and decisions the group

FIGURE 2-5
First version of Anna Trompkin's report

Anna Trompkin, M.S.W.
1818 34th Place
Edmonds, Washington 98111

February 3, 1985

George MacDonald, Administrator
ParkView Nursing Home
Carnation, Washington 98222

Dear Mr. MacDonald:

Thank you for the opportunity to serve ParkView Nursing Home during the month of January. I visited the home on January 4th and January 17th and on each occasion I stayed for 4½ hours.

On the fourth, Mrs. Smith, your assistant administrator, introduced me to six patients who were admitted since my last visit. All of these patients appeared to be in good spirits; I don't think any special counseling or intervention will be necessary in their cases. Before lunch, I met with the physical therapist to discuss our common patients; and after lunch I met with my own group of eight long-term patients. We discussed problems in coping with visiting grandchildren. No special problems came up during the staff meeting, although I agreed to give a short course on intervention strategies annually.

On the seventeenth there were no new patients at the home, although Mrs. Bryan in Room 27 died on the 15th. Several staff members reported to me how understanding and warm the family was. I met with my long-term group for two hours and we agreed to propose that the medical clinic waiting room be used for quiet games during off hours, as the main lounge is often too noisy for bridge or other table games.

Two other important issues need to be dealt with fairly soon. I spoke with Mrs. Bryan's roommate, Mrs. Williams. The staff is being very helpful with her grieving, but I think special efforts will be needed here before a new patient is moved into Mrs. Bryan's area. I think Mrs. Williams might become disruptive if this transition is too abrupt. This situation reflects the need for an advanced in-service course on grieving and intensive nursing home management of grief. The other matter concerns the patients in the nonambulatory ward, who are often anxious. There has been some talk of patient abuse in the ward, and while I have seen no documented evidence of abuse, I think the problem needs attention. After talking with some of the patients, I reached the conclusion—which is shared by the assistant administrator and the ward chief—that there has been an outbreak of fear of abuse rather than any actual incidents. There should be a special in-service seminar on recognizing and allaying patient fears and avoiding modes of speech and action that encourage these reactions.

Thank you for contracting for my counseling services. If you have any questions, please call.

Sincerely,

Anna Trompkin, M.S.W.

has reached. When you find yourself writing a report that has more than one purpose, decide which purpose is most important and deal with it first. Then address your other purposes in separate sections. As far as possible, try to keep your purposes separate so that they do not become confused.

We can see how important it is to assign priorities in the example in Figures 2-5 and 2-6, a report that had three important objectives: to provide a record, to make recommendations, and to inform. Anna Trompkin, a social worker, provided psychological counseling services to fifteen nursing homes that were too small to afford their own full-time counselors. On an average, she visited each institution twice each month. During her visits, she trained the staff in counseling and intervention techniques, made policy recommendations, and offered limited group and individual counseling to the patients. While she made most of her recommendations orally to the administrators of the various homes, she also wrote a monthly report for each home. Ms. Trompkin also knew that when her contracts came up for annual review, her reports would provide a record of her hours and her accomplishments. The report in Figure 2-5 was written for one January.

As you can see, this report was organized chronologically. Issues raised at Ms. Trompkin's first January visit are discussed before issues raised during her second visit that month. Notice that none of her purposes is well served, although the three purposes are probably achieved in this order: to record the visit, to inform about work done, and to make important recommendations. The report is chatty and wandering, and significant recommendations appear at the end, where they are unlikely to be noticed, because many readers simply give up reading if they do not find important information at the beginning of a report. Because her recommendations were not being acted upon, Ms. Trompkin consulted an editor, who suggested that she decide on her purposes and put the most important information first. Figure 2-6 is her January report as it was rewritten.

FIGURE 2-6
Second version of Anna Trompkin's January report

Anna Trompkin, M.S.W.
1818 34th Place
Edmonds, Washington 98111

February 3, 1985

George MacDonald, Administrator
ParkView Nursing Home
Carnation, Washington 98222

REPORT FOR JANUARY 1984: VISIT DATES: JANUARY 4 AND 17.

CRITICAL NEEDS

1. Mrs. Bryan in Room 27 died on January 15. Her roommate, Mrs. Williams, needs immediate and continuing grief therapy. No new patient should be placed in this room until Mrs. Williams' condition has stabilized.

2. Mrs. Williams' case points up the need for an advanced in-service course on grieving. With your approval, I propose to begin this series in March. The staff supported this recommendation at our meeting on the 17th.

3. Several patients in the nonambulatory ward are spreading rumors about patient abuse by staff members. While there is no evidence that any abuse has occurred, there is a clear need for an in-service seminar on allaying the fears of nonambulatory patients. The staff also supported this proposal at our meeting on the 17th. Both of the proposed seminars fall within the provisions of my current contract with ParkView home. In the meantime, I held counseling sessions with the patients who were most concerned.

NEW PROGRAM RECOMMENDED

The long-term counseling group suggested that the clinic waiting room be used for quiet games when the clinic is closed. The lounge is often too noisy for bridge or board games. I support their recommendation and agreed to bring it to your attention and to the staff meeting.

STAFF NEEDS

1. At the staff's request, I agreed to offer an annual in-service refresher course on intervention strategies.
2. I also met with the physical therapist to discuss our common patients.

NEW PATIENTS

I interviewed six new patients on the 4th; there were no new patients to see on the 17th. None of the new patients presented any critical counseling problems.

REGULAR COUNSELING

I met each time with my regular group of long-term patients. We talked about two issues: the clinic lounge recommendation and the problems raised by visiting grandchildren. Visits by children are not a significant problem at this time; there is no need for additional guidelines or policies.

TOTAL HOURS
I spent a total of 9 hours at the home during January.

If you have any questions, please call me at 555-7000. I will submit my next monthly report by March 10.

Sincerely,

Anna Trompkin, M.S.W.

Notice the complete reorganization of the report, as well as the use of subtitles, boldfacing, and the numbering of points. In the revised report, the purposes are given priority in this order: to make recommendations, to inform the director of staff decisions, and to provide a record for the annual review. From studying this example, you can see how much more effective reports can be when the purposes are clearly defined and given clear priorities. The second version of Ms. Trompkin's report is not perfect—you can probably think of a few changes that would improve it—but it is vastly better than the first version, which achieved none of its purposes.

SUMMARY

1. Audience, purpose, and organization work together. You must determine both the audience and purpose of a report before choosing a pattern of organization.

2. The seven common purposes of technical documents are to direct or initiate action, to coordinate, to propose, to recommend, to provide a record, to inform, and to entertain. A report will seldom be purely informational.

3. Most reports will have more than one purpose. At a minimum, any report will need to do its major job and also serve as a record.

EXERCISES

1. Identify the primary and secondary purposes of Excerpts 2-1, 2-2, and 2-3. Start by reading each document through. Then read the document again, asking yourself what purposes the writers and designers intended to achieve. In a short paragraph, explain your analysis. If you can identify a purpose but conclude that it has not been achieved effectively, discuss your finding. Use quotations to identify the sections to which you refer.

Excerpt 2-1
From a Texas Instruments Calculator Manual. Courtesy of Texas Instruments Incorporated.

II. BASIC OPERATIONS

The keys have been selectively positioned on the keyboard to provide for efficient calculator operation. Although many of the operations may be obvious, the following instructions and examples can help you develop skill and confidence in your problem solving routine.

INITIAL OPERATION

The fast-charge, nickel-cadmium battery pack furnished with your calculator was fully charged at the factory before shipping. However, due to shelf-life discharging, it may require charging before initial operation. If initially or during portable operation the display becomes dim or erratic, the battery pack needs to be charged.

Under normal conditions, a fully charged battery pack provides typically 2-3 hours of continuous operation.

With the battery pack properly installed, charging is accomplished by plugging the AC Adapter/Charger AC9132 into a convenient 115V/60 Hz outlet and connecting the attached cord to the calculator socket. About 4 hours of charging restores full charge with the power switch off or 10 hours if the calculator is in use. CAUTION: The battery pack will not charge if not properly installed in the calculator.

Sliding the ON/OFF switch to the right applies power to the calculator and sliding it to the left removes power. The power-on condition is indicated by a lighted display.

Your calculator is designed to be energy efficient. After about 1 to 3 minutes of non-use, the display will shut down to a single decimal point traveling in the display. To restore the display at any time, just proceed with a calculation, or press the [2nd] key twice.

STANDARD DISPLAY

In addition to power-on indication, the display provides numerical information complete with negative sign and decimal point and flashes on and off for an overflow, underflow or error condition. An entry can contain as many as 8 digits. All digits entered after the eighth are ignored.

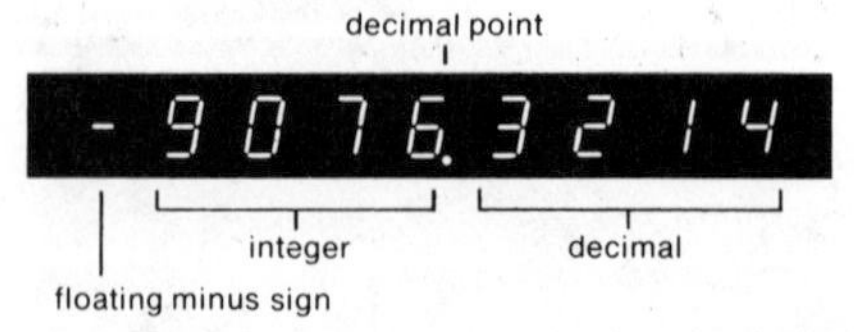

Any negative number is displayed with a minus sign immediately to the left of the number.

See Appendix C for the accuracy of the displayed result.

DATA ENTRY KEYS

[0] **through** [9] **Digit Keys**—Enters the numbers 0 through 9.

⊡ **Decimal Point Key**—Enters Decimal Point. The decimal point can be entered wherever needed. If no decimal point is entered, it is assumed to be to the right of the number, and will appear when a function key is pressed. A zero will precede the decimal point for numbers less than 1. Trailing zeros on the decimal portion of a number are not normally displayed. Only the first decimal point entered is accepted, all others are ignored.

π **Pi Key**—Enters the value of pi (π) to 11 significant digits (3 14 15926536) for calculations: display indicates the rounded value to 8 digits.

Excerpt 2-2
From a brochure describing the Mariner Seagoing Kayak. Courtesy of Mariner Kayaks.

Result of Efforts

The boat we had been searching for, the ultimate sea cruising kayak, the Mariner. A kayak with the following characteristics:

SPEED AND PADDLING EASE

The Mariner is considerably faster than all of the touring kayaks we've tested, including the highly regarded English expedition kayaks. The Mariner is as fast as the fastest wildwater racing boats, and the only kayaks we've come across capable of outdistancing it are Olympic flatwater racers.

The Mariner is an easily paddled boat that can cover more miles with less effort than conventional touring kayaks. Carrying a heavy gear load it has less resistance at cruising speed than most kayaks have empty. There are several reasons for this:

- ☐ The narrow beam means less wave-making resistance as well as less wetted surface for a given load and therefore *less frictional resistance.*
- ☐ Its distribution of buoyancy (prismatic coefficient) minimizes wave-making resistance at cruising speed.
- ☐ Its waterline shape remains very much the same through a range of 3 inches difference in draft. This range covers a paddler in an empty boat to a paddler with 250 extra pounds.
- ☐ Its bow shape promotes laminar flow over the forebody. Laminar flow has *four times* less frictional resistance than the alternative, turbulent flow.
- ☐ The stern minimizes eddy-making resistance. Eddy-making causes much greater resistance than even turbulent flow.
- ☐ The swede-form shape (greater underwater volume *behind* the midpoint) has less resistance moving at the water's surface than either a fish-form shape (its opposite) or a symmetrical hull. This is due to less wave-making resistance and a longer area of laminar flow. (Note: A fish-form shape is fastest through the air or underwater where there is no wave-making resistance, but *not* on the surface, where kayaks normally travel.)

Paddlers have commented on the ease with which the Mariner catches and surfs even small waves. This is due not only to its quick acceleration and high top speed, but also the planing shape (slight V-keel and hard chines) of the midsections. This shape lifts the Mariner more out of the water at surfing speeds, the hard chines spit water away to the sides further reducing wetted surface. Also, the stern configuration allows the waves to push the Mariner along in a more horizontal attitude, helping to prevent the bow from digging in and adding friction.

TRACKING AND MANEUVERABILITY

The Mariner is easily kept on course in all kinds of conditions thanks to its hull design, windage balance, and sliding seat trimming capability. This is *not* at the expense of maneuverability, however. Unlike some "tracking" boats, the Mariner is *not* so stiff that turning it becomes a chore. (Some fin keel expedition kayaks are so difficult to turn at cruising speed that even making the small course corrections necessitated by wave action is a major effort.) At cruising speed the Mariner can be turned up to 30 degrees with one sweep stroke and a slight lean to the stroke side to allow the V-keel to more easily skid.

The Mariner's long V-keel and hard chine aft not only improve tracking ability, but also help minimize the tendency to broach in quartering and following seas. Combined with the ability to "sit back" into following waves by sliding the seat back, this results in excellent control in these conditions. In fact, even when caught sideways in a large foaming broken wave, or "soup", the Mariner can be turned to point down the wave allowing the paddler to back off if need be. This was an unexpected, and pleasant, surprise on testing. Our experience had been that once sideways in soups a kayak would be held there until stopping on the beach, or on whatever the wave was pushing it towards.

BEAM WINDS

UNDERSTANDING THE PROBLEM AND SOLUTIONS.

In beam winds a kayak is subject to two opposing forces: windage, the wind acting on the exposed area above the waterline, and lateral resistance, the resistance of the area below the waterline to being pushed sideways by the wind. Both forces have centers that will vary with the design of the boat. If the center of windage is at the same point as the center of lateral resistance the boat will be pushed at 90 degrees to the wind. However, the center of lateral resistance *moves forward* as the boat gains speed and the kayak wants to turn into the wind.

Now, if you have a kayak that weathercocks like this (and most do) you must paddle constantly on one side with tiring turning strokes to stay on course. This is frustrating work, and gives rise to exasperation and profanity.

One solution is to add a rudder. This enables you to maintain course paddling with both blades, but with the considerable added drag of towing an angled rudder through the water. And then you've got a delicate device on your boat that requires wires that can foul up and break, has small parts that can corrode, come loose or get lost, gets tangled in towlines, and is subject to damage in surf, beaching, during rescues, while towing, carrying, or just being tripped over in camp.

There is another solution, the Mariner solution.

The Mariner is designed to have a balanced "wind-water couple" *at cruising speed*. In addition, the paddler's position can be adjusted instantly, at sea, by simply sliding the seat/footbrace unit forward or back. This changes the centers of windage and lateral resistance to fine tune the Mariner to suit conditions. This is advantageous because the wind-water couple changes with different gear loads and at different angles to the wind. The seat can be adjusted at speed without so much as putting the paddle down, yet once positioned it stays firmly in place until the next adjustment is desired.

EXCERPT 2-3
From the City of Seattle's Requirements for Electrical Service Connection

12.0 *INTERCONNECTION REQUIREMENTS FOR COGENERATION/SMALL POWER PRODUCERS*

Small Power Producers and cogenerators who wish to interconnect with and supply electrical energy to the City Light Department of Seattle must meet the following requirements.

12.1 *Approval*

All specifications and drawings of the complete C/SPP installation shall be submitted to the Utility's Technical Advisory Service Unit for review and approval before the system is energized. City Light reserves the right to inspect the system to verify compliance with all design requirements. The final decision as to the requirements for each installation will be made according to several criteria, including the type and capacity of the generator, the magnitude of other loads and generation on the Utility system, the safety of Utility personnel, the effect on other Utility customers, and other factors.

12.2 *Point of Interconnection*

12.2.1 The point of interconnection between the Cogenerator/Small Power Producer and the City Light Department distribution system shall be:

(a) the service entrance point for generation delivered through an existing service, or
(b) a separate service entrance point serving only the C/SPP generation, or
(c) one or more disconnecting devices for generation delivered at the Utility's primary distribution voltage.

12.2.2 *Division of Responsibility*

The Utility will maintain the metering transformers, meters, and all equipment on the Utility side of the interconnection point. All other equipment shall be maintained by the C/SPP.

12.3 *Frequency, Phase and Voltage*

12.3.1 *Frequency*

The frequency of the C/SPP system shall be 60 Hertz. Frequency tolerances are discussed in detail below.

12.3.2 *Voltage*

12.3.2.1 The delivery voltage of the C/SPP system at the point of interconnection shall be specified by the Utility.

12.3.2.2 *Secondary.* In general, generation of 100 KW or less will be accepted at the existing secondary service voltage: 240 or 480 volts single phase, or 208Y/120 or 480Y/277 volts three phase.

12.3.2.3 *Primary.* Generation exceeding 100 KW shall be three phase at the existing primary distribution voltage.

12.3.3 *Three Phase Generation in Single Phase Areas*

For installations requiring *three phase* service in an area served only single phase by the Utility, the C/SPP shall pay the cost for the Utility to extend three phase service to the point of interconnection.

12.4 *Transformers*

Transformers required to match the generator voltage to the specified utility primary voltage shall be furnished, installed, and maintained by the C/SPP. When generation under 100 KW capacity is accepted at the existing secondary voltage, the Utility will furnish and install the necessary transformers.

12.5 *Switchgear and Control Equipment*

12.5.1 *General*

12.5.1.1 Switchgear, circuit breakers, visible-break disconnect switches, and other equipment required to control the generator output shall be furnished, installed, and maintained by the C/SPP. All equipment shall be capable of withstanding and interrupting the maximum fault currents specified by the Utility.

2. Rewrite Anna Trompkin's recommendation for the clinic lounge as a short proposal. How does the shift from *suggesting* to *asking*—from reporting the concern of her patients to proposing a change on their behalf—change the language and attitude she must adopt? Here are some additional facts to use (if you wish) in your own one-paragraph proposal. (Assume that this paragraph might be inserted in the monthly report.) There will be no extra costs for janitorial service because the night janitor does not get to the clinic area before 10 PM, when the new game room would be closed. The nursing staff lounge is across the hall from the clinic lounge, so that nurses would be available in that wing in case of emergencies. Four folding card tables would need to be moved to the lounge, where there is storage space in an unused coat closet. The clinic itself is locked off from the lounge, so that patients would not have access to drug cabinets or other medical supplies.

3. Write section 2 of the Community Health report as it should have been written, using Professor Gordon's outline (p. 27). Purpose: to inform the professor about the parts of the health delivery system and provide background for the analysis and recommendations. Here is some additional information about the Chinese-American Downtown Lunch Program. The trustees of the Chinese-American Nursing Home operated the lunch program in a Chinese Baptist Church. Community volunteers talked with the clients, taught crafts,

***Figure* 2-7**
Chinese-American downtown hot lunch program: a community health care system providing hot lunches to elderly Chinese-Americans who live in downtown apartments

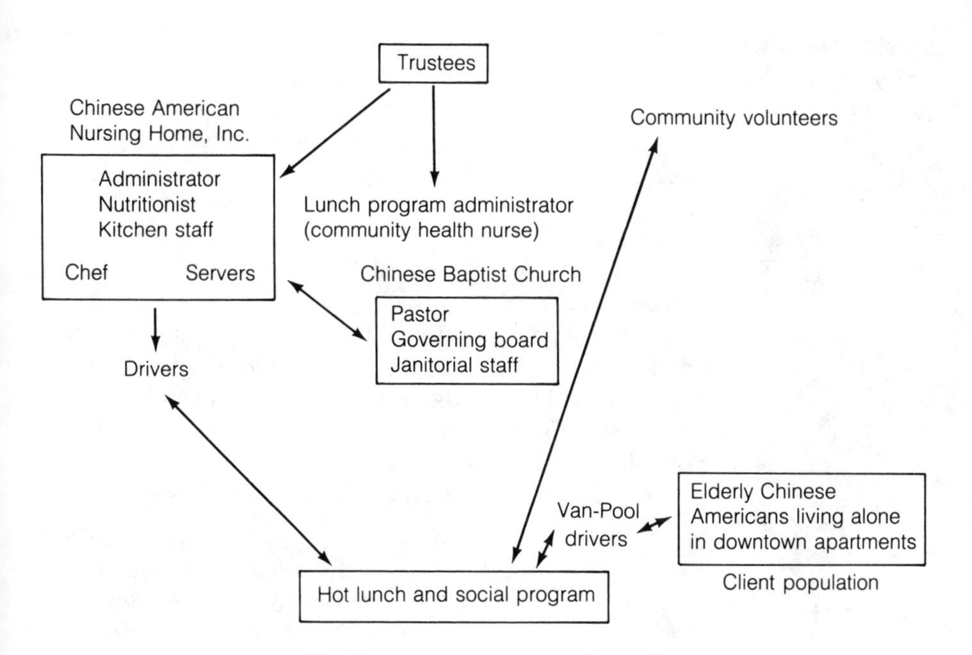

initiated games, and so on. The lunches were cooked at the nursing home and brought to the church by nursing home employees, who also drove the vans that picked up the clients. In turn, the church provided space, tables, and janitorial services. To manage the program, which was free to the clients, the trustees hired a part-time community health nurse. Figure 2-7 shows an outline of the organization.

4. Write three short incident reports about the same experience. Find a place where you can spend an hour or more observing. You might select a workplace connected with your major or job, a public agency, or a restaurant. You will need to collect some basic information about the address and location of the site, the time of day, the people present, and descriptions of conflicts or interactions that take place. After evaluating your experience, write one short report that provides a record, another report that makes a recommendation, and a third intended to entertain. Pay close attention to differences in the selection and organization of data.

Purpose Case 1

The Supermarket Scanners

You are a systems analyst for Frontier Incorporated, a local supermarket chain. Frontier has owned a chain of twenty grocery stores in your home town for over thirty years, and they've been very successful. Lately, however, sales seem to be slipping. A marketing survey reveals part of the problem: Frontier is perceived as a somewhat stodgy, old-fashioned firm. Younger shoppers seem to be taking their business to the newer Total Markets, a national chain.

The company has decided to modernize its stores, but as yet no concrete plans have been made. One change that occurs to you is the idea of using automated checkout stands. You know these systems have been used in other cities with some success; you think they could be used here as well.

You do some research and compile the following data:

Definition: Optical scanning systems have scanners that read the Universal Product Code on an item and use computerized price lists to enter prices automatically at the register.
Practice: Checker runs item across glass screen; scanner under screen reads number, refers to computer memory, and supplies price to cash register.
Checkout tape prints price, item description, and department for each item.
Price is displayed on terminal. Everything happens simultaneously.
Universal Product Codes (UPCs) appear on most drugs, frozen, and canned foods.
Total checkout time reduced. Illegible or missing price stickers make no difference if UPC is used.
Electronic scales for produce can be added to system: Clerk keys in number identifying produce, and computer supplies price for weighed amount automatically.
Credit verification for checks can also be part of system.
Estimated minimum savings in checkout productivity: 0.65 percent of gross sales.
Fewer errors in price occur with scanner system, also fewer discrepancies with scale readings.
Estimated accuracy savings: 0.2 to 0.6 percent of gross sales.
Scanner system can also store information on sales: inventory, checkout speed, number of customers per register, etc. Printout would be available to management.

Possible savings in administrative costs (i.e., writing reports, record keeping): 0.15 percent of gross sales.
Possible savings in personnel use: Shelf items need not be individually priced. Only one price per shelf needed.
Altering prices means only altering price on computer program and shelf listing; individual items wouldn't need to be repriced.
Poor consumer acceptance for dropping individual pricing; some states have passed laws against it.
System components: in-store processor and memory backup (to take over if processor goes down); terminals with keyboard, cash drawer, customer display screen, receipt printer; manager's terminal to give manager access to data without closing down lane; hookup with central computer at store headquarters (maintains price files and provides data printout for all stores).

ASSIGNMENT 1

Write an informational report about the automated checkout system for the store management. Remember, at this point you're not concerned with recommending the system; you're only trying to describe it.

ASSIGNMENT 2

Write a recommendation report about the automated checkout system for the store management. Here you are informing, but your ultimate purpose is to persuade. Notice how your emphasis changes.

Purpose Case 2

HERBAL MEDICINE

As a midwife with five years of experience, you have become increasingly interested in the use of herbs in pregnancy and delivery. Part of your interest is personal: As a self-styled "natural food nut" you're concerned with the amount of artificial and highly processed substances most people ingest. But part of your interest is strictly professional: Midwives are not allowed to use drugs in home deliveries; if a birth develops complications, herbal remedies are the only recourse you have.

After some conversations with your colleagues, you conclude that others are as concerned about this topic as you are. You decide to do some research on the subject for your local professional newsletter to see what you can find.

You discover these facts about herbs:

Raspberry—Contains frangine, muscle relaxant, relaxes uterus in tests with lab animals. Vitamins A and D; can be taken as tea. Easing labor; maybe for preventing miscarriage (no research data on this). Inducing labor in that it relaxes cervical muscles so they can work efficiently.
Peppermint—Good for nausea; traditional for indigestion. Sometimes with raspberry.
Blue Cohosh—Can induce labor. May elevate fetal heartrate; use **only** after labor begins, otherwise may bring on premature labor. Contains oxytocin (an octapeptide); natural labor inducer. Also lupine methyl cystine; induces strong contractions in lab animals. Can also be used to expel placenta postpartum; will not overstimulate cervix like synthetic oxytocin. Can restore cervix that isn't contracting properly postpartum.
Squaw Vine—Can be used to ease labor with blue or black cohosh. Not to be taken before term; can cause premature labor.
Lobelia—Good for cases of swollen cervix, can slow labor. Also good for primary inertia or incoordinate contractions. Antinausea. May be able to prevent miscarriage either with raspberry or alone (no research data).
Motherwort—Contains leonurine and Alkaloid A. Stimulates uterine contractions. Good for postpartum hemorrhage. Antinausea.
Wild Yam—Good for nausea.
Alfalfa—High vitamin K content (blood clotter). Also organic salt, calcium, magnesium, phosphorous, potassium. Best for postpartum hemorrhage. Use only in tincture form, not tea (K not water soluble). If given during pregnancy may prevent hemorrhagic disease of newborn (crosses placenta and enters baby's bloodstream).
Camomile—Sedative.
Shepherd's Purse—Vasoconstrictor (also high in K). Best for postpartum. High vitamin C content. Use as tincture. For postpartum hemorrhage. Will restore cervix not contracting properly.
Cayenne Pepper—Works on nonchronic hypertension.
Comfrey—External use only to heal perineum and for engorged breasts. May be carcinogenic if used internally.
Hops—Sedative.
Bethroot—Labor inducer; works on postpartum hemorrhage. Can cause premature labor if taken before term.

Skullcap—Sedative.
Bayberry—For postpartum hemorrhage.
White Oak—For postpartum hemorrhage.
Baddies (to be avoided)—goldenseal, pennyroyal, mugwort, tansy, yarrow; all cause premature labor if taken before term.
Black cohosh—vertigo, vomiting, tremors, lowered pulse in overdose.
American mandrake, senna, aloe, ephedra, valerian—may harm fetus.
Sage—may dry up breast milk.

You conclude that the effectiveness of the herbs you studied is very hard to evaluate. The plants are affected by the way they're grown and the methods used to preserve them. Sometimes they're hard to measure accurately because of differences in potency. Dosage is difficult to specify. In general the rule seems to be to take the herb until the desired effect is achieved since there seem to be few ill effects from overdoses (with the exceptions noted in your research).

ASSIGNMENT

Write a report on herbal remedies for the local newsletter. You have two purposes in this report: You want to give your readers information about herbal remedies, and you also want to make some recommendations about using them. Assign priorities to these two purposes and then organize your report so that both purposes are met.

PART

Gathering Material

CHAPTER • 3

Gathering Information Inside the Library

INTRODUCTION

Where does information come from? Research data can come from your own experiences and experiments, from libraries and archives, from professional associations, from foundations and university laboratories, from think tanks, and from surveys and interviews. As you take jobs with more responsibility, you will spend increasing amounts of time learning and collecting new information. You will read professional journals in your field; attend conferences, short courses, and workshops; commission searches of library holdings; and gather data by means of telephone calls and surveys. Library research, surveys, and basic research are required in product development, market research, and financial planning, to name a few areas where professionals use research techniques. This chapter will introduce you to some basic and some very sophisticated techniques for locating information in libraries.

After defining the difference between primary and secondary sources of data, this chapter presents three progressively more sophisticated methods for finding information inside libraries: traditional library catalog and index searching, patterned library searches, and automated library searching. The chapter closes with guidelines for evaluating the quality of the sources you find.

The approaches to information gathering described here are based on the actual practices of professionals. You will learn techniques for gathering data from a variety of printed, organizational, and human sources, and you will learn how to keep current in your chosen field. The assignments at the end of the chapter will help you practice research skills, gain an understanding of what is going on in your discipline, learn about sources in your field, and develop expertise in a subject area related to your career choice.

Primary and Secondary Research

Your objective as a researcher is to arrive at an informed judgment based on the best available information. That information could come from your own original experiments, analyses, memories, or surveys, or it could be based on original work done by others. Original work done by you or others is called primary research. The materials that record the results of that work—notes, laboratory records, data analyses, and so on—are primary research data. When you draw on the work of others, whether published or derived from interviews, you are doing secondary research. Most professional research and analysis requires a combination of primary and secondary research. Library work is generally limited to secondary research, unless you are doing certain types of historical research.

Libraries

Your college library and your college faculty are the two most accessible sources of information on your campus. To use your library effectively, you should be aware of its strengths and limitations. Libraries offer good access to printed materials that are (1) widely published and (2) made available to libraries by publishers and donors. That observation may seem simpleminded; it isn't. Many documents, such as some trade journals, internal corporate reports, and membership lists, are published only for limited distribution. These published documents are not found in most libraries. No matter how large your college library may be—and most college libraries are fairly small—there will be some limits on its collection. A library's budget usually limits the range of materials it can acquire, and many libraries respond to this limitation by trying to be very good in some areas at the expense of other fields.

The most important limitation on the library, however, results from its emphasis on printed materials. Almost everything you find in your library will be dated. By the time an article gets into print and on the library shelf, it will have been through the hands of many editors, printers, and distributors. The original idea or research may have been developed months or even years before. Once you realize that these limits exist, you have an incentive to learn techniques for finding out now—if you can—the ideas that will get into print six months from today. Because of its limitations the library must be used with intelligence and care. We will present three ways to find information in the library: traditional approaches, patterned approaches, and automated searching.

Traditional Approaches to Finding Information

The Reference Staff

A library's reference staff is trained to use bibliographies, data bases, and other reference tools. Even in smaller libraries, you may discover area specialists who have highly developed skills in the use of reference materials in engineering, the social sciences, or the natural sciences. Reference librarians can save you hours of time and frustration by helping you find the indexes and guides best suited to your search. Before you ask for help, however, take some time to define your research question, and be prepared to explain what sources you have already tried. Librarians are professionals; you should prepare for a consultation with a reference specialist just as you would prepare for an interview with any other expert. Unfortunately, librarians are frequently assaulted with requests like: I need something for a paper. Do you have anything on nuclear power? If you ask a vague question, you can only expect to receive very general help, which in many cases is an introduction to the subject cards in the card catalog.

The Card Catalog

The card catalog is a guide to the location of materials in the library. The card catalog is not designed to be a primary reference tool. Students who use the catalog as their main reference often end up writing papers that are (1) out of date and (2) based on unrelated sources. There are two specific reasons to avoid using the card catalog as a reference tool:

1. The catalog generally lists books, not journal articles. In many scientific and technical fields, books are not compiled or written about new subjects until a large body of experimental literature has already been published and reviewed in journal articles. As a consequence, books in the sciences are often out of date. Therefore, articles are everything.

2. Most libraries have the time and personnel to produce only three or four cards for each book: a title card, an author card (sometimes more in the case of multiple authorship), and one or two subject cards. If a book is published by a government agency, however, the official title or author may be the name of the agency or the title of the series in which the booklet is included. You might miss works by an individual expert that were indexed under an agency author.

   It can also be difficult to locate the subject classification for your topic. For example, you might try to find a book on the bubonic plague under *plague, bubonic,* or *epidemic* only to find later that the heading currently used by libraries is *pandemic*. Most libraries follow the Directory of Library of Congress Subject Classifications or the equivalent Dewey schedules, and library staffs try to insert cross-reference cards to help you. Still, in a large catalog it is easy to miss the subject cards for the books you need and to find, instead, a collection of unrelated, out-of-date titles.

As a rule, it is best to use the catalog only to discover whether your library owns a book and to find its classification numbers to locate it on the shelves. Once you do find a good book on your subject, however, you can use any of its catalog cards to identify the subject classifications under which you will find other studies of the same topic. The subject classifications are assigned arabic numbers and are listed toward the bottom of the card. Remember, though, that in most technical fields, articles and interviews will provide you with more current information.

Reference Books

The heart of any library is the reference collection, which offers you access to most of the available information about your hypothesis or topic, including information not in your own library's collection. There are several levels of reference books, including guides to other references, guides to specific fields, and indexes of periodicals and abstracts within a single field. This section will list only a few guides from each category to give you an idea of what is available in most fields.

Encyclopedias and Field Encyclopedias

If you need general background information on a new subject, a general or field encyclopedia can be useful. The best general encyclopedia is the *Encyclopaedia Britannica*, followed closely by the *Encyclopedia Americana*, which is less sophisticated but stronger on American subjects. Encyclopedias are massive editorial projects, and they are always dated and often include inaccurate information. In many fields, you will find specialized encyclopedias, such as *Kempe's Engineering Year Book* or *The Encyclopedia of Chemistry.* These can be found by consulting your reference librarian or the guides listed below.

Guides to Reference Books

There are two basic, selective guides to reference books:

Malinowsky, Harold R., et al. *Science and Engineering Literature: A Guide to Current Reference Sources.* 3d ed. Littleton, Colo.: Libraries Unlimited, 1980.

Sheehy, Gordon. *Guide to Reference Books.* 9th ed. & supp. Chicago: American Library Association, 1976, 1980.

These works will provide you with a two- or three-page overview of the encyclopedias, handbooks, and other specialized references in your field. There are also many less-comprehensive books of this type, such as:

Johnson, H. Webster, et al. *How to Use the Business Library With Sources of Information.* Cincinnati, Ohio: South-Western Publishing Co., 1984.

Indexes to Periodicals

Periodical indexes provide access to journal articles. There are dozens of specialized and general indexes in print, including *The Engineering Index*, the *General Science Index*, the *Business Periodicals Index*, the *Marketing Information Guide*, and the *Applied Science and Technology Index*. Kujoth's *Subject Guide to Periodical Indexes* lists most of them. While general indexes of this type are highly useful places for starting an information search, they have three limitations that are worth keeping in mind:

1. Although each index keeps track of the articles published in a specific list of journals, which can range from a few hundred to a few thousand titles, there are literally tens of thousands of journals in print.
2. Most indexes list only periodicals (journals or magazines), not government documents or books.
3. Each index has its own system of organization, and some are notorious for their complexity.

What do these limitations mean to you? Generally speaking, you can't expect to use only one reference source to find books and articles, although a few indexes, like the *Pandex Current Index of Scientific and Technical Literature*, lists periodical articles, books, and documents. Moreover, new journals or new fields of study are likely to be underrepresented for a time. With an understanding of these limitations you can start your exploration of indexes by getting to know just one: the main one in your field.

If you have not used a major index or abstract before, you should start by spending about twenty minutes studying the introduction and organization of the principal index in your field. Engineering students should have a look at *Engineering Index*, premed students should scan the *Index Medicus*, and so on. Ask your librarian to show you a major index in your field. If you can learn to use one index well and understand its limitations and eccentricities, you can quickly learn to use others. On the other hand, if you attempt to scan half a dozen indexes quickly, you might never learn to use any of them properly. The chapter exercises include some questions that will help you get to know one index in your own field of study.

A word of warning: In the past you have probably used the *Reader's Guide to Periodical Literature*. *Reader's Guide* is a fine general index to popular, nontechnical periodicals, but it is of very limited use in the development of technical reports.

Professionals sometimes avoid using these indexes and at the same time stay reasonably current in their fields by using the volume of *Current Contents* for their field. Each *Current Contents* prints the tables of contents pages from the major journals in a single field. There are many editions of *Current Contents* covering everything from the arts to medicine. Libraries that already subscribe to the individual journals are unlikely to buy the same tables of contents again, but corporations and research centers often buy *Current Contents* to provide employees with a quick and convenient overview of recent publications.

Abstracts

An abstracting index is an index to publications that includes a 60- to 200-word description of each article's contents. By using a volume of abstracts, you can find out more about an article than its title, journal, date, authorship, and page numbers: the information you get from an index. Abstracts can save you the trouble of looking up articles that are actually far from your topic.

Over 1000 abstracting services are in print, including specialized abstracts in engineering, applied mechanics, aeronautical engineering, chemistry, biology, economics, environmental studies, agriculture, and medicine. As you will see later, both indexes and abstracts are also widely available on line through computer data systems. Consult your reference librarian about the availability of appropriate titles in your field.

Government Documents

The U.S. Government Printing Office and the Department of Commerce's National Technical Information Service publish an extraordinary number of documents monthly. These documents include unclassified reports generated by government agencies, as well as many reports authored by individuals and private agencies working under government contracts. Add to these the publishing efforts of other federal agencies, state agencies, local jurisdictions, and regional governments, and you have an enormous volume of technical, economic, social, and political information that is generally not indexed in periodical indexes or abstracts. Beginners who need to gain access to this material often find sufficient title, author, and series information listed in the bibliographies of articles and books. Specially trained government document librarians can also help you find the report you want if your campus or city library maintains a collection of government reports. Four reference books offer a good start to your own researches in this area:

Andriot, John L., ed. *Guide to U.S. Government Publications.* 2 Vols. McLean, VA: Documents Index, 1982.

Schmeckebier, Laurence F., and Roy B. Eastin. *Government Publications and Their Use.* 2d ed. Washington, D.C.: Brookings Institution, 1969.

Monthly Catalogue of U.S. Government Publications. Washington, D.C.: Government Printing Office.

Monthly Checklist of State Publications. Washington, D.C.: Government Printing Office.

Chapter 4 will offer some advice on more rapid methods for finding government documents and other hard-to-locate information.

You can't always escape from indexes and abstracts, but there are easier and more organized methods for finding information that can help you focus on a single issue or debate within a subject area. We call these methods patterned approaches to information gathering.

PATTERNED APPROACHES TO FINDING INFORMATION

If you are like many students, at least once in your educational career you have compiled a bibliography of ten or twenty articles only to find that they addressed completely different aspects of your research question. Faced with such a disheartening result, many writers cut and paste together a paper that forms no coherent whole. The underlying difficulty is that for every highly technical article published, there are only a few readers who are trying to answer precisely the same question. It follows that most of the time you will locate sources that address questions tangential to your own interests. Chapter 4 explains how you can get experts to address *your* questions rather than their own.

Sometimes, however, you need to write a research paper on some topic of interest other than the one that interests you most. In those cases, a patterned approach to research will get you rapid results. Patterned library research proves particularly useful when you want to gather a list of books and articles that refer to each other and are part of a single controversy or a single exploration of new scientific territory. This section of the chapter introduces three such research strategies: one based on *Bibliographic Index, Public Affairs Information Service,* and book review collections; one based on citation indexes; and one based on annual reviews. In each case, the strategy is to use a few reference books in a search that is broad, deep, and relatively rapid.

The Bibliographic Index/*Book Review Method*

The *Bibliographic Index*/book review method combines a general index with a book review guide. The index will often lead you to a single source bibliography, while the book review allows you to check whether your single source of information is respected. *Bibliographic Index* differs from other indexes in that it lists both books and articles—on the condition that those books and articles include bibliographies—yet it is as easy to use as the *Reader's Guide to Periodical Literature.* If you find a listing for your subject in a recent volume of *Bibliographic Index*, you are guaranteed to find a bibliography that will probably be extensive enough for most term projects. It is always dangerous, though, to rely heavily on one source of information for a paper. If that source is not trustworthy, or if it represents only one side of a controversy, its bibliography might also be flawed or biased. Should you locate the title of a recent book on your general subject, you can check its quality by looking it up in one of the reference books that list or excerpt book reviews. These publications include:

Technical Book Review Index
Book Review Index
Book Review Digest. (This includes excerpts from reviews.)

To check the value of journal articles, you should use either a citation index, discussed below, or scan major indexes to see whether the journal itself is frequently indexed or abstracted.

While this two-step method will not generally permit you to work on the most up-to-date topics, it can provide a fairly well focused and reliable starting bibliography in a limited amount of time. The *Public Affairs Information Service Bulletin (PAIS)* is another good place to start a general search. Unlike *Bibliographic Index, PAIS* will not always lead you to a source that includes a bibliography, yet *PAIS* does index materials in many fields, including some government documents, and it is relatively easy to use.

The Citation Index Method

The citation index method focuses on the names and careers of researchers rather than on your topic. The object is to identify, in one step, both the publications of a researcher and the response of the scientific community to those publications. Briefly, a citation index lists references that other researchers make to a particular article by one writer. The index is organized alphabetically by the names of authors, and each annual volume lists all the citations that have been made in the preceding year to every author listed.

Consider this situation. While watching *NOVA* on television, you see a fascinating program about the death of the dinosaurs, an asteroid colliding with the earth, and the creation of Iceland. A scientist at Berkeley has connected all of these events by doing sophisticated research that involves measuring the age of a rare ion. You would like to write a term paper that explores some aspect of this controversial theory, but where do you start? *Geological Abstracts? Geophysical Abstracts? Astronomy and Astrophysics? Nuclear Science Abstracts?* A general science index, or any one of half a dozen other indexes? Fortunately you remember the last name of the principal investigator, Alvarez, who was interviewed on the program. Using the researcher's last name alone, you can find many articles on this hypothesis by turning to the *Science Citation Index.*

In this case you start your search without the researcher's first name, so you need to scan the journal titles under each Alvarez until you find something that looks right. In the 1981 *Science Citation Index,* under L. W. Alvarez, you find a 1980 article in *Science* followed by twenty-one citations to that article, all dated 1981 (the year of the volume you tried first). The twenty-one responses to Alvarez's article appear in journals in the fields of biology, paleobiology, astronomy, nuclear science, medicine, fluid mechanics, and general science. Under Walter Alvarez there are also several citations to 1980 articles appearing in geology journals. As it turns out, Luis W. Alvarez and Walter

Alvarez are two of the four authors of the major 1980 *Science* article, "Extraterrestrial Cause for the Cretaceous-Tertiary Extinction."

In *one step* the citation index has provided a bibliography of well over 40 recent articles, including half a dozen by the two Alvarezes and over 30 responses by other scholars. More important than the sheer numbers, all of these articles are about the *same* controversy. What a wealth of interconnected material for a research report!

To bring your search up to date, you might check the most recent issues of the *Science Citation Index*, as well as indexes in biology, paleontology, nuclear science, or astronomy. Once you have located a few of the titles you will be in a better position to judge whether the controversy is being waged primarily by astrophysicists or by paleontologists, which would help you decide which additional index you might want to consult. Citation indexes are also available in some nontechnical fields. Unfortunately, small libraries sometimes do not subscribe to citation indexes, because they are among the more expensive references on the market.

The Review Article Method

A third patterned approach focuses on review articles that summarize recent research in a field. Annual reviews, which typically have titles like *Annual Review of Anthropology* or *Recent Advances in Photochemistry*, contain article length assessments of recent research and publications in a discipline. Ulrich's *Irregular Serials and Annuals* lists these review volumes. Journals also publish review articles. Some of these, like the annual reviews, assess a year's progress within a specialty, while others may review past and current work on a single research problem. While these essays can be gold mines, they do have drawbacks. They can be too technical for easy understanding, and they can place an undue emphasis on the author's own opinions. If you feel a review article would help, the next section explains how a computerized data base search can be organized to select one.

Computerized Information Searches

There are two ways to use computers to search library collections: data base searches and library system searches. Data base searches provide access to the material in indexes and abstracts. Library system searches allow you to identify what books are available in a single library or a group of libraries that have joined a national or international catalog system. Most indexes and abstracts and many large library collections can now be searched through computer systems. This section and the following section introduce the principal U.S. data bases and explain how to use them. We also list the main library networks and discuss some of the strengths, weaknesses, and costs of using automated information retrieval systems.

Three "master" data base systems offer direct access to a huge range of computerized periodical indexes and abstracts. Your college or city library probably subscribes to one or more of these systems. If you have your own computer and modem, you might already have access to one of them yourself. Lockheed's DIALOG system (Lockheed Information Systems, 3460 Hillview Avenue, Palo Alto, CA 94304) offers 110 data bases (at the time of this writing). BRS (Bibliographic Retrieval Service, Corporation Park, Building 702, Scotia, NY 12302) is similar in scope and size to DIALOG. Systems Development Corporation of Santa Monica (2500 Colorado Avenue, Santa Monica, CA 90406) operates ORBIT, a system that specializes in business and science data bases. Each of the data bases within these systems is the computerized equivalent of one or more indexes. For example, COMPENDEX is the data base form of *Engineering Index*, while ELCOM corresponds to two indexes, *Electronics and Communications Abstracts* and *Computer and Information System Abstracts*. Many libraries also have direct access to systems such as the National Technical Information Service's data base and other government and private information networks. This field is developing faster than references can be printed. Two general lists of available services are:

Directory of Online Databases, with quarterly updates (Cuadra Associates, Santa Monica, CA)

The Library and Information Manager's Guide to Online Services (Knowledge Industry Publications, White Plains, NY)

Using the Systems

You can arrange two types of access to these systems: on-line and off-line. In either case, unless you personally subscribe to the systems in your own home or office, the keyboard work will be done by a specialist who has been schooled by the company that markets the system.

On-line searches are carried out through direct telephone access to the data bank. You pay a fee for each minute of time, as well as a set fee for each full citation or abstract that is printed out on your local terminal. Off-line search requests are telexed or mailed to a central location for processing, and the finished search is then mailed back to the library. On-line data bases often allow for off-line access at a lower cost. Typically, off-line searches are performed late at night when the data base is not being used by many on-line operators. On-line searches can sometimes be done immediately (although an appointment may be necessary) particularly if you only need basic bibliographic information: authors, titles, journal names, dates, pages. Printing abstracts takes up to a week. In either case, you will be asked to fill out a search request form, which helps you describe your needs, topic, or idea in considerable detail. The search operator may ask you questions that will help in the selection of an appropriate data base and search vocabulary. All of this information is then used to generate only those entries that will be of interest to you.

Computer searches are now interactive. As the operator reads the responses that are generated by each search command, he or she can modify the search until it produces only articles that will be relevant to your topic. This way, you don't pay for citations until it is clear that the system will generate citations that will be useful to you.

An example of an interactive search performed on MEDLINE (the data base equivalent of *Index Medicus, International Nursing Index,* and *Index to Dental Literature*) is printed in Figure 3-1. This search concerns articles about a specific complication of cesarean section. After generating a single entry that was of no interest to the client, the programmer asked the data base to report the number of citations it held for various combinations of *cesarean section, infection, postoperative complications,* and *incision.* At one point the computer reported a staggering 18,139 entries for the general subject postoperative complications. As the search draws to a close, note how the operator progressively limits the field by adding new constraints or definitions: *human, English* (language), and finally *review* (for review articles only). When the system responds that the field is reduced to just six articles, the operator finally requests titles (TI) and sources (SO), which correspond to the journal name, date, pages, and so on.

At its best, computerized information retrieval can provide you with extremely rapid access to a few relevant articles and review articles, saving you hours of hand searching through indexes and abstracts. At their worst, computerized data searches can leave you with pounds of useless computer paper. The best uses of these systems are discussed later in this chapter.

Figure 3-1
Example of a MEDLINE search

PLEASE ENTER/LOGIN

PLEASE ENTER USER ID/PASSWORD OR LOGON

NLM TIME 19:33 DATE 82:011 LINE 136

HELLO FROM ELHILL AT NLM [National Library of Medicine]

USER:

CESAREAN SECTION [programmer's second field entry]

PROG:

SS(2) PSTG (673) [field 2, 673 articles on cesarean section]

USER:

POSTOPERATIVE COMPLICATIONS OR ALL INFECTION

PROG:

SS(3) PSTG (18139)

USER:

2 AND 3 [Programmer instructs the computer to find the union of the previous two fields, which have been assigned subject numbers SS2 and SS3]

PROG:

SS(4) PSTG (98)

USER:

HUMAN

PROG:

SS(6) [field 6 because a few steps have been cut out] **PSTG (97)**

USER:

(LA) ENG [language: English]

PROG:

SS(7) PSTG (63)

USER:

SS(7) AND REVIEW

PROG:

SS(8) PSTG (6)

USER: PRT TI, SO [Programmer instructs system to print bibliographic data for the six review articles.]

Computerized Library Networks

Limited access to book titles is provided by computerized library networks, which are data bases that contain the contents of member libraries' card catalogs. The Western Library Network (previously the Washington Library Network), based in Olympia, Washington, offers subject searches of books acquired by the Library of Congress since 1968, as well as access to the collections of member libraries. Western Library Network is the most technically sophisticated of the currently operating library networks, but it is not as widely available as the Ohio College Library Center system (OCLC), based in Columbus, Ohio. While book subject searches are possible on these systems, their primary function is to allow member libraries to search each other's collections to speed interlibrary loans and cataloging procedures. These systems are changing and growing rapidly. (You can also obtain information about books in other libraries by hand searching the *National Union Catalogue* and other reference works.)

Strengths and Limitations of Computerized Information Retrieval

Data-base searches are highly useful when (1) you want to coordinate two sets of subjects, (2) the subject is so new that it might not appear as a subject heading in the printed indexes, (3) the topic is so narrow that very little may have been published about it, or (4) a subject appears under so many subject headings that manual searching could take forever.

One significant limitation of data-base searching is that it does not generally offer access to information more than ten years old. As time passes, however, more indexes and older information are being entered into the computer files. Cost is another limitation. Searches cost from $1.50 to $3.50 for each minute and 10¢ to 30¢ for each citation printed. Many data bases are so organized that you have access to increments of two or three years at a time, so that additional costs will be involved in a search that goes back five or six years. On the average, a complete search will cost from $5.00 to $40.00, although some will cost more and, at times, free searches and special student rates are available.

While it is exciting to contemplate the idea of almost instant access to millions of pages of citations, in the end automated bibliographic searching is simply a sophisticated way to get somewhat dated print information. If you want to answer *your* question or to find out what will be happening in your field next year, not last year, indexes (printed or computerized) are only a partial solution to your need. Using these sources in your library will give you several sources of information; other sources of material will be covered in Chapter 4. Before you use what you find, however, you should systematically evaluate the data you have gathered.

EVALUATING INFORMATION

Before you use your sources in a report or presentation, it is important to assess them for coherence, currency, and quality.

Coherence

You have two criteria for assessing the coherence of sources.

1. Are they part of the same debate or discussion? and
2. Do they contain the same level of detail?

For example, if you have found five reports on the general economics of electric automobiles in the U.S. market and one report on the economic feasibility of long-range batteries, you might drop the article on batteries because one article will not give you enough coverage of that subtopic. Similarly, if you have collected several general economic analyses of marketing non-traditional vehicles in the American economy, then you must be sensitive to the differences in level of detail in your sources. Your final report should be balanced in all respects.

Currency

You should always assess the currency of your sources. Older sources should be discarded unless there is a good reason to use them. When you do use older sources, you should carefully identify their age and specify their current value or lasting importance.

Quality

Finally, assess the general quality of your sources. Do they come from within the field? Or are some of them drawn from outside your field—from newsmagazines, for example? Knowing the issues in a field, as well as the names of key individuals and groups, will help you avoid sources that are not part of a discipline at all: sources that are not consulted by researchers within a field and consequently are not subjected to expert review and criticism. Earlier in this chapter, we warned against using the *Reader's Guide to Periodical Literature.* After you have considered the types of information channels that are typically used in a technical field, it will be easy to understand why it would not normally be appropriate to cite an article in *Reader's Digest, Newsweek,* or *Popular Mechanics* side by side with articles from a corporation's archives, an academic research center, or a government agency. The *Reader's Digest* and *Newsweek* articles might be factually accurate, but they would not be written at the same level of precision. At best, they would be someone else's simplification of material that you should read and interpret first hand. All the citations in the *Science Citation Index* to Luis Alvarez's article on the extinction of the dinosaurs come from highly technical scientific journals. None of the citations is from the popular press.

Some sources outside a discipline may be highly useful and relevant, but you should be able to specify why a particular source is relevant. From your study of a field, you should begin to notice which journals and scholars are frequently cited, which organizations are trustworthy, which research centers are well known for what. You should also be aware of which sources are mentioned in reviewing media such as technical book review indexes, review articles, or citation indexes. Chapter 4 discusses the structure of fields of study in detail.

SUMMARY

The information in Table 3-1 is a handy summary of the uses and limitations of the sources discussed in this chapter.

EXERCISES

1. Pick a subject you know something about. Look up that topic in three recent general or technical encyclopedias. Write a brief report comparing the three sources in which you discuss the differences in tone and level of coverage and any differences in facts or interpretation of facts. Do you think the information is up to date?

2. Consult a recent issue of *Scientific American, High Technology, Byte,* or any other technical journal that includes bibliographies of sources or further reading with major articles. Pick an article of interest to you. Try to find something current on that subject by using your college library's card catalog. When you have spent fifteen minutes with the catalog, turn to the bibliography in the periodical. In the catalog did you find any of the sources or readings listed in the magazine? Write a long paragraph describing and analyzing what you found or failed to find and what you learned about the uses of the catalog.

3. After several quarters of basic courses in your major, you may wonder when you will begin to learn about what is happening today in your field. Choose a major journal in your field and write a summary of one of the articles to be given to your classmates for analysis and discussion. Because your classmates come from many different majors, you must consider them a general audience. In a class discussion of your summary, explain the effect of any simplification or interpretation or other changes you were required to make to fit

TABLE 3-1 ***Uses and Limitations of Basic Reference Tools***

Source	*Uses/Strengths*	*Limitations*
Encyclopedias	Overview information	Dated, not detailed, often inaccurate
Card catalogs	Location of books	Dated
Guides to reference books	Overview of basic references	Infrequent updating
Indexes	Periodical articles	Often do not list documents, weak on new subjects, limited to specific list of journals
Abstracts	Summaries of articles	Same as for indexes
Bibliographic Index	Books and articles that include bibliographies	Limited depth of coverage
Public Affairs Information Service Bulletin	Documents, books, and articles	Limited subjects
Citation indexes	Access by names to groups of related articles	Limited availability
Annual reviews	Overview of recent work in a field	Limited topics, highly technical
Computerized searches	Quick access to review articles, new topics, topics that combine disciplines	High cost, sometimes dated, unselective
Library networks	Access to rare books subject search ability	Limited public access, complexity

the summary to the understanding of a general audience. For example, was it necessary to delete equations? Simplify examples? Leave out discussion of complications or theoretical questions? Generalize more than would be appropriate if you were summarizing for a professor in your major?

4. Do Exercise 3 for an annual review essay or a review article in your major.

5. With the reference librarian's help, choose a major index or abstract in your field of study. Spend twenty minutes to half an hour reading the introductory material in the index and familiarizing yourself with the system of organization. Write a brief report in which you describe the coverage, limitations, organization, and publication frequency of the reference. Specifically, how many journals does it index? What are its subject limitations?

6. Select the name of a living scientist, business leader, inventor, or some other person you have heard about in your classes. Look up this person in an appropriate citation index and locate one major article by that author and two articles that cite the article. Write a

short report summarizing the original article and the ways in which it was cited or reviewed or responded to by the other two.

7. Write a short report on the computerized search services available on your campus. Does the campus library tie into DIALOG, BRS, WLN, OCLC? Is the library's catalog on-line? Are any special services available to students?

8. If you have access to bulletin board services and/or data bases, write a brief report on a bulletin board or service that is useful in your own field.

CHAPTER • 4

Beyond the Library

NETWORKING AND INTERVIEWING

INTRODUCTION

This chapter explains how you can use three questions to find the information you need: Who is likely to know about this? Where is this information likely to be found? and Who would pay for this information? These questions place an emphasis on people for it is always people who have insights; develop ideas; and write reports, books, and articles. Yet research is not always carried out by investigators who work alone. Throughout the world, research and reporting activities are performed by groups of people organized into institutions, centers, academic departments, public interest organizations, corporations, think tanks, and so on. As a result one way to find information about a topic is to learn the names of the organizations that work in a field. In many cases you can contact interested organizations directly and interview experts or receive copies of publications that are not widely distributed and that you would be unlikely to find in a college, municipal, or corporate library.

The main purpose of this chapter is to give you an overview of the information structure of a typical research field. The word *field* can be taken broadly or narrowly; it could cover something as broad as environmental studies or as (relatively) narrow as research in electric vehicles or enhanced oil recovery. Once you know about the range of organizations in the United States, you are in a far better position to gather and evaluate information.

Knowing the Facts Behind the Facts

Suppose you are writing a paper on nuclear power plants and you find a single source that provides an impressive array of facts about the outputs, cost-efficiency, safety records, and accident probabilities for all the nuclear power plants operating in North America. While you make use of several other sources in your paper, you constantly quote this one report as you write your essay and develop your own conclusions about the future of nuclear power. Instead of the praise you think you deserve, your professor writes you a barbed note about using "one-sided" sources. Later you find out that the report you relied on was funded by a lobbying organization and written by a well-known nuclear industry advocate.

The moral of this story is that it is not enough to have the facts, you need to know where the facts come from and who is interpreting them. In order to do this, you need to know about the issues in a field and the sides that are being taken. Knowing about the structure of debate within a field—who's who and who is doing what—will help you interpret and assess the information you collect.

WHERE DOES INFORMATION COME FROM?

Where *does* information come from? Many types of organizations produce information and release it in a variety of forms. For each type of organization and publication, there are appropriate methods for gaining access. Your objective in reading this section should be to expand your awareness of the extraordinary variety of information sources around you, although you probably know this lesson already, at least subconsciously. Faced with a difficult information-gathering task, most people turn instinctively to associates, teachers, or fellow students for help. Like many instincts, the impulse is correct but could use some training. After discussing sources of information in a general way, this section concludes with a presentation on information sources in electrical vehicle research—an example that shows how these approaches to information gathering work.

Before you consider who develops and publishes information, however, you should consider another important issue: funding sources. Funding sources are important because the people who pay for information usually have access, or know how to get access to it. If you can answer the question: Who is likely to pay for this information to exist? you have a key to the other two questions: Who is likely to know about this? and Where is this information likely to be found?

Funding: Who Would Pay for This Information?

If you know who paid for a particular piece of research—or what companies or agencies would be *likely* to pay for information in a given field—you have gone a long way toward getting the facts you seek. If the private sector has invested in a project, the investors or owners have two strong reasons to make information available to you: good will and later direct business with you or your employers. Foundations often publish the results of the research they commission, fund, or perform. Government agencies also often publish the results of funded research, and if the research is not classified you can generally get copies of the reports and findings. Because both foundations and government agencies depend on public support and interest they are often eager to help you; after all, you are part of their constituency.

In practice, it is relatively easy to use the question: Who would pay for this? to begin to look for information. If you are interested in the use of antibiotics in poultry farming, you can start with the question: Who would pay to know about the value, use, and dangers of antibiotics in poultry farming? Without a lot of thought, you can probably come up with several answers: government food inspection agencies, agricultural agencies, state extensions, manufacturers of antibiotics, agricultural colleges, and farmers' associations. From that point, you need about an hour's work in the library to come up with specific names and phone numbers, perhaps a few publications, and even some sense of the positions each of the organizations might take on the issue of feeding antibiotics to poultry.

Organizations: Who Knows and Where Are They?

This classification of organizations is presented to help you identify likely sources of data and to provide an overview of the complex political structure of fields of study and enterprise. Under "Methods of Access" later in this chapter you will learn how to find the names of individual experts based on what you know about organizations. Figure 4-1 is a list of organizational sources of information and their typical means of publication.

Inventors/Small Companies

You can find out the names of the small firms active in a field by consulting detailed industry directories, which are discussed on page 78. The research findings of these firms are occasionally noted in technical or financial journals and news reports. Otherwise, you must consult patents or rely on direct access to a firm's internal publications and reports.

FIGURE 4-1
Organizational sources of information

1. Inventors, Small Firms
    - Internal reports
    - News reports
    - Patents
2. Corporations
    - Annual reports
    - Internal reports
    - Patents
    - Proposals
    - Publications for sale
3. Small Publishers
    - Catalogs
    - Reference books
    - Specialty newsletters
4. Large Publishers
    - Books and journals
    - Internal publications
    - Trade lists
5. Lobbies, Professional Associations, Churches, Charities
    - Convention proceedings
    - Membership publications
    - Position papers
    - Trade journals
6. Foundations
    - Books and journals
    - Publication catalogs
    - Information offices
7. Think Tanks
    - Articles and books
    - Commissioned reports
    - Internal reports
8. University Departments
    - Articles and books
    - Funded research reports
    - Internal reports
9. University Research Centers
    - Commissioned reports
    - Internal reports
    - Journals
    - Monographs
10. Government: International, Multinational, Federal, State, Local, and Regional, Extra- and Quasi-Governmental Agencies
    - Reports
    - Speeches
    - Information offices
    - Libraries
    - Publications offices

Corporations

Larger corporations often maintain publication and public relations offices that make some internal reports available on request. Some companies even send out publications lists of hundreds of free titles. From their perspective, a report that was highly valuable, even secret, a few years ago now has public relations value. Companies that offer you reports show the quality of their work and the pride they take in their accomplishments.

Small Publishers

Specialized publishers produce reference books, catalogs, and newsletters in many fields. Operating out of the National Press Building in Washington, D.C., for example, are several information specialists that compile newsletters on state and federal developments in water treatment, energy research, transportation, and so on. One of these publishers alone prints 40 different newsletters each month. Because these newsletters are fairly expensive and are aimed at very limited audiences, general public and university libraries often do not subscribe to them. Knowing the field is the only way to discover these sources and locate special libraries, corporate offices, and research centers that may subscribe. "Methods of Access" includes a list of several publishers of reference books on this scale.

Large Publishers

Journals in scientific and technical fields are often published by major publishing houses (such as Sage, Elsevier, or Lippincott), which maintain editorial offices that specialize in single disciplines or groups of disciplines. Knowing these facts about publishing arrangements in your field can help you trace copies of books and journals and gather information about forthcoming books and journal articles.

Lobbies and Professional Associations

Lobbies and professional associations are major publishers and sources of information. Some professional associations, like the major engineering societies, maintain joint publications and information centers, print four or five journals, and maintain lists of hundreds of books, pamphlets, reports, convention proceedings, position papers, studies, films, cassettes, and so on. By contacting these organizations directly, you can often gain direct access to specialized materials that libraries might not acquire. Even if your library does maintain a standing order for all the publications from a major professional association, the easiest way for you to get an overview of what has been published may be to get a catalog of publications from the association itself.

Foundations

Like professional associations foundations often maintain information offices and publishing offices that produce both books and reports. These foundations include The Brookings Institution, the Ford Foundation, and other institutes that draw their staffs from universities, government agencies, and corporations. By this point, you should begin to appreciate the extraordinary diversity, and at the same time the striking interconnectedness of research efforts in America, where a researcher can work on the same problem in a university, an international foundation, for a government agency, a professional association, or a corporation.

Think Tanks

By *think tanks* we mean corporations that exist to do contract research, sometimes in a single area, sometimes in several disciplines. A beginning list of think tanks would include MITRE (a "private" research organization that was founded to do government research), the Rand Corporation, Battelle Institute, the Research Triangle Institute, and SRI (formerly Stanford Research Institute). Here the lines separating universities, government, and private corporations often blur. These centers generate books and reports, some of which are available either free or for a fee. Some of the nonclassified work these research centers produce for government agencies is also available through the publications offices of the sponsoring government bodies.

University Research Centers and Departments

While independent university researchers generally publish their work through journals and academic book publishers, some university centers publish their own journals or report series. Research center directories are discussed in "Methods of Access" later in this chapter.

Government Agencies

The main point to keep in mind about government information is the complexity of intergovernmental relations in this country, where executive, legislative, and judicial agencies often maintain information and research staffs that are working on the same issues at the same time. Federal, state, local, and regional government manuals and legislative directories provide guides to the basic structure of government but often leave out regional offices of federal agencies and/or fail to list the important extragovernmental associations that facilitate communication on a national and international basis, such as the Council of State Governments, the Conference of State Legislatures, and other associations of mayors, attorneys general, and other officials. Fortunately municipal and state libraries can help you find information in these areas. Because municipal government librarians spend much of their time working with government agencies on the local and state levels, you should consider using them in addition to your college or university library reference staff. The federal government also maintains two basic publishing agencies: the Government Printing Office and the National Technical Information Service.

Special Libraries

By now it should be clear that no single library can maintain a collection of all the references you might conceivably need at some time in your student or professional life. In addition to all the general city, state, and college libraries, there are many specialized libraries, some on university campuses, others privately maintained. In some fields, there are published directories of the special libraries. There are also many regional branches of the Special Libraries Association, as well as associations for medical librarians, law librarians, and so forth. Before this array of sources becomes too bewildering, consider an example.

PUTTING IT ALL TOGETHER

Our point is simple. In almost every field you can find examples of every one of these organizational sources of information. Professional market researchers use these organizations and their publications daily. A library will help you to find books and journal articles, but knowing the structure of your field or topic will help you get interviews with experts, copies of reports and publications that general libraries normally do not take, and insights into the future directions of the field. The chart in Figure 4-2 shows a beginning sketch of the informational structure of American research in electric vehicles. All of the information in this chart was drawn from a single directory, *Urban Mass Transit: A Guide to Organizations and Information Resources* (1979). As you can see, this chart is similar to Figure 4-1, with the names of organizations and their publications filled in. But how can you gather this much information about a topic area? The methods of access are less complicated than you may think.

This is not a comprehensive overview of the structure of electrical vehicle research in the United States, but if you can learn to view the production of information in this way, you will be able to develop the ability to tap an extraordinary range of information sources.

FIGURE 4-2
Informational structure of electric vehicle research in the United States

1. Inventors, Small Firms
   - Lyman Metal Products
   - Globe Union
   - Jet Industries
   - McFarland Design
   - Electric Auto Corporation
   - Electric Fuel Propulsion Corp.
   - Electric Passenger Cars, Inc.
   - U.S. Electricar
2. Corporations
   - General Motors
3. Small Publishers
   - Electric Vehicle News
   - Guideway
   - Energy Policy Newsletter
4. Large Publishers
   - Mass Transit
   - Metro
   - Transportation Engineering
   - (many others)
5. Lobbies, Professional Associations
   - Electric Vehicle Council
   - Society of Automotive Engineers
   - Modern Transit Society
   - American Public Transit Association
   - Motor Vehicle Manufacturers of America
6. Foundations
   - Public Technology, Inc.
7. Research Consultant
   - R. W. Bourke, Associates
8. University Research Centers
   - Purdue University Institute for Engineering Studies
   - University of South Florida
   - SUNY—Stonybrook, Department of Technology and Society
9. Government Agencies
   - LOCAL
   - Municipality of Metropolitan Seattle
   - STATE
   - Florida Department of Transportation
   - FEDERAL
   - U.S. Department of Transportation
   - Executive Office of Science and Technology
   - Department of Energy
   - Transportation Research Council
   - U.S. Congress, Office of Technology Assessment

REGIONAL
National League of Cities
10. Data Bases
TRISNET, HRIS
11. Indexes
Current Literature in Traffic and Transportation (Northwestern University)
12. Special Libraries
Directory of Transportation Libraries and Information Centers
Directory of Transportation Libraries in the U.S. and Canada
13. Directories of Associations
American Public Transit Association membership directory
Electric Vehicle News Annual Index of World Manufacturers
Metro Annual Directory
Motor Vehicle Manufacturer's Association Directory
14. Guides to Research in Progress
Innovations in Public Transportation (U.S. Department of Transportation, annual); *Directory of the Transportation Research Board of the National Research Council*

METHODS OF ACCESS

Before you can gain access to the information produced and published by organizations you need to know their names, so that you can telephone or write letters of inquiry and obtain interviews with experts to gather the information you need. Figure 4-3 lists methods and sources for gathering these names. Most organizations publish membership lists of individual, corporate, or institutional members; these lists are gold mines, but you often must borrow them from a member or a member institution's library or information office. In some fields there are directories of research in progress that include the names of people and agencies that are working on specific projects. On a larger scale Gale Research and similar companies publish directories of associations and research centers. Available in most college libraries, these volumes can be used to locate associations and university departments that are active in every area of research. The National Technical Information Service maintains archives of federally funded reports and will photocopy reports for a fee. A branch of the Department of Commerce, NTIS will also perform data-base searches in its own report files. The Library of Congress National Referral Service will put you in touch with other researchers who have registered themselves as having an interest in the same area or research question. Lists of special libraries (like the two directories of transportation libraries in Figure 4-2) will help you locate information centers dedicated to your field. Equally useful are directories to conventions, such as the *World Calendar of Forthcoming Meetings* and *World Meetings: U.S. and Canada,* and directories of convention proceedings, such as *Proceedings in Print* and the *Directory of Published Proceedings.*

Industry and Field Directories

All of these sources can be used to help you identify individuals and groups you can contact for assistance. Special industry directories and telephone books are also available in many fields, just as government agencies publish special phone books for the use of state and local government employees. Many states and state Chambers of Commerce publish frequently updated lists of licensed manufacturers and corporations. Equally useful are the field directories put out by publishers and foundations, such as G.K. Hall, Bowker, Marquis Who's Who, and the California Institute of Public Affairs, and by professional associations and industry lobbies. The Motor Vehicle Association of the United States, for example, publishes an extensive guide to businesses in that field, and similar guides are available to many subjects, including the world food crisis, mass transportation, energy, international environmental organizations, nuclear power, women's issues, civil rights, human rights, and the world water shortage. A few guides to organizations are listed in Figure 4-4, solely to indicate the range of available titles. You will find that even small college libraries carry an impressive range of these reference works.

In the end, your most important resource is people; one of the most valuable resources on campus is the faculty. Once you contact a useful and helpful person, interviews and referrals can do much of your research on most topics in most medium to large American cities.

Applying the Methods

You might well ask: How can all of this information help me to find anything? Now that you have an overview of what can be found inside and outside the library, you are prepared to orga-

FIGURE 4-3
Methods and sources for gathering information

1. Patent Office Publications
2. Letters of Inquiry
3. Interviews
   - Telephone
   - In person
4. Surveys
   - Telephone
   - Mail
5. Membership Lists
   - Separately bound
   - Published in journals
6. Directories of Research in Progress
   - Special directories
   - Notes in journals
   - Announcements of funding
7. Directories of Associations
   - Independently published
   - Published by groups within the field
8. Directories of Research Centers
   - Independently compiled
   - Compiled by members of field
9. Bibliographies and Indexes
   - (standard library tools)
10. Publications Catalogs
    - From publishers, foundations, associations
11. Citation Indexes
12. Directories of Persons
    - Telephone books—public
    - Telephone books—government or corporate
    - Special field directories
13. National Technical Information Service
14. National Referral Service, Library of Congress
15. Professional Networks
    - Access through conventions
16. Government Manuals
    - Local, regional, state, federal
17. Industry Directories
    - In journals
    - Separately published
18. Telephone Books
19. Directories of Conventions and Proceedings
20. Indexes of Presentations
    - In-house speeches and presentations at large corporations
21. Data Bases
    - Public: Dialog, Orbit, BRS, etc.
    - Marketing
    - Government: NTIS and others (public access)
    - Library systems: WLN, OCLC
22. Directories of Special Libraries

FIGURE 4-4
Sample list of industry and field directories

A Directory of Information Resources in the U.S.: Physical Sciences and Engineering. Washington, D.C.: Library of Congress National Referral Center, Science and Technology Division, 1971.

Encyclopedia of Governmental Advisory Organizations. Detroit, Mich.: Gale Research Co., 1975.

The Energy Directory. New York: Environmental Information Center, 1974.

Gale Research Centers Directory. Detroit, Mich.: Gale Research Co., periodically updated, supplemented by the Gale *New Research Centers.*

How to Find Information about Companies. Washington, D.C.: Washington Researchers, annual.

The Media Encyclopedia: The Working Press of the Nation. 5 vols. Chicago: National Research Bureau, annual.

Moody Manuals. A series of manuals describing the history, indebtedness, and bond ratings of governments, banks, industrial firms, utilities, transportation agencies. Annual.

National Trade and Professional Associations of the U.S. and Canada. Washington, D.C.: Columbia Books, annual.

The Nuclear Power Issue: A Guide to Who's Doing What in the U.S. and Abroad. Claremont, Calif.: California Institute of Public Affairs, 1981.

Population: An International Directory of Organizations and Information Resources. Claremont, Calif.: California Institute of Public Affairs, 1976.

Thomas' Register of American Manufacturers.

Washington V. Washington, D.C.: Potomac Books, irregular. Includes an annotated list of federal contractors.

World Food Crisis: An International Directory of Organizations and Information Resources. Claremont, Calif.: California Institute of Public Affairs, 1977.

nize your information searches more effectively. Three important points to remember are:

1. Most printed information is at least slightly out of date.
2. Some articles will puzzle you until an expert gives you their context, which might be an ongoing controversy.
3. Only part of the information you want is in any library.

It follows that an effective search will have two parts: a standard library search for printed books and articles (and any other media the library may have, such as films, videotape, microfilm) and a nonstandard search through directories for the names of groups and people who can help you identify the positions that are being taken on controversies and their advocates. Once you locate these groups and individuals, you can use interviews and letters of inquiry to gather information.

People are the final object of any comprehensive search for information.

For example, turn again to the chart of information sources in the electrical vehicles field (Figure 4-2). Suppose you were working on a report on the economic feasibility of electric automobiles. Immediately, you face difficulties in choosing an index. Would articles on the economics of new technologies appear in business journals, engineering journals, or is it possible that the published information on this topic would be restricted to limited distribution reports for investors? It would be tempting at this juncture to turn to *Reader's Guide* and find an article in *Popular Mechanics*, when in fact you should be doing what a *Popular Mechanics* author might do: find an expert in the field and interview him or her.

In this case, a trip to the library will get you to the guide to urban transit organizations. A brief glance at that book (or at Figure 4-2) suggests a short list of groups worth contacting: electric car manufacturers themselves, *Electric Vehicle* magazine, the Electric Vehicle Council, the Motor Vehicle Manufacturers of America, and the Transportation Research Board. A college reference librarian could help you find a copy of one of the directories of transportation libraries and *Current Literature in Traffic and Transportation* and a library that is tied in to TRISNET, the national transportation information data base. All this work might seem difficult; in fact, it is often far more efficient to spend a few dollars on letters and long distance phone calls than to spend time and money searching indexes, ordering data-base searches, and photocopying articles that are not exactly on your subject.

Another fruitful approach can also be used. Faculty members often subscribe to publications that are not taken by the library, and of course faculty who are active in research or consulting activities generally have contacts throughout their disciplines as well as broad general knowledge of their fields. A brief interview with a faculty member in your own school or another nearby college might help you choose the best approach to gathering data. In the case of electric vehicles, you could start by asking if your college engineering department participates in any of the national electric vehicle or high gas mileage races.

Interviews with faculty should be treated like any other interview. You need to have questions ready and be prepared to show how much groundwork you have completed. The next section explains how to prepare interviews and letters of inquiry, but before we introduce those subjects, consider one final example of a full-scale information search.

Several years ago, a researcher posed the following question: Would it be technically feasible to develop a computerized simulator of the human musculoskeletal system that could be used to test experimental surgical procedures? Automated data searches and hand searching of new indexes produced no results, even though it was clear from the new medical machines on the market that researchers were working on related issues. In short, this was a case where the people who were publishing wrote about their interests, not that of the researcher, and yet many people in the field could obviously shed a great deal of light on the problem of surgical simulation. The lead researcher on the surgical project turned to the telephone, and within two weeks he gathered a list of names of experts who could contribute to the solution of the problem, as well as information about the probable cost of development and the probable organizational sources of funding and support. Everyone who was contacted was both interested and helpful. In this case, personal contacts provided information and advice that simply was not available in published form.

INTERVIEWS AND LETTERS OF INQUIRY

Whether you choose to interview an expert or to write a letter of inquiry, the basic principles are the same. Respect your contact's time; show that you have done your homework; ask specific questions; and record the answers. Chapter 13 contains a more detailed discussion of the letter of inquiry; here we focus on the basic principles.

Respecting Your Contact's Time

Make an appointment for an interview or, if you plan to do the interview by phone, ask if you have called at a convenient time. If you write, keep your letter under two pages. Be sure to write a note of thanks for the time your interviewer or correspondent spent helping you.

Showing What You Know

When you identify yourself as a student, you will often be treated as if you know very little about a subject. People will be nice and helpful, but they will tend to give you very basic advice, such as a recommendation to try the library card catalog. To get past this barrier, you should present a brief, coherent, and well-organized review of the work you have already done. Mention specific ideas, articles, references, researchers' names, organizations, and contacts. Then ask for help getting further with your research and information gathering.

Asking Specific Questions

Keep your letter or conversation on track by focusing clearly on your main point. If you want information about the economic feasibility of electric cars, don't ask questions that elicit information about the technical feasibility of new battery designs.

Some people behave like tape recorders when they are interviewed: A chance question will trigger a long tape of personal reminiscences and information about favorite topics. You can maintain a clear focus by carefully limiting the number and type of questions you ask. First, limit the number of questions, and in a letter assign each question a letter or number and a separate line or paragraph. Second, word your questions to get essay answers, not a simple yes or no. It helps to word questions neutrally, so that your respondent does not feel pressed to take sides.

Recording the Answers

Ask permission to use a tape recorder in the interview, or take careful written notes.

Evaluating Organizational Sources of Information

At the end of Chapter 3, we offered three criteria for evaluating print sources of information: coherence, currency, and quality. There are some additional criteria you can use to evaluate both print information and the organizations that publish that information.

Scientists and technicians, like other experts, strive for objectivity, yet there is no perfect perspective from which anyone can observe present events. The information you find is always part of a continuing search for facts and truths that involves a process of testing, debate, case building, and occasionally both overstatement and subjectivity. The advantage of working directly with people is that you can ask *directly* whether you are hearing one side of a case. Most people are very honest about who their technical and theoretical opponents are and where you can find out about their positions.

The following are some questions you should ask yourself as a reader and your contacts during interviews:

1. Who are the key people, organizations, and theories?
2. What are the classic studies on this issue?
3. What are the best journals in the field?
4. What are the key reference works?
5. Where do you think this field is headed in the next five years?
6. What are the controversies in this field?

7. What are the most reliable sources on both sides of the issue, and who is the responsible opposition?

These are tough questions. You can't expect every expert to know the answers to all of them. Most professionals, though, will be able to give you some of the answers, and it is a mark of your own growth as a member of your chosen profession that you can probably already give at least some partial answers.

SUMMARY

1. A professional search for information has four goals. The information must be current; it must be of high quality; it must be produced by researchers in the appropriate fields; and it should be relevant to the question under investigation.
2. To meet these four goals, a full-scale search for information includes a review of published information, contact with experts in the field, and contact with organizations that are also in the appropriate fields.
3. Library research (covered in Chapter 3) can be performed in three ways: by using computerized data bases, by using indexes and abstracts, or by relying on references, such as annual reviews and citation indexes, that provide quick access to highly focused bibliographies.
4. Gathering information from experts and organizations can be done efficiently with the assistance of government directories, directories of associations, and guides to special fields or by contacting local professionals and asking their help in developing contacts and bibliographies.
5. All of your data and sources should be systematically evaluated.
6. Your own mind and experience are a primary source of information. The answer may lie within you, not in a book. People are always the final source of information.

EXERCISES

1. Using a directory of associations or one of the other references cited in this chapter, report on three national organizations in your field of study. Can you identify and name the major organizations in

your field? Are there regional offices or regional chapters of the major organizations? Where is the nearest office?

2. Using any references or methods you choose, identify a local organization in your field. List the name, address, and telephone number of this organization. Also find the name of a local or regional journal in your field. If possible, obtain a copy and bring it to class. List the name, address, telephone number, and frequency of publication.

3. Undertake a major research project, completing the following steps:

   Step 1. Formulate a question. Examine your present situation as a student or professional by brainstorming or doing freewriting to identify information that would be useful to *you* at this time. You might start by formulating a question. For example, should you major in industrial engineering? How could you make use of your work experience last summer? Or you could formulate a working hypothesis by drawing together lessons and facts you have learned in two different classes. Could a statistical method you just learned be applied to your marketing class? How are computer-modeling techniques being applied in your field?

   Step 2. Do primary research: personal. Once you have uncovered a question, reflect on it further, refine your purpose in answering the question, and define an audience that might be interested in your findings.

   Step 3. Develop a hypothesis. Develop a hunch about what the answer to your question might be. Make a list of what you need to learn, what you need to know.

   Step 4. Develop a research strategy. Ask yourself the following questions: Who is likely to know about this? Start by thinking of people you know, including faculty members, other professionals, friends. Compile a list of experts. They can probably help you answer the next two questions. Where does information on this subject come from? What organizations or institutions would focus on this area of research or interest? Consult directories of organizations and ask your contacts for information about who is doing what in the field. Who is likely to pay for this information to be in print? Again, consult directories of government agencies, professional associations, publications, volunteer groups, and lobbies. Save time by talking to experts.

   Step 5. Gather information. Do not neglect to use library sources, but remember that many print sources may not be available in the library for various reasons. As you use the library, rely on review articles, recent publications that include annotated bibliographies, and citation indexes. In addition, be sure to interview people who can provide information, bibliographies, and publications that are not available in libraries, and clues to the debates that are going on in the field. Ask for references to other people and build a network of contacts. Write out questions before you interview, and be prepared to explain your purpose and the work you have done so far.

Step 6. Write a report. Present what you have learned about sources, the availability of information, and the issues current in your field that relate to your hypothesis. Discuss your hypothesis if you unearthed sufficient information in the time your instructor allows.

Research Case 1

THE FROZEN EMBRYOS

To earn money during your summer vacation you go to work for your older brother on his cattle ranch in the western part of the state. Frankly, it's not much of a ranch; your brother has only worked it for the last two years, and his herd is still small. But he has high hopes for the future.

One night at dinner your brother is discussing ways of improving his herd. He would like to introduce better bloodlines, but the stud fees he's been quoted seem much too expensive. Then he tells you he's heard that a process exists whereby average cows can become "surrogate mothers," carrying embryos produced by crosses between superior strains. The embryos are frozen and purchased like frozen bull sperm.

To you the whole thing sounds like something out of *Brave New World,* but the more you think about it the more intrigued you become. After some more discussions with your brother, you agree to research the whole thing for him.

ASSIGNMENT

Find the answers to the questions below as efficiently as you can, but be prepared to defend the accuracy and reliability of your sources. Some of the information will be available in journals; some will need to come from nonconventional sources. Consider what organizations, both public and private, would be interested in this process and how you would find the information they have.

1. How is the frozen embryo transfer carried out? What is the process by which the embryos are obtained, stored, and implanted?
2. Who performs this service? What companies or government agencies are involved?

3. How much would it cost for your brother to use the process? Are there any tax breaks involved?
4. What would his chances of success be?
5. What are the advantages and disadvantages?
6. Would you recommend this process to him?

Research Case 2

PHASE CHANGE MATERIALS

As your senior project in civil engineering, you've been investigating several types of energy-efficient heating and cooling systems. So far you've studied solar, wind, and even geothermal systems. But just recently you've come across something you have never heard of before: phase change materials (PCMs). As nearly as you can determine, this material changes form at a certain temperature, either giving off or absorbing heat in the process. Apparently, this material would be used in wall coverings, ceiling tile, and flooring along with a passive solar system; it could also be used on the solar collectors to store heat in active systems.

You discuss this idea with some of your professors. Some of them are vaguely familiar with the process, but no one seems too definite. You do pick up a few more facts, however. The most frequently mentioned substances are sodium sulfate decahydrate and calcium chloride hexahydrate. Paraffin and polyethylene glycol are also mentioned. Dr. Maria Telkes seems to have done much of the work on sodium sulfate; work has also been done at the University of Delaware and MIT. Dow Chemical Company has done some work with calcium chloride. The Solar Research Institute has worked with paraffin. But you still need hard facts on PCMs and what they do.

ASSIGNMENT

Write an informative report for your civil engineering professor on phase change materials. Consider the following questions as well as others.

1. Do PCMs work and, if so, how?
2. Are they economical?
3. Are they safe?

4. Who is doing the research?
5. What are the prospects for the future?
6. What material seems most promising?

Assume that your reader knows nothing about PCMs, but would be interested in them from a practical standpoint. Again, some material will be available in journals, but some will have to be found in other sources.

P · A · R · T

Preparing Your Solution

CHAPTER • 5

Organization

INTRODUCTION

After you have defined your audience and your purpose and gathered your information, your next step is to put that information into some order. Many beginning writers fall into the trap of organizing all reports chronologically, beginning with the first task they undertook on the project and progressing through each subsequent task in turn. Yet a quick review of both your purpose and audience should convince you that this is seldom the most effective method of ordering your material: your audience is usually most interested in what you accomplished, rather than how you did it.

In this chapter we will discuss the most common types of order in technical writing—orders based upon the needs of your audience. Technical reports contain both summary sections for general audiences and detailed discussions for those directly involved with your project. We will discuss the various sections of your report: the opening, the discussion, and the conclusion. We will also suggest ways to generate information and ways to order that information once you have generated it. Finally we will explain how to convey your organization to your audience, since all your work in ordering your material will be useless unless that order is clear to your readers.

THE THREE PARTS OF TECHNICAL REPORTS

There is a common axiom about order in technical writing: Tell your audience what you're going to tell them, tell them, and then tell them what you told them. This is a simplified outline of basic technical organization. In most cases you begin with a "forecasting beginning," which lets your readers know what you'll be covering in your report. In the middle "discussion" section you present your data, the material on which you're basing your conclusions. Your final summary is a recapitulation of the major points you want your readers to remember about your report.

This three-part sequence may seem repetitious, but this is functional repetition. Much technical writing deals with complex information, so complex that your audience may simply not absorb it in one explanation. Repetition plays a role similar to the redundant systems in many technical designs. If one system fails, another system can take over its functions; similarly, if one statement fails to convey your meaning entirely, a repetition later in the text may do the trick.

Three-part formats respond to the special demands of the various technical audiences. As you learned in Chapter 1, each audience has different needs and different requirements. Some will want all the details of your ideas and the material behind them. For them you'll want to include in-depth explanations, examples drawn from your data, definitions of terms, and graphic support. But other members of your audience, particularly those who are not working on the same project, may want only a general idea of your conclusions and the facts leading to them. This group will be looking for a clear summary without detailed explanations.

You can solve the demands of these audiences by designing your reports so that everyone can find what they want. The opening section will give your general audience an overview of the major points of the report and the conclusions you draw from them. The second part or discussion will be directed toward that part of your audience concerned with specifics. Here you'll give the detailed discussion of your data. The third part or conclusion will present conclusions or recommendations or another summary of your data, depending on the subject of the report.

The Opening

Your opening can include a foreword or preface stating the exact problem or problems your report addresses, as well as the context in which it was prepared (departments and persons involved, the original assignment). The preface can also include the purpose of the report and, perhaps, an explanation of the technical questions dealt with in the report.

The opening should include a summary (sometimes called an informative abstract or executive summary). This is a concise statement of the key points you make in your report. After reading it, the audience should be able to understand how your conclusions were derived from the data you present. Your summary may include any of the following: a statement of method, the results of your research, your conclusions, your recommendations, and a summary of the costs involved. A summary of this chapter could read:

> Chapter 5 discusses technical organization. It describes the three standard sections of technical reports, methods of generating information for the report discussion, the grouping and ordering process, and outlines. Finally, the chapter discusses methods of conveying organizational information to readers through forecasting and transitional material. A detailed illustration is followed throughout.

For more information on summaries and abstracts, see Chapter 7.

Your reader should be able to understand this opening segment without referring to the rest of your report. Many of your readers will use this segment as a substitute for reading the report in its entirety. Consequently there are several things to be avoided in your summary:

- References to graphics or equations in the body of the report
- Terms that are given special definitions in the report
- Any material that is not immediately clear from the summary itself

The Discussion

Like any piece of writing, your discussion should be a whole, with a beginning, a middle, and an end. The beginning (or introduction) and the end (or conclusion) will be relatively short; the middle (or body) section can be much longer, depending on the amount of material to be included.

The Introduction

The introduction to your discussion should give your audience any information needed to understand the rest of the discussion. It should begin with a clear statement of both the subject of your report, the problem to which the report is addressed, and the purpose of the report, what the report is going to do with the subject.

Your introduction should also include what is called a forecasting statement or plan of development. This is simply a state-

FIGURE 5-1
A typical forecasting statement. (Courtesy of Keuffel & Esser Company, Parsippany, New Jersey)

Scope of Manual
This manual is divided into five main sections. Section 1 contains descriptive data for the VECTRON Surveying System and illustrates and explains the functions of the operating controls of the VECTRON Surveying instrument. Section 2 provides instructions for visual inspection and performance checks of the equipment. Section 3 provides detailed operating instructions, and Section 4 provides instructions for operator maintenance and adjustments. Section 5 lists technical data. Descriptive data and operating instructions for the AUTORANGER EDM instrument and VECTRON Field Computer are furnished with those instruments.

ment laying out the structure of the discussion: what will be covered and in what order. Notice in Figure 5-1 how the structure of the material is clearly indicated before the reader ever begins the discussion proper.

Finally, your introduction can include any necessary background information the reader should have before tackling the detailed discussion. This could include a history of the project, the basic theories involved, or definitions of vital terms. But make sure that this is really necessary information; don't fall into the trap of using your introduction as a dumping ground for material you don't know how to use in the discussion. Your introduction should be relatively short, no more than a few paragraphs. Any extended discussions should be placed in the body of the report.

The Body

We will discuss the body of your discussion in more detail in the following sections where we consider generating and organizing information. The body of the discussion presents all the data necessary to make your case, together with the necessary supporting material: tables, figures, case studies, and examples.

The Conclusion

The ending of your report will generally restate your objective or purpose and summarize your major points and conclusions. In purely informative reports, which do not lead to recommendations, a summary of your main points may be enough.

In reports that work to conclusions and recommendations, you will state them following your discussion. Both conclusions and recommendations are based on the evidence presented in your discussion, and your final summary should lead to them logically. Conclusions can be followed by recommendations, which are statements about actions to be taken or avoided based on your conclusions. You can also put your conclusions and recommendations in a separate section placed near your opening summary so that your reader can refer to them immediately. For more about writing conclusions and recommendations see Chapter 11.

The important point to remember about your conclusion is that you must have one. Don't leave your readers hanging, and don't rely on them to deduce accurately the points you want to make. Be sure to direct your readers to the significant facts you want them to take away from your report, and also let them know you have finished what you wanted to say.

GENERATING IDEAS

When you actually begin putting your material together, your first step is to generate the ideas for your report. You may feel that you have assembled a great deal of data already in your research process; it may seem a little late to be "generating ideas." Yet although you have this data, you may not know exactly what you want to say about it. You may know the conclusion you want to present, but you may not be sure how to derive that conclusion directly from your data. When you generate ideas during the organization process you will probably not generate any more data about your subject. Instead, you will generate ideas about the data you have already collected: conclusions, supporting generalizations, and related evidence. After your research you have collected a body of facts. Now you can proceed to consider what those facts mean.

There are two methods by which you can generate ideas about your data: subdividing your main points and brainstorming.

Subdividing Points

Sometimes when you begin to organize your material you already know what your major points will be. If you were basing a recommendation report on three or four criteria, for example, you might simply organize the report around those criteria, defining each one and explaining how each of your alternatives fulfills each particular criterion. Or you might be reporting on an inspection tour of a project site, in which case you might begin with the most important discovery you made (from your reader's point of view) and work your way to the least crucial.

If you know what main points you want to make, you can simply subdivide them. You would consider what subpoints support each main point, then what further subpoints support each subpoint. You continue this subdivision until you cannot subdivide any further and then introduce the actual experimental data that supports your subpoints.

Suppose you are recommending the purchase of ceiling fans to cut down on air-conditioning bills in your office. One of your points might be that ceiling fans are more efficient than ventilating fans. You could subdivide this point into operation and cost, then you could further subdivide operation into the amount of air moved and the extent to which each fan was able to lower the temperature in a room. At this point you would introduce the data to support each statement (e.g., the ceiling fan moves almost twice as many cubic feet of air as the ventilating fan). Cost could be similarly subdivided.

Ideally, at the end of this subdivision process, you should have several major points, supported by subpoints and appropriate data. Then you need only order the points to have a clear organization.

But frequently you have no very clear idea of the points you want to make. You may know the conclusion you want to come to, and you may have some fragments of information you want to include, but you may not be sure how to go about putting it all together. If this is the case, you can use our other strategy for generating ideas: brainstorming.

Brainstorming

In the brainstorming process you list all the ideas and facts that you have about your subject. Don't worry, at this point, about the relevance or importance of what you produce. Try not to be critical. Simply list your material in the order it occurs to you. You can reread your notes, consult your memory, even discuss the subject with others, anything that will get your mind working. The entire sequence may take many attempts. The information flow may seem to start and stop. Just stick with it and allow yourself time to think of as much material as you can.

When you have what seems to be a sizable list of material, return to your audience and purpose. Consider what material will be of most interest to your audience and what material will most clearly serve your purpose. Then eliminate any material on your list that doesn't seem to fit either of those categories. This should still leave you with a useful amount of material; if it doesn't, try some more brainstorming with your clarified sense of audience and purpose.

We can look at this process of generating ideas in an example. A large military hospital adopted a national heart disease prevention program called the Coronary Artery Risk Evaluation (CARE). During routine physical examinations, the hospital staff

***Figure* 5-2**
Typical information sheet

Data: Physical exam results
Lab results
Cardiac risk evaluation

File for each person

Possibilities:

Series 2000:
- Personnel familiarity
- Fast (relatively)
- 200 K/development
- 420 K/ installation
- 95 K/ personnel
- 2 K/ maintenance
- 95 K/ equipment
- 200 K/ computer time

URISOL 2300:
- Centralized, but keypunching done at computer center
- Reports mailed
- Is centralization necessary/desirable?
- 375 K/ development
- 95 K/ personnel
- 720 K/ new equipment
- 750 K/ maintenance
- 3.5 K/ installation
- 190 K/ computer time
- 300 K/ mailing (!)
- Training—no one at hospital has used the thing

Microcomputers:
- FAST, 2-3 second turnaround
- How many people know how to use?
- Training necessary (probably)
- 300 K/ equipment
- 475 K/ personnel
- 350 K/ development
- 200 K/ time
- Could we make a deal with a computer shop?
- Probably couldn't get them for less than Series.

Manual (?!):
- SLOW, files could get lost, misplaced
- 7 new clerks—14 K/ year each
- 135 K/ Filing cabinets, space, etc.—per year

1500–2900 victims/year, counting death and disability
$50 million/year—benefits, damage, medical expenses, accidents, etc.
CARE—Identify, counsel, prevent coronary disease.

would collect data regarding exercise, diet, and physical condition of base personnel. The data would be stored and evaluated, and personnel who were identified as having a high risk of heart disease would be given counseling and therapy. An assistant administrator at the hospital had to decide on the best method of processing this data at the lowest cost and present her conclusion in a report to the hospital administration.

She first considered the kind of information the program produced: files containing the results from physical exams and lab tests. Looking at alternative data-processing methods, she came up with four possibilities:

1. Using an existing computer system at the hospital (the Series 2000)
2. Using the central base computer (the URISOL 2300)
3. Processing the data on five microcomputers at the hospital
4. Processing the data manually by using seven file clerks.

The administrator researched all four alternatives in terms of cost, retrievability, and training required.

During her research, the administrator also uncovered some interesting related facts. Between 1500 and 2900 military personnel suffer from coronary disease annually, including both death and disability. Fifty million dollars per year are lost because of accidents, medical expenses, equipment damage, disability, and death benefits. The CARE program would, for the first time, identify people who might suffer coronary disease and try to prevent it.

Through her research and her brainstorming, the administrator generated the information sheet shown in Figure 5-2. This information sheet gave her an overview of her data. Notice that some of the items on her list are statements of fact, such as the costs for each of the systems, but some (like the question "Is centralization necessary?" under the URISOL section) are speculations produced as the administrator studied her data. The administrator now had a list of information, some of which she could go on to order for her report.

ORGANIZING INFORMATION

After you have brainstormed or subdivided your material, your next step is to organize what you have produced. Organization can be broken down into a three-step process: grouping your material, ordering groups, and producing an outline.

Grouping

Grouping is a way to begin bringing order to your list of information. You first study your list, looking for common denominators, general topics, or features that some of your information fragments have in common. Then you try to put these similar pieces of information together, writing a generalized statement for each group that explains what the fragments have in common.

After you've done one round of grouping and generalizing of the material on your brainstormed list, you can try grouping the generalizations that you've just made. You should consider whether any of them seem to fit together naturally, then try to develop a new generalization that covers the two or three smaller generalizations. You continue this process of grouping and generalizing until you reach the highest level of generalization that makes sense; usually these generalizations will be the major points you want to make about your subject.

In our hospital example, the hospital administrator had already grouped her material somewhat on her information sheet under the headings Data, Series 2000, and so forth. She now tried to generalize about these groups, and she produced these statements:

> The data to be stored consists of examination and evaluation results for all active duty personnel.
>
> The Series 2000 is lowest in expense, moderately fast, and requires little training for hospital personnel.
>
> The URISOL 2300 is centralized, but more expensive and would require personnel training.
>
> The microcomputers are extremely fast, but more expensive than the Series 2000 and would require light personnel training.
>
> Manual processing is lower in expense, but extremely slow with high risk of error, and would require personnel training.

Generalizing again, the administrator produced another statement based on these generalizations:

> The Series 2000 is best in terms of cost and training, and acceptable in terms of retrievability.

Ordering

Whether you begin with your major points and subdivide or begin with fragments of information and generalize, you will generate several groups of information. Your next step after generating this information will be to order these groups in some meaningful and effective way.

To begin, return to your original audience and purpose analysis. Your order, like the other aspects of your report, should be based on the needs of your audience. Consider what information will be most important to them and what information they will find most interesting. Also consider what order is most likely to achieve your purpose.

Several orders are possible for most reports.

Order of Importance

For a management report it is common to begin with the most important information and move to the least important. Since many managers want to read only the most important points in your report, this order allows them to find those points easily. It may also ensure that they will read those points, since they come early in the report: You run less risk of losing your reader's interest before they get the points you want to emphasize.

Order of Acceptance

If you are dealing with controversial material, it is usually easier to begin with the most widely accepted points in your report before moving to the points that seem more likely to inspire disagreement. In this way you can build up some common ground and establish your credibility before challenging your audience's assumptions.

Chronological Order

Chronology is particularly useful in giving directions or describing a process (see Chapter 12); you simply begin with the first step and follow through to the last. Be careful not to overuse this order, however; do not fall into the habit of describing your research step by step when your audience is really only interested in your conclusions.

Spatial Order

The traditional order for descriptions is left to right, top to bottom, north to south, etc. To use it, pick a particular point on the object you're describing and go in a uniform direction to complete the description. Make sure your direction is uniform, however; if you begin by going left to right and switch to top to bottom half way through, your reader will be hopelessly confused.

Order of Complexity

If you are dealing with a particularly complex subject, it is often best to begin at the simplest, most basic point and move gradually toward the more complex information. This order also operates with audience familiarity; you can begin with the material most familiar to your audience and then work gradually to the material that is unfamiliar to them.

Logical Order

If you are presenting an argument for or against a point, you can use the traditional orders of formal logic. *Deductive order* begins with the conclusions and then presents the evidence that supports them, usually in order of importance. *Inductive order* presents the evidence point by point in a logical progression, leading inevitably to the conclusion. Deductive order is more common in technical writing because of the preference in many audiences for conclusions presented at the beginning of the report; you should only use inductive order if you are convinced that your audience needs to see your entire reasoning process to understand your conclusion. Even so, you should be prepared for the fact that many readers will read the conclusion first no matter where it is placed in the report.

Consider our hospital example once again. The administrator had several groups of data she needed to order for her report. Before proceeding any further, she considered the needs of her audience. Her report would be addressed to her immediate superior, the deputy director of the hospital, but she knew that eventually it would be passed on to the hospital governing board. The administrator asked herself what these people would be most interested in. Cost would obviously be a paramount factor; the hospital budget was already strained. Efficiency would be important, too; the governing board wanted a system that made the information readily available for use. Along with this they would probably prefer a system that required minimal training and minimal hiring of new personnel, a system that could fit into the existing structure. These points—cost, efficiency, and use of personnel—became the criteria for the administrator's decision.

The administrator's purpose depended largely on her conclusion. After studying the data she had collected, she decided that the current hospital computer, the Series 2000, would best fulfill the needs of the system. Her purpose thus became to recommend: to convince the hospital administration that the Series 2000 was the best choice they could make for processing the CARE data.

The administrator had two possibilities for ordering her data to achieve her purpose: She could order her data around the four alternative processing methods or she could organize it around the three criteria. The administrator decided, after considering both of these choices, that ordering around the criteria was best. This kind of organization avoided unnecessary repetition; it also stressed the audience's concerns—the criteria that were important to them—rather than the writer's concerns—the alternatives that she had researched.

The administrator now had the main points for her report: cost, training, and retrieval time. Her ordering of these points was relatively simple: since she was writing a recommendation report she decided to use order of importance: cost, retrieval time, and training. But how would she order the alternatives around each criterion?

She could have arbitrarily begun with one of the alternatives, the Series 2000, for instance, and then followed it with the others, the URISOL, the microcomputers, and the manual system, for each criterion. Or she could have begun with the most effective alternative each time and worked through the three others to the least effective. The administrator chose the latter order so that she could emphasize the comparison for her readers, pointing out the most effective choice in each case, rather than forcing her readers to make the comparison for themselves.

Outlines

The whole process of generating information, grouping, generalizing, and ordering will produce a type of scratch outline. This will be a rough blueprint for your report and frequently this will be enough for a short, informal version, or for a memo. However, some reports require more fully developed outlines. Formal reports, for example, usually require an outline written in phrases or sentences rather than single words to convey completely the points you want to make. Likewise, any report done as part of a group requires a fully developed outline to ensure that no topics are overlooked and that no topics are repeated.

Many writers don't like outlines, finding them too restrictive and troublesome to prepare, but outlines have several advantages. A fully developed phrase outline, for example, can serve both as a table of contents and as a source for your headings and subheadings. An outline can also allow you to work out large sections of your report in advance; when the time comes to write, there will be less for you to do because much of the thinking will already have been done.

Multiple Outlines

One way to make outlining easier is to work with multiple outlines. Your beginning scratch outline—the result of your grouping and ordering—should provide you with a number of major points to consider. Now you can try creating a separate outline for each major topic. Begin by writing each major topic on a sheet of paper. Then list the pieces of information you have about the topic: subtopics, data, and points to consider. Essentially, you go through the same sequence that produced your scratch outline, but here you are concentrating on only one major point at a time. Finally you can group and sequence your material. You can repeat this process with each major topic, and do each one at a different time, even on a different day if you have the time. By spacing the entire outlining sequence over several days, you can keep the task from becoming too burdensome, and at the end of the process, you should find you have produced a multi-paged outline with little strain.

Outlining Hazards

There are some things to be careful of with any outline. First of all, you should decide whether you will be using a phrase, a sentence, or a complete paragraph format. Each of these alternatives is acceptable, although sentences and paragraphs will give you more material for the report and will force you to develop your ideas more fully. Once you have decided on a format, however, you must be consistent with it. Each entry in your outline is a cue for you to write something in your report. If these cues are out of balance—if one topic is expressed in phrases and another in sentences—the writing itself may be out of balance. You may wind up with some topics that are covered in depth and parallel topics that are covered in only a few sentences.

You should also make sure that each main topic you consider has at least two subtopics if it is divided. The reason for this is simple: Outlining is based on division, and it is impossible to divide a whole into fewer than two parts. If you find yourself with a main topic that cannot be divided into two parts, review the main topic itself. Could it actually be a subdivision of another main point? Could it be combined with another point to form a new main topic?

Finally, be careful not to leave out elements like your introduction, your conclusions, and even your transitional material. Some things, like the introduction, may actually be written after you outline the rest of the discussion; it may be easier to decide what to include in the introduction when you know what material will follow it and in what order. But if you omit these elements from your outline, you run the risk of omitting them altogether. Do not rely on your memory when you are writing your report; put everything you can on your working outline.

The entire organizational process is represented by the organizational flow chart in Figure 5-3.

Now return to the hospital example. Having decided on the order of her main points (her criteria) and her subpoints (her alternatives), the administrator created this scratch outline, organizing her alternatives in order of priority:

Conclusion: the Series 2000 is best in terms of cost and training, and acceptable in terms of retrievability.

1. Cost
Series 2000
Manual
Microcomputers
URISOL 2300
2. Retrievability
Microcomputers
Series 2000
URISOL 2300
Manual
3. Personnel
Series 2000
Microcomputers
URISOL 2300
Manual

Next she considered the other material she had uncovered: the statistics about the losses caused by coronary disease and the description of the CARE program. The administrator had to decide whether to include this material and, if so, where to put it. The statistics the administrator had found are interesting and would certainly be included if she were arguing for the adoption of the CARE program. However, the program had already been adopted; there was no real need for the administrator to argue for it. Interesting though they were, the statistics were not highly relevant to her purpose, and she discarded them.

The general description of the CARE program, however, did serve a purpose in the administrator's report. The program description provided an account of the data to be processed and thus helped to justify the administrator's criteria. Thus she decided to include this information as background in her introduction.

Finally the administrator constructed her complete sentence outline, and it looked like this:

Title: The Most Feasible Data-Processing System for CARE Information

1. Introduction
   - 1.1 The CARE program is a coronary disease prevention program that has been adopted by the base.
     - 1.1.1 It is a preventive coronary disease program.
     - 1.1.2 All base personnel will have coronary risk data added to their routine physical examination.

FIGURE 5-3
An organizational flow chart

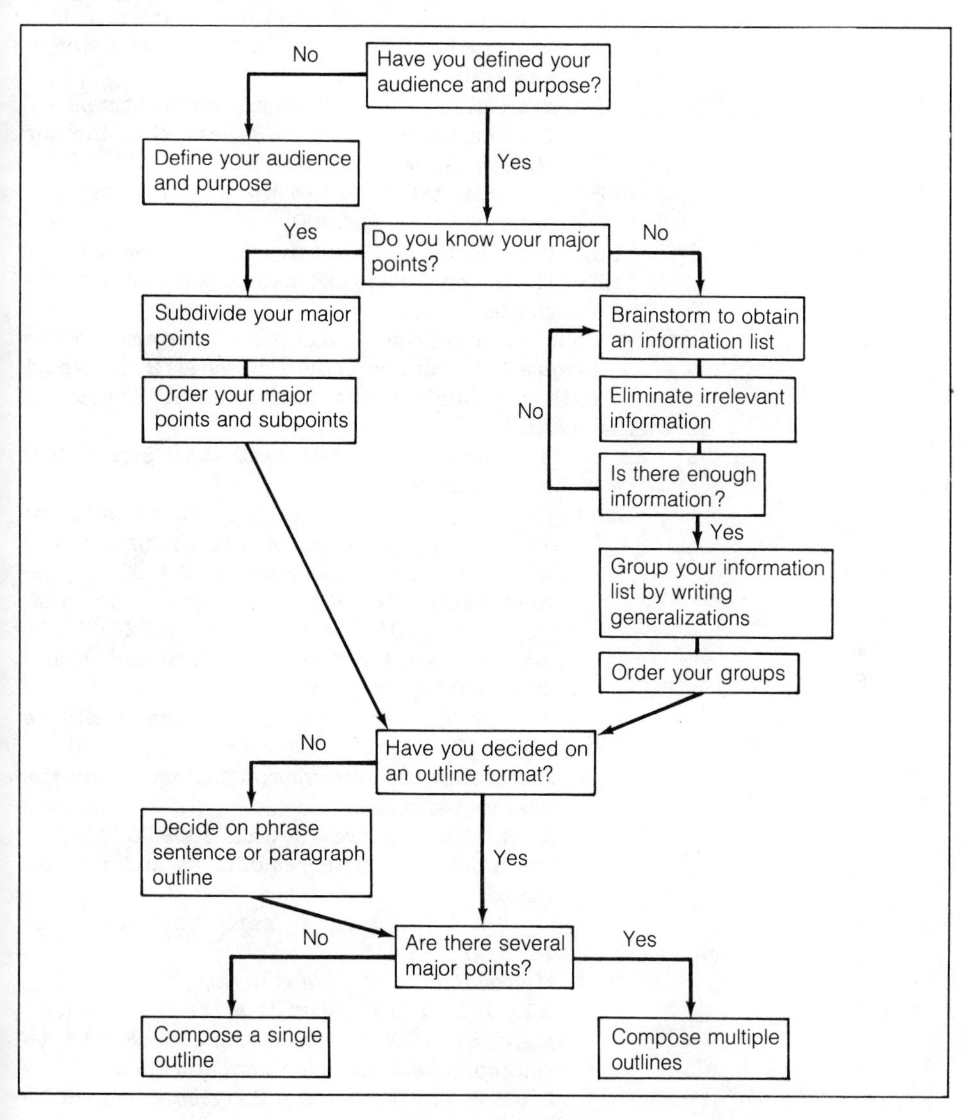

1.1.3 This data will be stored and evaluated.
1.1.4 Counseling and health care programs will be established for high-risk personnel.
1.1.5 The data-processing system chosen will store all CARE data on base personnel.

1.2 These are the criteria for the processing system to be chosen.

1.2.1 The system chosen should show low operating costs.

1.2.2 The system chosen should make the information easily available to the hospital staff.

1.2.3 The system chosen should require minimal training for current staff and minimal hiring of new staff.

1.3 These are alternatives for the processing system

1.3.1 The hospital could use the Series 2000 computer already in use.

1.3.2 The hospital could use the central base computer: the URISOL 2300.

1.3.3 The hospital could use five microcomputers.

1.3.4 The hospital could use manual processing by file clerks.

1.4 Conclusion: On the basis of these three criteria, the hospital's current computer, the Series 2000, would be the best choice for processing the CARE data.

2. Discussion

2.1 Cost comparison demonstrates that the Series 2000 would cost the least.

2.1.1 The Series 2000 computer shows the lowest cost for the nine-year period of the CARE program.

2.1.1.1 The cost breakdown is $200,000 for development; $420,000 for equipment installation; $95,000 for personnel; $2000 for maintenance; $95,000 for new equipment; and $200,000 for computer time.

2.1.1.2 The average yearly cost would be $112,333.

2.1.2 Processing the information manually has the next lowest cost.

2.1.2.1 The cost breakdown is $882,000 for personnel; $333,000 for equipment and maintenance.

2.1.2.2 The average yearly cost would be $135,000.

2.1.3 Microcomputers are next in cost.

2.1.3.1 Cost breakdown is $300,000 for equipment; $475,000 for personnel; $350,000 for development; $200,000 for computer time.

2.1.3.2 The average yearly cost would be $147,222.

2.1.4 The URISOL 2300 is the most expensive alternative.

2.1.4.1 Cost breakdown is $375,000 for development; $95,000 for personnel; $750,000 for maintenance; $3500 for installation; $190,000 for computer time; $300,000 for mailing.

2.1.4.2 The average yearly cost would be $190,388.

2.2 Microcomputers would provide the best retrievability, with the Series 2000 next best.
   2.2.1 The microcomputers would have the best retrievability since the terminals would be most accessible to hospital personnel.
   2.2.2 The Series 2000 would be next since the computer is in the hospital, although the data would have to be passed along to interested personnel.
   2.2.3 The URISOL 2300 would be next best.
      2.2.3.1 The data would be computerized.
      2.2.3.2 The data would have to be mailed from the computer center to the hospital.
   2.2.4 The manual system would be the slowest because files would have to be located and retrieved from filing cabinets.
2.3 The Series 2000 has the lowest training and personnel costs.
   2.3.1 The Series 2000 requires no training and no new personnel.
   2.3.2 The microcomputers and the URISOL 2300 both require light training but no new personnel.
   2.3.3 The manual system would require seven new personnel and light training.

3. Conclusions
   3.1 The Series 2000 is the cheapest alternative and makes the best use of hospital personnel.
   3.2 Although the retrieval time of the Series 2000 is not as fast as the microcomputers, it is acceptable.
   3.3 The series 2000 is the best choice for processing the data produced by the CARE program.

Notice how much of the administrator's report has already been written in her outline. The actual report writing should proceed smoothly since so much of the final version has been written in advance.

INDICATING YOUR ORGANIZATION

Your final step, after organizing your material to your own satisfaction, is to convey that organization to your readers. Sometimes writers do not realize that their organization, which seems straightforward to them, is actually not at all obvious to their readers. As a writer, you are aware of complex relationships between pieces of information in your reports; you see what points are subordinate to other points, what points are equal in importance, what the logical relationship is between various pieces of information. Your readers, on the other hand, are only aware of the sequence of your material: what comes first, second, third, and so on. Thus your readers may miss the more complex relationships in your information entirely unless you manage to convey them in your report.

Your headings and subheadings (and the way they are placed on the page) constitute one type of organizing material that can help convey these relationships to your readers. Numbering and lettering systems are another (see Chapter 7 for a discussion of both of these features). But you must also use prose indicators to communicate the relationships you see in your material. There are two principal types of prose indicators you can use: forecasting material that indicates the major divisions and subdivisions of your report, and transitional material that signals shifts from one topic to another. (The following discussion is based in part on research by Paul V. Anderson.)

Providing Forecasts

When you forecast your organization, you let your reader know what topics you include in your report and the order in which they are discussed. Thus the introduction includes a general forecasting statement that previews the topic of each of the major sections that follow. In turn each of these major sections begins with another forecasting statement that indicates the major subtopics the section covers. You could also begin these subsections with forecasts, indicating at each level of your report what material is to follow. Consider an example from an instructional manual. In the general introduction (Figure 5-4), readers are told what is included in the manual's major sections and appendix.

Then each individual section contains another forecast, like the one in Figure 5-5, indicating the general topics to be covered in that particular division.

Some of these forecasts may be repetitions of earlier material; your subsection forecasts may actually repeat the forecasts in your general introduction. But this is functional repetition; it will help your readers follow the general organization of your work.

Providing Transitions

Transitional material, too, can appear at all levels of your report. Transitions let your readers know when you are moving from one topic to another; thus at the end of each major section a transitional statement should inform your readers that you are moving on to the next major topic (e.g., "This data supports conclusion A, but other data indicates that conclusion B may also be tenable."). Similarly transitions within each section can let your reader know the relationship between the subsections. If the ceiling fan report cited earlier had included a forecast

FIGURE 5-4
General forecast. (Courtesy of Keuffel & Esser Company, Parsippany, New Jersey)

1.2 ORGANIZATION OF MANUAL

The Manual is organized into sections for operator convenience.

SECTION I contains an introduction to the System, simplified operating instructions and an explanation of how the techniques employed work. It also contains safety information, specifications of the equipment, and descriptions and functions of the System components, controls and indicators.

SECTION II takes you step-by-step through Registration Exposure, Processing and Toning of a SPECTRA Proof. It contains complete operating instructions for the SPECTRA Processor, including initial startup, routine cleaning, maintenance, and adjustment information to keep the Processor in optimum operating condition.

SECTION III details proper set-up procedures for the SPECTRA Laminator and takes you step-by-step through the Laminating and Mounting procedures to produce the final SPECTRA Proof on a printing stock of your choice.

SECTION IV This section contains routine maintenance and adjustment procedures to assure trouble-free operation and accurate register of all colors. It also contains a Troubleshooting Chart to help you with any minor difficulties you may encounter when making a SPECTRA Proof.

APPENDIX. We have included an assortment of backup technical data in the Appendix to which you may have occasion to refer, such as: Installation and Setup Instructions, Schematics, Parts Replacement Diagrams and other useful information.

FIGURE 5-5
Section forecast. (Courtesy of Keuffel & Esser Company, Parsippany, New Jersey)

3.1 SETUP OF THE *SPECTRA* LAMINATOR

The Laminator has been set up ready for operation by your Technical Representative. However, before attempting to use the Laminator, operators should become familiar with all controls, indicators and components, and review their description and function in SECTION I of this Manual. The operator should also become familiar with the procedures for installing register pins on the Image Drum. Register pins have been installed by a Technical Representative, but will have to be reinstalled whenever the Image Drum blanket is replaced. Of particular importance to the operator are the procedures for preparing and mounting the Receptor Sheet to the Receptor Drum. Operators will perform these procedures every time a SPECTRA proof is made.

such as, "Ceiling fans operate more efficiently than ventilating fans both in terms of the amount of air moved and the number of degrees the temperature can be lowered," the transition between the subsections on the amount of air moved might be something like, "Ceiling fans not only move twice as many cubic feet of air as ventilating fans, they also lower the temperature in test rooms an average of 3 degrees lower than ventilating fans." The section that followed the transition would present data about the lowered temperatures.

Forecasts and transitions help your readers focus on topics and find relevant material. A good forecast helps readers locate the topics they are most concerned with in each section. Similarly, a clear transition indicates to your readers the end of one topic and a shift to a new one. Transitions also help readers judge the relationships between what they have been reading and what is yet to come; a *however*, for example, will tell them that the following points are contrasts, while a *moreover* will indicate similarity. (See Chapter 6 for a list of common transitional words.)

SUMMARY

1. The organization of most technical reports is based on the needs of the technical audience.
2. Most reports can be divided into three parts: an opening segment, which responds to the needs of the general audience by providing summaries; a discussion segment, which responds to the needs of a specific audience with in-depth discussions of data; and a conclusion, which restates the major points.
3. To develop material for the body of the discussion, you can either subdivide major points or generate information by brainstorming.
4. The information generated can be grouped under generalizations and then ordered according to audience analysis and purpose.
5. After composing a scratch outline by grouping and ordering, you can compose a formal outline to guide composition of the final report.
6. Indicate your organization to your readers by means of forecasting and transitional material.

EXERCISES

1. Using a paper or report that you have written for this or another class, review the organization. Did you use a three-part format? Did you group similar material together? Did you use an effective order? Did you convey your organization through forecasting and transitions? Are there ways in which the organization of the paper could be improved?
2. Exchange papers with another member of the class and review the organization of that paper in the same way.
3. Locate a journal article in your field, and analyze the organization. Does the author use a summary? Forecasting? Transitions?
4. Write an outline of the reports assigned in either of the following cases: Supermarket Scanners p. 44, or The Eighth Grade Diet, p. 363. Assume that you're writing the report as a pamphlet rather than delivering it orally.
5. As part of a report on energy alternatives, you have collected the following material about solar heating; your purpose is to recommend a specific solar heating system for the average home. Group this material, devising generalizations to cover each group; eliminate any material that seems irrelevant. Then order the groups in the most effective sequence. Finally, outline your report on solar heating, using either a phrase or sentence format.

- Two main types: passive and active.
- Active most widely used.
- Two kinds of active: water and air.
- Passive uses design and mass of building; must be designed into house from beginning.
- Active uses flat plate collectors mounted toward south so that they collect maximum solar radiation all year.
- Sometimes collectors are adjustable to change with sun's position in winter and summer.
- Water system uses water as heat transfer medium; air uses air.
- Passive designs usually use large areas of glass so that sunlight can be used for heat.
- Water systems have collector, storage, piping, circulation pumps, and control system for transferring heat to house.
- Storage for active systems can be located anywhere (usually garage or basement); whatever is convenient.
- Air like water, but it uses air as transfer system.
- Congress increased solar energy tax credit in 1980.
- Water systems work well; circulation requires less energy than air circulation.
- Active storage terminals should be placed as close as possible to collector panels to minimize heat loss.
- Storage terminals for air systems usually made of 2-inch to 4-inch diameter stones.

- Air systems use ductwork rather than pipe systems used in water.
- Taxpayers can take a tax credit of up to 40 percent of the cost of solar systems.
- Credits are claimed on first $10,000 spent; up to $4000 can be claimed.
- Piping uses less floor space than ductwork and is easily installed.
- Leakage with air systems will not cause damage to walls or insulation.
- Air and water equally efficient overall, but air systems more complex and expensive.
- Water systems need antifreezing and anticorrosion additives.
- Air systems don't require heat exchange; there's no loss of temperature.
- Air leakage hard to detect; can lower system efficiency.
- Water leakage could contaminate domestic hot water supply.
- Air systems more complex; more expensive to install.
- Air systems use more energy for heat transfer because of lower thermal capacity of air.

Organization Case 1

The New Light Switches

You have just taken a technical writing job with a large electronics firm, Omni Electric. Omni manufactures everything from light bulbs to computer components, but most of their sales are in "consumer electronics": lighting fixtures and wiring accessories such as junction boxes and switch plates.

Lately Omni has developed some innovative alternatives to conventional electric switches for the home. As yet they are not economically feasible for the average house, but the company hopes to appeal to upscale buyers with money to spend on unusual accessories. The Omni marketing division has decided to try selling these switching systems to building contractors, concentrating on those who specialize in luxury homes. Your assignment is to write a description of the new systems that will be printed in a brochure and sent to the targeted customers.

To get details on the new systems you talk to one of the electrical engineers who helped design them. You take the following notes on his conversation.

Usual switching system involves 15A-120V AC toggle switches. Maybe 24 switches in a three-bedroom house. Installed in switchboxes mounted on studs.

Touch system: mounted next to door jamb inside. Enclosed in wall. Switch box. Metal strip under the door jamb—maybe halfway across room entry. Anybody touches strip turns on light. Switch works with existing wiring—could be added to house.

Sensor system: proximity system. Sensor projects field; any change in capacitance unbalances field, switches light on or off. Sensor electrode in hole in door jamb. Pass through door, activate system.

Computer system: Uses incandescent light sensors, infrared units, home computer. Beam of light projected at doorway (projector on one side of door, receiver on other). Break beam, switch activated. Sensors connected to computer; two sensors per doorway. Computer monitors number of people in room and whether entering or leaving (side-by-side sensors will record this). Computer will make sure lights stay on as long as someone's in the room.

All systems would provide some electricity savings since lights would be turned off automatically (not enough to offset cost of switches, though). Computer could also be used for other purposes.

ASSIGNMENT

Write the sales brochure. Organize the material and provide an introduction and conclusion, as well as transitional and forecasting material. Remember to consider your audience and purpose. What do you want to accomplish and how can you accomplish it with these readers?

Organization Case 2

The Carlyle Windsor House

As your first job after receiving your B.S. in civil engineering you go to work for your friend Bill Morrasco, who owns Vintage Restorations. Bill's company specializes in restoring older houses in the historical section of the city, and they've been very successful.

For your first assignment, however, Bill asks you to clear up the company's first disaster. Six months ago the company restored the Carlyle Windsor house, a large wood-frame Victorian on the city's west side. Extensive structural repairs were necessary, and these proceeded without problems. However, the final step was repainting the exterior. Something apparently went wrong; the new paint is now blistering and peeling. The owner of the house is understandably upset and demands an explanation, to say nothing of a new paint job. Bill asks you to investigate.

Your investigation produces the following facts (in no particular order):

The Carlyle Windsor house had had several coats of paint previously.

The choice of exterior paint was between latex and alkyd (oil base).

It had rained heavily two days before the house was painted.

Latex paints perform better than alkyd on damp surfaces because they allow moisture to pass through them readily.

The Carlyle Windsor house has a large elm tree on the north side; the south, west, and part of the east side are in full sunlight.

Three days before the house was painted, the exterior was sanded down to bare wood, then a primer coat was applied.

Alkyd paints perform better than latex on previously painted surfaces.

Moisture meters are used before painting to test the percentage of moisture by weight of actual water present in the wood, plaster, or concrete.

Paint applied in direct sunlight may blister because the surface paint dries before thinner can escape; as the sun vaporizes the remaining thinner, the paint blisters because the thinner cannot escape through the outer skin of dried paint.

The percentage of water permitted in wood that is to be painted is usually 15 percent to 18 percent. 20 percent is too high.

It was partly cloudy on the day the house was painted; the high temperature was 76 degrees.

The primer coat should be allowed to dry at least two days before the finish coat is applied.

The moisture percentage found in the walls on the day the house was painted was 19 percent.

The forecast for the week after the house was painted called for clear skies and unusually high temperatures.

The project supervisor selected alkyd paint as the most suitable for the Carlyle Windsor house.

Allowing too long an interval between the application of the primer coat and the finish coat can result in intercoat peeling.

The most common cause of blistering and peeling is moisture.

Study these facts and arrive at a decision about the cause of the paint failure at the Carlyle Windsor house.

Organize the material. Group and generalize, then order your main points for the following assignments. For each assignment, analyze both your audience and your purpose. What does each reader need to know, and how can you convey it?

ASSIGNMENT 1

Write a short report for Bill Morrasco in which you explain your findings and make recommendations.

ASSIGNMENT 2

Write a letter to the owner of the Carlyle Windsor house, Dr. Lois Aragon, in which you explain why her house has a peeling exterior. Incidentally, Bill has decided to repaint.

CHAPTER • 6

Writing and Revising

INTRODUCTION

Many people find writing difficult, but it doesn't have to be. This chapter is designed to help you write easily about practical subjects by setting priorities in your writing. We will review the way people actually read technical prose and present a functional approach to style. We will also review several ways of improving your writing and conclude with specific advice on how to revise your work. This chapter does not stand by itself, though. Writing and revising technical prose should be seen as part of a process that includes analyzing your audience, determining your purpose, and organizing your material. These subjects are covered in Chapters 1, 2, and 5. Chapter 15 on writing in a group presents further advice about revising your own work, as well as your coworkers' writing.

READING AND WRITING: THE BASIC SITUATION

Reading and writing are reciprocal activities. For this reason we will first look briefly at the way most people read before we consider how to make that reading easier through improved writing. The first point to be stressed about reading is a negative one: Few people have time to read.

Making Time to Read

Most of us must make time to read the material necessary for our work. You probably have to schedule time to read this book, because like every other busy person, you have other jobs to do. Although all of us read some things for pleasure, most of our reading is done because we must.

Reading under a deadline is different from reading for fun. Consider for a moment how you read most of your textbooks. Do you read every word? Every sentence? Every paragraph? Chances are you make some choices as you study. And the same thing will happen when others read your reports. Most of us read selectively when it comes to professional material.

Reading Selectively

Most readers skim. In fact some speed-reading specialists teach executives to read only the first and last sentences of each paragraph. Knowing this, you can understand why clear topic sentences and paragraph conclusions are so important.

Executives will also be selective in the parts of a report that they take time to read. In a classic study of professional reading habits, James W. Souther found that readers go straight to those sections of reports that help them make decisions (James W. Souther and Myron L. White, *Technical Report Writing*, 2d ed., New York: John Wiley & Sons, 1977, p. 20). For managers and executives, these are sections that explain conclusions, make recommendations, summarize content, and present problems.

As a writer you can help these readers choose what they will read by using the organizational material we cover in other chapters: forecasting, transitions, clear summaries, and headings. But you can also help readers by cultivating a clear, efficient style.

Remember, no matter how long your report turns out to be, you can plan on capturing no more than twenty minutes of your reader's time: enough time for a cup of coffee. Every element of your report should be designed to help your readers get what they need in that twenty minutes.

An Overview of the Reading Process

When people begin to read, they are already expecting to receive a specific kind of information; they are prepared to find familiar words and content. As they read each clause or sentence, they organize the material as rapidly as they can in their short-term memory. As soon as they know that a clause is complete and that a unit of meaning has been formed, they send that meaning from short-term memory to long-term memory for storage. For a writer, the most critical features of the reading process are the limitations of the short-term memory. If a reader receives too much information at one time, the short-term memory becomes overloaded.

If the short-term memory had an infinite capacity, people could probably read and understand anything, no matter how unclear, without becoming confused or needing to reread. As it is, there are two limits all readers (and writers) must live with: a limit on time and a limit on storage space.

Time Limitations

Each time readers begin to process a new clause, they have only a few seconds, first to organize the information into a meaningful form and then to connect the new meaning to the last sentence or clause processed. Under most circumstances, these few seconds are plenty of time for the job; yet there are many ways to bring the reading process to a grinding halt. For example, a writer may use obscure language or a long string of dependent clauses. Anything that makes the reader stop reading will exceed the time limit.

This same phenomenon applies to single words and words grouped into phrases. As people read a new sentence, they normally assemble the words into meaningful groups and pass them on for storage in long-term memory. The short-term memory stalls (or "goes down") whenever readers are confronted by a long string of words in which they can find no pattern, or when they are required to wait and wait for the missing word that makes sense out of a long sentence.

Space Limitations

The limit on storage space can be more troublesome than the limit on time. The short-term memory has about five to seven "boxes" for keeping information that is waiting to be organized. If a sentence requires readers to exceed this storage capacity, then they will begin to lose part of the meaning and will have to reread.

These are the basic limits: a few seconds and five to seven spaces for storing words (or phrases) that are waiting to be gathered into a meaning. Once you understand these limits, you are prepared to learn how to write prose that your readers can follow with ease.

Style: Writing for Efficient Reading

Effective writing can be understood in one reading. Good writing does not force readers to stop and reread because they are lost or confused. In other words, good style means efficient reading and efficient use of human time and energy. One version is better than another when it conveys the same meaning but demands less work on the part of the reader.

Reader-based Writing

Efficient writing is writing that meets the needs of readers. It does not force readers to stop and think in order to discover meaning or overtax the space and time limitations of the short-term memory. There are several ways you can make your writing more efficient. You can meet your reader's expectations, use precise and jargon-free language, employ connective words and parallel structures, and even practice strategic repetition. Along with a clear organization and an accurate audience analysis, this kind of writing should meet your readers' needs effectively.

Providing Clear Context and Expectations

Readers need, first of all, to understand the context of statements they read. Consider this sentence from a memo: "With reference to our last meeting, I think we all agreed on our first priority." Given a sentence such as this, readers would struggle vainly to find something in their memories that provides a context. With reference to what meeting? Who was there? What was discussed? What was given priority? In this case, actual reading stops while the reader searches for a connection. Giving the reader a clear context—"With reference to our January 12 meeting concerning a redesign of the UXB"—allows reading to continue without confusion.

Readers understand what you write when they know what to expect. As long as you tell your reader what you are going to say, and in what order, and then fulfill those expectations, you will have a reasonably satisfied audience. We covered this ground in our chapters on organizing and audience. Remember the importance of well-designed introductions (with forecasting), transitions, and summaries.

Eliminating Jargon

One way you can make technical writing needlessly difficult for readers is to use unnecessarily technical language or jargon. For example, several years ago a surgeon was attempting to gain attention for a new surgical procedure. He presented his research at conference after conference by reading one of his published articles, which overflowed with highly technical descriptions of procedures and the human anatomy. After several conferences, he realized that his method was not being adopted by other surgeons. At the next meeting, instead of reading an article, he brought two blocks of wood to the session and used them to demonstrate his principal point about the relationship between two major bones. In addition, he replaced the technical terms for positions (*anterior, posterior, distal, proximal*) with words such as *above* and *below*. Moreover, he began to use comparisons between anatomical processes and simple mechanisms, such as book covers and hinges. When the session ended, two members of the audience came forward to say that they wanted to use the new surgical approach but had not fully understood it until it was presented simply.

This story demonstrates two important points about reading and listening. First, it illustrates how we read long and difficult words. Each of us has an easily accessible vocabulary of words we recognize in reading but seldom or never use in our own speech or writing. When confronted with an unfamiliar word, we probably try these internal "dictionaries" before turning to a printed dictionary for help. Obviously, every time readers need to consult a printed dictionary the reading process stops completely. In the same way, each time readers pause briefly to remember a word they seldom use, precious time and energy—and the reader's attention—are lost.

The other point the story demonstrates is that the use of complex technical language is not always appropriate; you should always keep your readers' needs in mind. At a conference with all of its distractions, even specialists have difficulty following the technical vocabulary of their field. As a rule, technical vocabulary should be used only when there is no common word that will convey the same meaning. Remember, many readers may not share your technical background. Heavy use of jargon does not impress them; it confuses and loses them.

Using the Active Voice

In the active voice, the subject of the verb acts on something. "The president of the corporation toured the People's Republic of China," is a sentence in the active voice. The president performs an action: making a tour. In the passive voice, the subject of the verb is acted on by something. "The People's Republic of China was visited by the president of the corporation" is a sentence in the passive voice. Notice that the passive sentence has two additional words, *was* and *by*. Passive constructions may be preferable when a machine or process is the subject and an operator is not involved. For example, "the paint is dried in a dust-free room" is preferable to "a dust-free room dries the paint" when you want to emphasize the paint. Too many passive constructions, though, can bore your reader because of their additional length and complexity.

Using Precise Language

Words with many possible meanings can also confuse readers. Suppose your employer sends out a memo announcing that: "The new work program will be initiated on Monday, August 7." In this sentence, the word *initiate* has at least two possible meanings: "start" and "formally inaugurate." If your employer likes public celebrations, you might wonder whether the new work actually starts on Monday, or whether a ribbon-cutting ceremony is planned for Monday, with work starting on Tuesday. Your brain responds to this situation by putting the choice of meaning on hold (or in storage) until the context makes the intended meaning clear. Occasional longer words add variety, but imagine a report in which a writer constantly used words such as *terminate, procure,* or *requisition* rather than *end, buy,* or *ask.* Every time readers face one of those longer words, they need just a fraction of a second longer to decipher the intended meaning, and as those fractions add up, they come closer and closer to their memories' limits. Imagine the mind performing a lengthy computer process to select the right meaning from several possible meanings; that waste of time and energy can be prevented if you use simple words whenever they are appropriate.

Using Clear Modifiers

Some writers use too many adjectives in a row (*large, green, retrograde, transverse, refurbished, roofing shingles*), which severely strains the memory. Other writers use nouns to modify other nouns (*contract administration personnel cost analysis office*), which forces the reader to decide which nouns are being modified and which nouns are pretending to be adjectives. If you have these habits, consider: Could a specific noun be substituted for the adjective string plus general noun (e.g., *sedan* rather than *large, four-doored automobile*)? Could the adjectivized nouns be rephrased (*Office for the Analysis of Personnel Costs*)? Adjective strings and adjectivized nouns are characteristics of hastily written technical prose. A more subtle strain on the reading process can be caused by unclear prepositions.

Consider this example of vague preposition use: "The Blythe Corporation completed the project with the General Engineering Company." Here the preposition *with* can take on several different meanings, such as "assisted by General Engineering," "working in consort with General Engineering," "by delegating the final stages to General," and so on. Until the context clarifies the relationship between these two firms, the meaning of *with* cannot be established. Prepositions often serve as a writer's shorthand for full verbal phrases, and this shorthand can be confusing if the preposition has ambiguous meanings.

Vagueness can also be caused by stringing prepositional phrases together. Here is an example taken from a technical document. "The Nuclear Regulatory Commission has initiated safety studies as part *of* its safety program *under* the nuclear safety and engineering branch *of* the division *of* reactor development." In this passage the preposition *under* means "directed by," while *of the division of reactor development* means "which is administered by the reactor development division." You don't need to replace every preposition with the implied verbal phrase, but any time you notice that you are stringing together prepositional phrases, you should expand a few of them by writing out the "hidden" verb. Thus our previous example would become "The Blythe Corporation completed the project undertaken in partnership with the General Engineering Company."

Remember the reading process; every time your mind faces a vague preposition, it probably makes a short list of likely meanings (including possible implicit verbs) and then holds that list in short-term memory until the context clarifies the writer's intent. In other words, faced with a string of prepositional phrases, your mind may spend much of its time and storage space working out what the prepositions mean and little time organizing the meaning of the sentence.

Using Connecting Words and Transitions

We discussed transitional material in Chapter 5 when we suggested ways for you to signal the relationships between your ideas to your reader. Transitions can also be made between sentences by using connective words. Much technical writing uses short sentences of twenty words or less. Using concise sentences is one way to avoid overtaxing the time and space limitations of the short-term memory. Yet too many short sentences can drain the reader's energies if writers forget to explain how the ideas in them are connected. An insurance company field investigator wrote the following two paragraphs to be used by an underwriter as the basis for determining a policy premium. This is what the underwriter had to contend with:

Oxygen and acetylene and electric arc welding performed in the vehicle service area an average of two hours a month. The welding practices meet the majority of NFPA requirements and clearances to combustibles greater than 35 feet. Hazards well controlled. Gas cylinders chained upright on movable cart. Welding performed on concrete floor. Welding performed for short time period on monthly basis for cutting and loosening parts.

Oxygen and acetylene and electric arc welding performed on metal table in engine repair shop. See photo number 8. Main activity is fabricating small steps for helicopters. Per FAA requirements, insured not allowed to perform major welding on helicopters. Large welding jobs performed by outside shop. Welding deviates from NFPA requirements in that less than 35 feet clearances to combustible construction. Welding performed for short period of time on metal table on small parts. Concrete floor under table. Consider hazards limited and controlled.

Not only do these two paragraphs seem to contradict each other, but it is also almost impossible to tell which ideas are central and which are subordinate. The sentences badly need connecting words such as *because* or *even though* or *in addition*.

These connecting words can lace your sentences together by indicating the direction of your argument. Some common examples of these words are

Chronology: *until, before, after, previously, earlier, later, meanwhile*
Location: *above, below, to the right (left), nearby, in the distance, behind, before, adjacent to*
Addition: *furthermore, moreover, besides, likewise, too, again*
Cause and Effect: *therefore, consequently, accordingly, because, there, so*
Comparison: *likewise, similarly*
Contrast: *however, yet, but, still, nevertheless, on the other hand.*

If the field investigator had used connective words, as well as other organizing techniques, the insurance report might read:

Oxygen, acetylene, and electric arc welding are performed in the vehicle service area an average of two hours a month. These welding practices meet the majority of NFPA requirements and clearances to combustibles greater than 35 feet. Hazards are well controlled: gas cylinders are chained upright on a movable cart and welding is performed on a concrete floor. In addition this welding is performed for a short period on a monthly basis for cutting and loosening parts.

In the engine repair shop, oxygen, acetylene, and electric arc welding are performed on a metal table (see photo 8). The main activity of this engine repair shop is fabricating small steps for helicopters. Per FAA requirements, the insured is not allowed to perform major welding on helicopters; therefore, large welding jobs are performed by an outside shop. The limited welding done by the engine repair shop deviates from NFPA requirements in that clearances to combustible construction are less than 35 feet. However, welding is performed for only short periods of time on small parts. In addition, there is a concrete floor under the metal table used for this welding. For these reasons I consider the hazards limited and controlled.

Writing Effective Sentences

The key to writing effective sentences lies in variety. Using the basic subject-verb-object kernel of your sentence you can vary sentence type, order, structure, and length. As you revise your work, consider whether you're getting enough variation into your sentences to avoid monotony for your reader.

First of all, consider the length of your sentences. The average sentence length in technical writing is twenty words. But the fact that this is an average means that some of your sentences will be longer and some shorter. You should avoid an extended series of short sentences with no indication of the relationships among them; create longer sentences through subordination and coordination. However, you should use short sentences occasionally to emphasize certain facts and to make them memorable to your readers.

After length, consider order. English sentences use three standard orders: loose, periodic, and inverted. In a loose sentence, the main idea is completed at the beginning of the sentence, but may be expanded by modifiers in the latter part. For example,

> Gasohol seems to be a viable alternative to gasoline, considering the current data.

In a periodic sentence, the thought is completed at the end of the sentence. For example,

> It is possible that, having examined all the competing factors, gasohol may not prove practical.

In an inverted sentence, the standard subject-verb-complement order is reversed. For example,

> Never will gasohol be the only answer.

Although most of your sentences will use loose order, you can also use periodic order for occasional variety. You can also sometimes use inverted order, although this usage should be less common.

Along with length and order, you can also vary the way your sentences start. There are many ways to begin your sentences other than the traditional subject-verb-complement sequence. You can begin with a phrase, such as

> Of particular interest to the industry was the fact that 75 percent of the yield was recovered.

You can also begin with a subordinate clause.

> Although the filtering system was incomplete, some measurements could be made.

You could begin with an adverb or adverbial phrase.

> Significantly, our projections were within 10 percent of our results.

Or you could begin with a coordinating conjunction.

> Yet these results remain to be verified.

You can also vary your sentence structure. Changing from simple to more complicated structures is another way to vary sentence type and sentence openings. Simple sentences have a single main clause: Gasohol is the answer. Compound sentences include at least two main clauses: Gasohol is the answer, but more distilleries are needed. Complex sentences have a main clause and at least one subordinate clause. The subordinate clause cannot stand by itself as a sentence. Here is a complex sentence: As soon as more distilleries are ready, Gasohol will be the answer. Compound-complex sentences consist of at least two main clauses and at least a single subordinate clause: As soon as more distilleries are ready, Gasohol will be the answer; the demand for oil will slump.

Finally, you can even vary your sentence type occasionally. Along with the standard declarative sentences (sentences making a declaration), you can sometimes use exclamations, commands, or questions for emphasis.

> These results are significant in their short term applications. But what of the long term?

Although the majority of your sentences will be declarative, with loose order and medium length, you should consider the numerous possibilities that exist for varying these factors.

Using Parallel and Coordinate Structures

Another way you can make your sentences effective is to use parallel and coordinate structures. Parallel structures in sentences use the same grammatical forms for items that have the same function. Julius Caesar used parallel structure when he said "I came, I saw, I conquered" rather than "I came, I saw, and acts of conquest were carried out." People read grammatically parallel structures more rapidly than those that are not parallel. Notice the difference between the two paragraphs below:

1. Remove the air filter. Now you should look for the butterfly valve. When you find the valve, you can clean it by spraying cleaning solution into the intake tubes next to the valve.
2. Remove the air filter, locate the butterfly valve, and spray cleaning solution into the intake tubes next to the valve.

Parallel structures are similar to connective words in that they signal to the reader the relationship between pieces of information in your sentences. While connectives indicate complex relationships like chronology or location, parallel structures indicate that the pieces of information are of equal importance. Both connectives and parallel structures will simplify your reader's job by clarifying the relationship among ideas.

Repeating Main Ideas

Much of what you write will seem obvious and straightforward to you. But to your audience your conclusions and observations may be startling and unexpected. One of the most serious mistakes in professional writing is to assume that your reports will be "obvious" to your intended audiences. As you write you should repeat key ideas and terms, give your readers a chance to become accustomed to what you are saying, and avoid writing paragraphs that introduce four or five new concepts all at once. Contrary to what many writers have been taught, a little repetition, used correctly, is a good thing.

Finally, the best way to achieve a reader-based writing style is to develop your own sensitivity to good and poor writing. Most of us know when we are reading, or producing, poor work. We feel confused, mystified, even frustrated and angry. These "symptoms" are caused by writing that asks too much of its audience's memory capacity. You can write more efficiently by listening to what your brain tells you. When you revise your writing, you must learn to read your work as if you were a stranger. How to do that is the subject of our next section.

REVISING YOUR WORK FOR STYLE

When you revise your work for style, keep in mind the principles we have just discussed. Ideally, you will not need to do much editing for organization or content because you will plan and outline your work so carefully that the actual writing will go smoothly. (Outlining and organizing are covered in Chapter 5.) Whether you write with ease or difficulty, however, you will always need to revise, if only because no one produces perfection on the first effort.

Revision is a task that proceeds more smoothly and efficiently when it is well organized. In this section, we are chiefly concerned with how you can organize the revision process so that you can complete your writing assignments on time and with a sense of satisfaction. Organizing your revision helps you use your time efficiently, so that you do only as much work as you need to do.

Setting Limits

As a student you will probably spend more time revising your work than you will as a working writer. As a student you are still gaining experience in writing and have probably not yet achieved the level of effectiveness you would like. But working writers must learn to set limits to the amount of time they spend reworking their prose.

As a professional writer or a writing professional your task is to communicate effectively and work efficiently. Even though your job might require many hours of writing each week, you will have other important work to do. Fortunately not every report needs to be polished to the same degree.

In addition to the general limit on your writing effort set by the conditions of your job, you will discover a personal limit that changes as you gain more experience. When you start to work on a project, you progress rapidly, and soon produce an edited draft that is almost as polished as you can make it. As time goes by, improvements take longer to achieve, and gradually you reach the point where you have a completed rough draft. This represents your best effort, the maximum quality you can achieve after a considerable expenditure of time. From that stage you can spend hours or days working on your draft without making much headway, and further changes do not significantly improve the quality of your work.

Even as a student you can learn to recognize these stages in your own writing behavior and then aim for a version that is good, if a bit rough, or your best, depending on the occasion and the audience. We are not suggesting that you produce sloppy work or that you ignore the necessary process of revision. Don't let laziness convince you that substandard work is acceptable. But professional writers often aim for an acceptable rather than a perfect report, and they devote a large proportion of their writing time to prewriting activities such as organizing and audience analysis. Setting these limits saves them valuable time in the revising stage.

Revising: Step by Step

Once you understand that your goal as a writer is to produce competent, communicative prose, it is much easier to use an almost mechanical, step-by-step revision process. Keeping a distance from your work is the first step in revising. A Roman author recommended that aspiring writers keep their work in a drawer for nine years before showing it to anyone. If it still looked good after nine years, it was probably worth reading. Sometimes you won't have nine minutes, let alone nine hours or days, to revise a memo or a proposal. Still, you can achieve a sense of distance from your work by looking at it one feature at a time. Remember that your object is to judge and correct your work as impartially as you can, and you can do this best when you are not concerned about the overall quality of what you have written.

In the first pass through your draft, check the organization. Read the first and last sentences of paragraphs. Did you include transitional and introductory material? Have you summarized sections? If you are working on a long report, you should take a sheet of scratch paper and jot down the headings or make a short note about what each section contains. Many writers find that this "global" approach to revising produces the most results in the shortest time. You may find to your surprise that the final draft has deviated strangely from the outline. If this is the case, you will probably need to move paragraphs, pages, or sections around. You might discover that you made the same points in two widely separated sections of your report or that you have followed a tangent into territory far from your actual topic. If this is the case, you will need to throw out entire sections, and this is far easier if you start the process of discarding by simply drawing a line or two through some entries on the scratch outline you have made while reading your draft. Once you make that commitment on the scratch outline, it hurts less when you go back to the report itself and begin to throw page after page into the trash.

Finally, be sure that your introductory sentences repeat the content of your headings and subheadings. Remember, some readers use headings to skim and rest their eyes. A reader who is plowing straight through an entire report generally rests at each heading and consequently misses reading it. For that reason, you cannot expect headings to serve as transitional devices. Repeat the content of the heading in the first sentence of the new section.

During the second reading, which should take place after a break, check the content. Review your purpose in writing and your audience's purpose in reading your work. Have you written for the audience? Are you accomplishing your purpose? Are your arguments complete? Or have you left out premises and supporting data that seemed too "obvious" to include?

In a third reading, again after a break, scan for problems of grammar, usage, and style. By this time you should be so used to your own writing that your mind will stray at any vague passage. Be sensitive to these problems but remember that you are aiming for competence, not perfection.

Some writers use more elaborate systems, adding separate checks for spelling, punctuation, and graphics. However you choose to divide the editing process, remember that multiple pass editing works well because it helps you gain distance from your work by focusing your attention on one feature at a time. However you choose to proceed, remember that "global" revision is just as important as sentence-by-sentence stylistic revision. In fact global revision for organization, audience, and purpose often turns out to be far more important than sentence or paragraph editing.

Revising with Word Processors

The advent of word processors has made many aspects of the revision process much easier. The general routine is this: You enter your text into the computer and revise in either of two ways.

1. You can work with the report page by page on the processor screen, inserting revisions into the text.
2. You can generate a printed text ("hard copy") from the processor, mark your corrections on the text itself, and enter those corrections into the computer.

Whichever method you use, you can continue to make various changes in your text—adding, deleting, rearranging, or correcting—until you are satisfied. Then you can use your word-processing program to tell a computer printer to print your report. Depending on the printer available to you, the report will be generated at speeds varying from 12 characters per second (cps) to 160 cps or faster.

In addition to changes you decide upon yourself, there are various programs for checking the mechanics of your report. Spelling programs check your text against a dictionary in the computer's memory to find unfamiliar spellings. Other programs will monitor repetition and other aspects of grammar, usage, and syntax. Some programs will catch clichés and recommend stylistic revisions. If you use these programs, you need to find out whether you must reformat the entire report afterwards. For example, if your spelling program changes the length of a word when it corrects a spelling or typing error, the length of the line may change and, in some cases, push part of a word beyond the margins that are set in the computer's memory. If you printed the report before rejustifying the margins, you would either get rough margins or lose parts of words. The importance of this issue depends on the sophistication and design of the word-processing program you use; read the documentation carefully and ask questions.

Some writers also use processors for "incremental" report writing. They write up parts of their research while the project is in progress, and at the project's completion they recall the sections they have already completed and rework them into a final report.

Processors obviously offer many advantages, but they also have drawbacks. Many writers find it difficult to revise directly on the processor screen without printed copy. Because most microcomputers allow you to see only a portion of a page at a time (a third of a page is common), it is difficult to study the entire text. Moreover, even simple changes such as replacing one word with another may require a series of commands to the computer. Without a printed copy, it is possible to lose track of the change that you wanted to make during the time that it takes to enter commands.

These problems can be offset somewhat if you work with the printed text rather than revising directly on the screen. However, generating a printed text is not always possible, and some writers feel it is an inefficient use of processor capabilities. This particular problem can be resolved if you use a computer printer that has a large buffer. A buffer is a separate memory unit that receives data from a computer and then feeds it to the printer. By using a buffer you can use your computer for writing or other work while the printer is operating. Without a buffer, your computer will need to feed data to the printer while you stand idle or find other tasks to do that do not require the computer.

There are also limitations in the processing programs. Spelling programs can recognize only words that are misspelled as nonwords. If you misspell a word so that it becomes another existing word—if you type *form* rather than *from*, for example—the spelling checker program will not detect the error. For the same reason, these programs will not distinguish between misspelled homophones (*their, they're,* and *there*, for example). Obviously, these programs can supplement, but not replace, your own proofreading.

A more serious problem is that some writers using a word processor become more concerned with minor, cosmetic changes than with substantive ones. The computer makes changes in wording and phrasing quite simple, but changes in organization and development—additional explanations, say, or changes in the order of paragraphs—are more difficult. Thus some writers are tempted to "polish" language and mechanics, rather than deal with more necessary and substantial revisions.

Finally, if you use a computer to write, make backup copies of your work. On most computers, you are writing to a memory chip and then saving that data as magnetic signals on some form of disk. Power shortages, lightning, dust, sloppy handling, and many other causes can erase the chip memory or the disk. Backup disks are also important when you are first learning a new system or using a laboratory computer that may have seen rough service. On some systems it is easy to touch a button that wipes out all your work. Many writers keep a minimum of one backup file on the same disk *and* a backup disk that is stored in a separate location.

You should also be careful not to overload your disks. On a single-sided, 5¼-inch floppy disk, you theoretically have space for 196 kilobytes of data, which is roughly 100 pages of double-spaced text. Many word-processing programs, however, will use up to two thirds of that space for their own operations. For example, with a simple form of WordStar, one of the most widely used word-processing programs, you are limited to about thirty-five pages of writing on a single-sided disk, because the program will automatically use up another thirty-five pages of space for a backup file and reserve yet another thirty-five pages of space for juggling data when you tell the program to move paragraphs or perform other large-scale editing operations. Some programs will notify you when you are approaching the limit of your actual usable space, others will tell you nothing until you suddenly discover that you must erase page after page before the program will allow you to exit, or print, or do anything else.

A comparison of word-processing programs and their features is beyond the scope of this book. If you are just beginning to use a computer for writing, you can find many guides on the market. As with other machines and routines, if you learn one system and one word-processing program well, you will find it easy to learn others later, because the basic features of most systems are similar. For a good comparison of medium-priced microcomputers and the software they support, consult *Word Processing for the IBM PC and PCjr and Compatible Computers,* by Carole Matthews and Martin Matthews (New York: McGraw-Hill, 1985). This book does not review lower-priced systems, such as the smaller Kaypro and Apple Computers, but it does introduce fourteen micros in the $2000 to $5000 range, and it reviews sixteen of the more common word-processing programs and their accessory programs for spelling, grammar, math, mailing, and so on.

SUMMARY

1. Most readers must make time to read; therefore, they read quickly and selectively, concentrating on material that meets their needs.
2. Good writing style makes efficient use of readers' time and can be understood in one reading.
3. The reading process is dependent on the short-term memory, which has two limits: time (a few seconds) and space (five to seven items).
4. Writing that recognizes the limits of the short-term memory provides clear contexts, eliminates jargon, and uses precise language, clear modifiers, appropriate connectives, parallel structures, and strategic repetition.
5. Writers should set limits on the revision process, striving for the level of effectiveness appropriate to the assignment.
6. The revision process should be handled in at least three stages: one reading for organization; another for content; and another for grammar, usage, and style.
7. Word processors can be valuable tools in the revision process if they are used carefully.

EXERCISES

1. Revise the following sentences for clarity.
   a. We will be in telephonic contact with you.
   b. No difficulties of an operational nature were discovered during that period.

c. After a cooling down period of around twenty minutes, the solution was seen to be red in color.
d. Turn the switch to off, the power cord should be disconnected, and check the left L.C.D.
e. We inspected the find. The position of the funerary display suggested high social position. There were no funerary vestments.
f. Utilization of recording devices is strictly prohibited within the facility.
g. The duties of the safety officer include inspection of work areas, education of staff, and he should file monthly reports.
h. Decompression becomes obligatory when diving personnel ascend from depths of over fifty feet down.
i. The roadbed was examined. The portions without shielding showed 8% erosion. The portion using geotextiles showed 2% erosion.
j. Advanced preparation greatly facilitated the utilization of hydrolysis apparatus.

2. Rewrite the following passage to improve its clarity.

 The agency's regulations regarding tax payment relief, effective 30 July 1974 (including Rule 18, in regard to applications for bankruptcy relief; Rule 42, in regard to elderly tax credits; and Rule 25, in regard to deferred payments) stipulate that all matters which individuals or their representatives wish to bring to the agency's attention which are not already covered by the stipulations relating to late tax payment or local property tax credits, shall be presented to the agency through a motion, indicating in full the situation in question, or the action or relief required, along with sufficient supporting materials including affidavits, depositions, and financial records, when questions of fact are involved, or citations of authorities, when the situation involves legal questions. The stipulation is also made that any neglect in completing in detail the forms required, or in furnishing other documentation which may be specified, shall be considered as adequate grounds for dismissal of the motion.

Writing and Revising Case 1

The Instruction Booklet

You have just been given your first assignment in your new job as a technical writer for Flash Electronics. Flash has recently purchased and begun marketing a radio/cassette player manufactured by another company, and you are working on a revision of the instruction booklet. The radio is a high-quality product, but Flash has been unsuccessful in cracking the market so far. After you read the original instruction booklet you begin to have an idea why. This is the section on safety instructions.

Safety Instructions

Read carefully this manual before unit operations. Retain this forever. It helps you depend on when operation is troublesome. To operate correctly follow this. Attention paid to warnings printed on this and unit.

Heat, Water, Air

- Do not place unit where is hot
  Like stove, appliances (amplifiers), hot registers, radiation, etc.
- Place unit away from water
  Like tub, sink, laundry, lavatory, swimming, etc.
- Place where is proper ventilation
  Not on bed, rug, cushion where is blocking openings

Cord

- Connect unit with power cord is proper
- Cord should not be walked on
- Objects should not pinch
- Recommended power supply only, pay attention to plug at convenience receptacles
- Unplug when DC power supply use
- Unit not use for two weeks should unplug

Internal Shorting Prevent

- Foreign objects should not enter accidentally
- Foreign objects or spilling may result in shocking
- Do not drop or push in ventilation any foreign objects

Service

- Unit should not service except as in instruction manual
- Unit should be service at Flash service station
- Unit should service when:
  - cord or plug damage
  - Unit has rain
  - Unit has change performance
  - Objects have fallen on or spilled on
  - Unit has drop

Pack unit for moving as same as when receive. Clean with cloth soak in synthetic detergent only. Ground if need as in instruction manual.

ASSIGNMENT

Rewrite and reorganize the safety instructions so that they will be clear to the purchasers of the radio. Remember that your audience may range from high school students to electrical engineers.

Writing and Revising Case 2

THE SIKES CHALK TREND

After your graduation with a degree in geology you take a job with a geological research company, Rogers Exploration. Rogers is interested chiefly in enhanced oil recovery, working with already drilled wells and using various methods to recover oil that cannot be reached by conventional means.

For your first assignment, you join a project already in progress: a field owned by Carolina Petroleum in the southern part of the state. Rogers is investigating the possibility of using downhole steam generation to recover oil from the field, and they have already completed several months of work.

To your disappointment, you discover that your first job will be to write a report rather than actually to work on the field. The owner of the field expects a progress report on the project every three months and the first three-month deadline is fast approaching.

Although you know very little about the project, you have something to work with. The former project director, now working on another project in Venezuela, dictated some notes on this project before he left. His secretary has made an exact transcript of the dictation for you; you read it with a sinking heart.

Transcript

This is a description of what we've done so far on this project, let's see, the Sikes Chalk Trend, Field 82. What we're talking about here, essentially, is EOR, I mean enhanced oil recovery, only what we're using is downhole steam generation instead of something like pumping down chemicals or gases. They use chemicals and gases now sometimes, but we're not going to use them here. Actually, they don't use chemicals and gases nearly as much as DSG, but they may use them more in the next few years because the technology's getting a lot better. But DSG is really where it's at right now.

Where was I? OK. What we had to do on this was do a study of the geological makeup of this field and see whether DSG would give them much of a return on their dollar. Because the field's not doing too hot right now. So, all right, we've done most of the analysis. What we've got here is, well, chalk, you know, like the name says. Sikes Chalk Trend and all. Except it's not chalk like blackboard chalk—you better explain that. What we've got is calcite, what do the profs call it? Lithified limestone mud. Only it's in layers with shale, interbedded with it, thin shale.

Do you think maybe we should explain DSG in case somebody at Carolina Pete doesn't know what it is? OK. See, they used to generate all the steam on the surface and then pump it down into the well, but that meant some heat loss because the steam had to travel so far. So sometimes you ended up injecting hot water instead of steam, which was a real pain. But some DOE* boys down at Sandia Labs came up with this thing, what would you call it, generator, I guess,-that you can put down the well bore all the way to the oil layer. Then you just pump down the water and the fuel and this thing makes steam and you put that steam right into that oil layer. Real slick. You can get these things, these downhole steam generators, from Tenneco or Chevron. These things work best when you can use them on a whole field instead of on a well at a time. Say you've got your hexagonal pattern of well bores—typical, right? Well, you just slip generators down all the peripheral wells and then the steam acts like a collar or a tightening band sort of and it forces all the oil in the reservoir up the collection bore in the middle of the field. So, OK, we've got the right kind of field here and we're trying to figure out whether the geological structure is OK for DSG.

Now you've got your chalk and shale here, but unfortunately the permeability isn't worth much. I mean it's less than 1 md.** So the movement of the oil to the bore isn't so hot. But there are a lot of fractures and faults in the rock, you got a lot of little local oil accumulations and flow channels. We figure they can enlarge these with some hydraulic fracturing so it shouldn't be a big problem. They've already done that higher up in the field. Porosity is good—from 4 percent up to 14 percent with a 6 percent average. That's a lot of available space in that rock for oil. If you took the whole trend and not just this field, we're talking at least five billion barrels.

There's a problem with calcite for DSG, though. It goes into solution real easy and that's going to really slow down

*Department of Energy
**millidarcy

that oil movement. They're going to have to keep that calcite cool—your steam's pumping in there at 100 degrees Celsius and calcite dissolves, begins to, at 110 degrees Celsius.

The viscosity of the oil is about 29 degrees, which is pretty close to normal. It'd be better if it was high because that steam is going to lower the viscosity. But it's OK. The average output for the wells here is around 40,000 barrels over 10 years which is about the lifetime of the well. That's pretty low, so they'd better be careful about evaluating EOR, make sure it's cost effective and all.

So, OK, we've finished all the field analysis, checked the geology of the field and the quality of the oil, so what have we got left? We still got to do the economic side of it. How much would it cost to bring in the DSG, how much more we could get out of the wells with it, how much the oil is going to bring per barrel (who knows, right?), and so on. They're going to want to know whether they can get enough extra with DSG to make it worth their while. Right now, with the prices they're getting, I'd say no, but give the finance boys a chance to tinker with it and see what happens. I mean don't say I said no in the report, OK?

ASSIGNMENT

Organize this material into an effective progress report (see Chapter 9 for progress report formats). First, decide on the information you'll include, then group and generalize as you learned to do in Chapter 5. Write an outline. Then write the report, using headings, forecasting, and transitional material to convey your organization.

CHAPTER · 7

Report Supplements

INTRODUCTION

Thus far we have covered the preparation of the most important part of your report: the text itself. This task will require most of your time. But formal reports may also have supplemental elements. Some, like the table of contents, are easy to prepare; others, like the abstract, may require more effort. You might use only two or three of these elements on a short report and you might never use all of them on a single report. Yet each contains a particular type of information that may be necessary to your audience, and it is necessary for you to be familiar with all of them. In this chapter we will discuss these supplements in the order in which they occur in a report. We will begin with supplemental material placed before the report: the title, the abstract, the letter of transmittal, and the table of contents and list of illustrations. Then we will discuss supplemental material that goes within the report: the headings and subheadings and the reference list. Then we will consider supplements that are placed at the end of the report: the glossary and the appendixes. We will close with a discussion of the general appearance of your report, which, although not a "supplement," plays an important role in its acceptance or rejection.

Before the Report

There are four supplements that you may use at the beginning of your report: the title, the abstract, the letter of transmittal, and the table of contents and list of illustrations.

Titles and Title Pages

Your title is, of course, the shortest part of your report, but it is also one of the most important parts. It is, after all, the only section of your report that every one of your potential readers will look at. It can cause your report either to be used or to be misfiled and forgotten.

Writing Your Title

There are several ways you can make your title helpful to your readers. First, your title should be brief, but informative. Titles are seldom written as complete sentences; they are phrases. Describe your subject and your focus with specific words. Be specific when referring to the topic of your report. A title such as "Studies of Certain Bacteria" will not allow your readers to make an informed decision about the report's usefulness to them; on the other hand a title like "Microscopic Studies of *E. coli*" will enable them to be more discriminating. Instead of "Analysis of Inorganic Compounds," use "Laser Raman Spectroscopic Studies of Fe(II) Compounds."

An effective title will indicate the scope and objective of the report and, if possible, attract the reader's attention. It should also include "key words" that will ensure that the report is indexed under the right topic. These are the words that indicate the subject of your report, the main topic. Computerized indexes use key words to classify your report by its contents; in the last title above, for example, the key words are "Fe(II) Compounds" and "Laser Raman Spectroscopic." The wrong key words in your title can doom it to oblivion under the wrong index heading.

Finally, be aware of your audience. Do not use technical terms that may be unfamiliar to them; similarly, avoid jargon that will be unknown to at least part of your audience. Remember, your title should be as informative as you can make it.

As an example consider a civil engineer interested in suggesting a new stadium design to a city planning department. The new stadium would use an air-supported fabric roof to reduce construction and maintenance costs. If the engineer were to title the report "A Proposal for a New Stadium Design," readers would have no idea what kind of design they were being asked to consider; their interest would probably be minimal. The title "A Proposal for a New Stadium Design Utilizing Structo-Fab Poly-Amalgamated Material," would be more specific but might confuse the audience by using unfamiliar terms. The best solution would be to title the report something like "A Proposal for a Stadium Design Using an Air-Supported Fabric Roof," which is both specific and audience-oriented. Chances are the report would find its way to the appropriate readers.

The Title Page

Most organizations require the use of a title page; sometimes, particularly in government reports, a special printed form is used for this purpose. You should follow the format of your organization, but most requirements for title pages include the title of the report and any subtitles, the name of the author or authors and their division or organization, the date of the report, and any identifying material common to your organization (e.g., report numbers, availability, type of report). The order in which this material is placed on the page will vary with the organization.

Abstracts

An abstract is a condensed summary of your report's contents. It is placed at the beginning of the report for the convenience of your readers; the reader can study the report abstract to decide whether to read the entire report or only a particular part. Occasionally a reader may use an informative abstract as a substitute for reading the report as a whole.

Abstract Types

There are two principal types of abstracts: descriptive and informative. The **descriptive abstract** is quite brief and describes the content of the report without giving any of the conclusions or recommendations. It is actually a table of contents in prose, a listing of the major topics covered in the report (see Figure 7-1), and it serves the same purpose as the table of contents: to give a brief overview of the report's major points. A descriptive abstract cannot serve as a substitute for reading a

FIGURE 7-1
A descriptive abstract

Abstract

This report discusses a stadium design for Tucson using an air-supported fabric roof. Disadvantages of current designs are outlined along with specifications for the fabric roof. Savings in construction costs are analyzed, as well as lowered construction time and greater space utilization. Maintenance and energy costs are also considered.

report; it can aid the reader only in deciding whether to read the entire report or only a particular section. Descriptive abstracts frequently appear in abstracting journals; they are usually written in passive voice.

Informative abstracts are more extended summaries. They include the report's conclusions and recommendations along with the major facts that support them (see Figure 7-2). The purpose of the informative abstract is to give the reader a capsule version of the report's major points and conclusions without the supporting data. These abstracts are usually included in a formal report or a journal article; they can be substituted for reading the report as a whole. If you are asked to supply an abstract of your report, you are usually expected to write an informative abstract.

Occasionally, in addition to or instead of an abstract, you will be asked to provide an **executive summary** or even just a **summary**. This is similar to an informative abstract, but usually somewhat longer (up to 1000 words). Executive summaries are provided at the beginning of a report for management readers; they frequently serve as a substitute for reading the report as a whole.

Writing Your Abstract

You should, of course, write your abstract after the report is completed, and your first step is always to reread the report. Look at your headings and subheadings to get some idea of your major topics and subtopics; it may be possible to base your abstract on your table of contents, particularly if you're writing a descriptive abstract. You might also work from your original outline.

For an informative abstract begin with a clear statement of the purpose and conclusion of your report; this will serve as the topic sentence of your abstract paragraph. Then list the major points in the report that support the central statement, either ordering them from the most to the least important or following the order you use in the report itself. Confine yourself to major points. Your abstract should not present your supporting data; that is the function of the report body. When you have finished, your abstract should include all the major points and conclusions of your report, but none of the details; for those your reader can consult the report proper. Obviously, your abstract should never include information that is not included in your report.

Abstract style should be very clear and simple; remember, the abstract is the first thing your reader will probably read after the title. Use complete sentences and don't leave out articles (*a,*

FIGURE 7-2
An informative abstract

Abstract

An air-supported fabric roof stadium design would meet all demands currently being considered for a new civic stadium in Tucson. This report recommends it be adopted as a design criterion.

Stadium designs have been evaluated with the purpose of constructing an attractive stadium that seats the largest number of people for the lowest construction and maintenance costs. Conventional stadium designs studied proved to be costly and difficult to finance. Designs using an air-supported fabric roof to cover the stadium were found to cut costs dramatically in other cities.

The properties of such structures make them ideally suited to Tucson's moderate climate. Air-supported roofs can allow a variety of innovative and attractive stadium designs. The stadium structures have lower initial building and maintenance costs, satisfy all safety codes, and allow greater flexibility in design. The fabric roof design produces a lower cost per unit area as the size of the structure increases. Stadiums with steel and concrete roofs show the opposite tendency.

an, the). The abstract should be self-contained, that is, your audience should be able to understand it without reference to the report. For this reason you should omit references to figures or tables and avoid technical terms and jargon that the audience will not understand until after they have read the report. You should also include the important key words that indicate the subject of your report; abstracts, like titles, are used by computer indexing services to index reports by contents. If you can, study the style used in abstracts written for other reports in your particular organization or journal, and use them as models for your own work.

When you write an executive summary, follow the general rules for writing an abstract. The summary should be self-contained, using simple style and complete sentences. However, executive summaries place special emphasis on the conclusions and recommendations of your report, since those are the primary interest for your audience. Begin with a statement of your conclusions and recommendations as the topic sentence of your summary, then list your major supporting points so that the executive can see how you arrived at your conclusions. Again, confine yourself to major points and support; if your readers want details, they can consult the report itself.

Letter of Transmittal

When you are directing your report to a particular individual or organization, it is customary to include a letter of transmittal to introduce your report and place it in context for the reader. The letter reminds the readers of why the report was written in the first place and helps them decide how to treat it: to read it immediately, pass it on to another reader, or file it for reference.

Because your letter is directed to a particular audience, it should be constructed with that audience in mind. Consider what they need to know before reading the report, and what concerns are likely to be uppermost for them. Keep your tone personal, though respectful; consider how you would address your readers if you were speaking to them in person.

The letter of transmittal is occasionally bound within the report, usually directly after the title page. Frequently, however, it is sent separately. If you do so, be sure to indicate to your reader when and how the report is being sent: regular mail, private carrier, or hand delivery.

Writing Your Letter of Transmittal

Usually a letter of transmittal is brief, around three to four paragraphs. The first paragraph is your introduction. Here you remind your reader of the occasion for the report: the subject and purpose of the report, why it was written, and significant dates in the project. Be sure to include the title of the report, especially if the letter is to be sent separately. Your second paragraph is the body of your letter. Here you mention the material of most interest to the particular reader. You can call attention to features of the report that might be of special interest to the reader or you can explain anything unusual or unexpected (if, for example, you have changed your emphasis in the report since you last contacted your reader). You can also acknowledge help you had in preparing the report and, for persuasive reports like proposals, you can add reasons why your recommendations should be accepted. Occasionally you can summarize the report's contents; however, you will probably also have a summary in the report itself, so an additional one in the letter of transmittal may not be necessary, particularly if the letter is bound within the report. Your third paragraph is your conclusion. Offer to be available for consultation or discussion and tell your reader where and how you can be reached. You may also express the hope that your report will be satisfactory to your reader, although some authorities advise against this, since it implies the possibility that the report might *not* be acceptable. At any rate, close positively and with an indication of your willingness to be of further help.

For an example of a letter of transmittal for the report about the stadium design, see Figure 7-3.

***Figure* 7-3**
A letter of transmittal

Jonas Berghoff
Berghoff Associates
128 Pinon
Tucson, Arizona

March 24, 1984

Dr. Harriet Tobin
City Planner
Tucson Department of Public Works
Tucson, Arizona

Dear Dr. Tobin:

The enclosed report, A Proposal for a Stadium Design Using an Air-Supported Fabric Roof, is the final product of the Stadium Design Committee organized on June 19, 1983. It discusses in detail the conclusions I described to you in our telephone conversation of February 14.

I think you will be particularly interested in the section on construction costs beginning on page 10; these turned out to be even lower than we anticipated. The section on energy costs beginning on page 17 has been considerably expanded since our conversations in September, and I feel you'll be pleased with the result.

I will be happy to answer any questions you or your staff might have about our work; I can be reached at 555-7621, although I will be out of town from March 31 to April 5. Jane Colby can also provide information at 555-7628.

Sincerely,

Jonas Berghoff
Consulting Engineer

JB:msw
encl.

Table of Contents and List of Illustrations

A **table of contents** is a convenience to your reader, and it allows your report to be read more efficiently. The table of contents gives your readers an overview of the report, indicating the relationship between the parts, and it allows your readers to select the most useful sections for their purposes. You should include a table of contents for any report of ten pages or more; they are usually not included in short, informal reports or in journal articles.

If you have more than five or six illustrations, both tables and figures, you must list them separately in a **list of illustrations** immediately after your table of contents (the only exception to this rule occurs when all illustrations are placed at the end of a report rather than in the text; then the illustrations are included in the table of contents). Sometimes these tables, figures, photographs, and other graphic aids are called exhibits, sometimes illustrations or figures and tables. You should follow the style of your organization (for more on graphic titles, see Chapter 8). If you have only tables or only figures you can simply call your list "List of Tables" or "List of Figures."

Writing Your Table of Contents and List of Illustrations

If you have constructed an effective outline, you can take your table of contents directly from that (for suggestions on constructing an outline, see Chapter 5); the same rules apply to both elements. You should make clear the subordinate and coordinate relationships between your various entries; subsections should be indented below major sections and sections of parallel importance should have parallel placement on the page.

Any entry in your table of contents should also appear in your text as either a heading or a subheading; make sure that the wording and any numbering system that you use is the same in both places. Your table of contents should include all major headings in your text and at least the first rank of subheads; three levels of headings and subheadings is usually considered enough. Be careful to be inclusive but avoid cluttering your table of contents with too many items. Your goal is to enable your reader to find individual items easily; too many entries will defeat that goal. Figure 7-4 is an example of a table of contents for the fabric roof report.

When writing your list of illustrations, always list figures and tables separately (anything not a table or a photograph is considered a figure). Photographs may have a separate listing or be included and numbered as figures. List illustrations in

Figure 7-4
A table of contents

TABLE OF CONTENTS

their order of appearance; include the illustration number, its title, and its page. Again, be sure to use the same title and number in the text that you use in the list.

Within the Report

Two supplemental elements may be necessary within the report itself: headings and subheadings and a reference list.

Headings

Headings help your audience follow the organization of your report. Your forecasting and transitions work within the text itself to indicate the relationship between sections, but headings allow your readers to understand at a glance what the major and minor subdivisions are. Your headings and subheadings call attention to your topics and subtopics; they also signal changes of subject. In general your headings should help your readers find the material they need more quickly; they will also divide the report into easily understood sections, making it easier to follow.

You should make your headings brief but informative, like your title. Try to make them specifically describe the material that follows; although you can sometimes use standard titles such as *Introduction* or *Conclusions*, remember that these tell your reader very little about the contents of the section that follows. In the engineer's report about the fabric roof, a heading like *Costs* will not tell readers much. *Construction Costs for the Fabric Roof* would be more helpful to them. Occasionally a scholarly journal will require you to use certain standard headings for a report of research, but in most cases the more particularized your headings are, the more effective they will be.

Writing Your Headings

If you write an outline when you put your report together, you can frequently take your headings directly from that, particularly if you have used a phrase outline (for more on outlines see Chapter 5). Headings should always be phrases rather than sentences. As with your outline, these phrases should be parallel grammatically; all of your major topic headings should be parallel with one another, and your subtopics should be parallel with other subtopics in a section. For example, if one of the engineer's main headings were *Construction Costs for the Fabric Roof*, the next could not be *How the Fabric Roof Is Maintained*; that would not be parallel. The parallel heading would be *Maintenance Procedures for the Fabric Roof.*

The number of headings you use will differ with each report. You should use enough headings to indicate your topics and subtopics clearly, but not so many that the report is fragmented and hard to read. Remember, not all sections need to be divided into subtopics, just those where such a division is appropriate and helpful. Also, be sure that your table of contents uses the same wording as the headings; otherwise your reader will be confused.

There is no standard format for headings; usage will differ from organization to organization. Usually three levels of headings and subheadings are enough, but for long reports up to five may be possible. Journals, however, usually accept only two.

There are several possibilities for making headings prominent and subheads less prominent, even with conventional typing (typesetting will allow even more possibilities). First of all there is the size of the letters: CAPITAL LETTERS appear more prominent than lowercase letters. You can also vary the placement of your headings on the page; a centered heading appears more prominent than one placed against the left margin; likewise, a heading on a separate line will appear more prominent than one on the same line as the text. Underlining is another way to emphasize a heading; you may even use double underlining.

Using all of these options you could create the following hierarchy: the first level headings that you use for your major sections could be centered, capitalized, and underlined. The second level headings, subheadings, could be centered and capitalized, but not underlined. Your third level headings could be on the left margin, capitalized and underlined. The fourth level could be on the left margin, underlined and the first letters capitalized; and the fifth level could be on the margin, on the same line as the text, and underlined with the first letter capitalized. Such a hierarchy of headings would look like this:

FIRST LEVEL HEADING
SECOND LEVEL HEADING
THIRD LEVEL HEADING
Fourth Level Heading
Fifth level heading. Followed by text.

This is only one possibility; you can think of others based on the needs of your report.

No matter what format you use for your headings, consistency should be a major concern. Headings of the same level should all have the same position, typeface, and underlining. Any inconsistency is liable to confuse your readers, making them wonder whether this section is actually a subsection or a major division. Likewise, you should be careful not to use the same typeface and position for any other elements of your report, such as titles of lists or instructions. Make sure there is no confusion as to what is a heading and what is something else.

A final concern is the relationship between your text and your headings. The first sentence in a section should not refer to the heading as if it were part of the text. For example, if the engineer's heading were Construction Costs for the Fabric Roof, the first sentence beneath that heading should not be, "These were lower than expected." Like a title, the heading is an independent unit and not part of the paragraph that follows it. Moreover, the first sentence of the paragraph below a heading should repeat the material in the heading; many people who read the report straight through will skip over the headings or use them strictly for eye relief. You cannot rely on people to read the heading itself. Thus the engineer should begin the paragraph beneath Construction Costs for the Fabric Roof with a sentence like, "The construction costs of a typical fabric roof are consistently lower than the costs of conventional designs."

The Reference List

There are both ethical and practical reasons for documenting your sources. From an ethical standpoint you need to indicate what material in your report is the result of your own research and what material is taken from work done by other people. Not only does documentation pay proper tribute to other workers in your field, but also it indicates that you've done the necessary literature search and know the important research that has already been done in your area. By citing previous research, you place your own work in context and show how it fits into the ongoing research process.

But documentation is also a service to your readers. It allows them to find more material on the same subject from other researchers. You are giving them a list of reputable sources for further reading.

In most cases you should cite the source of your data whenever you're referring to information you have taken from another researcher. You should cite the sources of any direct quotations or paraphrases that you use and the source of any facts that you feel are likely to be controversial. However, you need not cite sources for data that can be found in any source you consult; this data is referred to as "common knowledge." For example, you would not need to give a source citation for the information that oxygen and hydrogen unite to form H_2O; this qualifies as common knowledge. However, if you were referring to an unusual or not widely accepted theory about how this union takes place (usually a theory based on the research of only one or two people), you would need to give a source citation. Give source citations only for material actually used in your report; if you read other material and did not use it, or used it only as general background, you would not include it in your documentation.

Reference List Types

No one format is universally accepted for source citations. According to Robert Day (in *How to Write and Publish a Scientific Paper*, Philadelphia: ISI Press, 1985), a recent study revealed that in fifty-two scientific publications, there were thirty-three methods of documenting sources. However, no matter what format you use, be careful to record complete bibliographic material about each of your sources (i.e., author's name, title, publisher, and place and date of publication) as you do your literature search. By noting this material you save yourself the bother of going back to look up a page number or a place of publication after your research is completed. In the next sections we will describe four common reference formats.

Footnotes Some organizations still use the footnote or endnote system with which you may be familiar. In this system your citations are numbered consecutively using superscript numbers (numbers above the text) placed next to the material you're citing. This number refers the reader to a note placed either at the bottom of the page (footnote) or in a list of notes at the end of the report (endnotes). This system has the virtue of familiarity since most people have used it at some time; however, it can be cumbersome for both the reader and the writer since it requires a separate note for each citation. It is also expensive for publishers since footnotes create page makeup problems, and endnotes can require many pages.

Scientific and technical writers have developed alternative documentation systems that use a single list of references at the end of the report or chapter rather than footnotes or endnotes (although footnotes may still be used for short definitions or commentary). All citation numbers then refer to this reference list. There are three major reference list systems: author's name and year of publication, citation order, and alphabetical order.

Author's Name and Year of Publication Using the author's name system you put the last name of your source and the year in which the source was published in parentheses next to the material being cited in the text; it looks like this (Smith 1971). You may also include the page number of the material being cited by putting it after a colon following the date (Smith 1971:38). If you are using more than one work published by the same author in the same year, you can distinguish them with letters (Smith 1971a). The reader can then refer to the reference list at the end of your report to get the complete title and publication data for the source. This is a convenient system for a writer, since citations can be easily added and deleted during revision.

Citation Order With the citation order system, you list your references on a numbered reference list in the order that they occur in the text. Then each time you cite a particular work, you refer to the number of the reference. For example, if your first reference is to a work by Smith, Smith is number 1 on your reference list; then each time you cite that particular work by Smith you write a 1 in parentheses or a superscript 1, even if it is the last reference in your report (see Figure 7-5). This system is easy for readers to use unless the report is a long one with many references; then it may be hard to keep track of which reference is which.

Alphabetical Order With alphabetical order the reference list is set up alphabetically, using the authors' last names. As with citation order, you cite a work by referring to its number on the reference list. Thus if your first reference is to a work by Smith and that work is number 8 on your reference list, the first reference number in your text will be 8. This is an increasingly popular system since it is convenient for both author and reader; obviously, the reference list must be constructed before the numbers can be inserted in the text. As with the citation order system, the reference numbers can be either superscript or in parentheses.

Writing Your Reference List

There are many different formats for reference list entries; be sure to follow any directions you are given by your organization or by journal editors. Here is one format developed by the American National Standards Institute (see Figure 7-6 for an example of a reference list).

For book entries you include these items in this order:

1. The author's name, last name first; if there is more than one author, separate the names with a semicolon. Follow the author's name with a period.
2. The book's title, capitalizing the first word and any proper nouns; follow the title with a period.
3. The series title, if the book is part of a series, followed by a period.
4. The edition, if the book is other than a first edition, followed by a period.
5. The city of publication, followed by a colon.
6. The publisher, followed by a semicolon.
7. The year of publication, followed by a period.
8. The volume number, if there is one, followed by a period.

The entry should look something like this:

> Smith, Hortense; Jones, Vincent. A terribly technical book about Hoover Dam. Dam Books 5. Fourth edition. Los Angeles: Obscure Publishing Company; 1986. Vol. 6.

FIGURE 7-5
Text with reference numbers

THE AC GAS-PLASMA DISPLAY

The gas-plasma display cell consists of a glass envelope filled with a gas (frequently neon) which is at a low pressure (3). Conductors surround the gas, and a spacer is used to keep the glass envelope walls separated because of the low pressure needed for the gas-plasma cell. When this gas is subjected to a high voltage (the threshold voltage), the electrons of the gas become excited and release energy in the form of a bright, orange-red light (5).

FIGURE 7-6
Alphabetized reference list

REFERENCES

1. Adkisson, Gerald. Electronic design. New York: Beta Press; 1981.
2. Apt, Charles; Ketchum, John. Flat panels. Computerworld. 19(4):ID11–16; 28 July 1985.
3. Display devices. Electronic Engineering. 57(700): 104–135; April 1985.
4. Hector, Gary. The race to perfect the flat screen. Fortune. 109(11): 97–106; 28 May 1984.
5. Shuford, Richard S. Two flat-display technologies. Byte. 10(3): 130–136; March 1985.

For periodical entries you include these items in this order:

1. The author or authors, last name first, followed by a period.
2. The title of the article with the first word and any proper nouns capitalized, also followed by a period.
3. The name of the periodical, each major word capitalized, followed by a period.
4. The volume number.
5. The issue number in parentheses, followed by a colon.
6. The page numbers, followed by a semicolon.
7. The date of the periodical, followed by a period.

The entry should look something like this:

> Jones, Aloyisius. A terribly technical article. Technical Jargon. 22 (10): 1–32; 1 October 1985.

For technical reports published by the government or by other organizations, you can use the following format:

1. The author's name, last name first, followed by a period. Should there be a corporate author (a government agency, for example), simply put it in the author's position, capitalizing each major word.
2. The complete title, first word and any proper nouns capitalized, followed by a period.
3. The year of publication, followed by a semicolon.
4. Where the report can be obtained.

The entry should look something like this:

> U.S. Department of the Interior. A terribly technical report. 1986; Government Printing Office.

This is only one format and may not be the one which your organization prefers; in this respect as in other parts of your report, let yourself be guided by the standards your organization has developed.

After the Report

At the end of the report you may use two supplemental sections: a glossary and one or more appendixes.

Glossary

A glossary is an alphabetized list of terms and definitions usually placed at the end of the report, although it may occasionally appear at the beginning. Glossaries are particularly useful for mixed audiences; by using a glossary you can provide information for general readers without cluttering the text with definitions an expert would already know.

You should use a glossary when there are several terms (five or more) in a formal report that need defining. In this way you can avoid breaking the flow of the discussion with definitions. When you use a glossary, you should mention it in the body of the report and direct readers to its location. Indicate in your text which terms are included in the glossary, perhaps by marking each one with an asterisk the first time it appears. Sometimes you may want to include a working definition of the term in the text if it is particularly important to your discussion; then you can give a more extended definition in the glossary for those who need more detail (for more on writing definitions, see Chapter 12). Remember, although glossaries are useful, they do not relieve you of the need to adapt your terminology to your audience. If you find that your glossary contains so many terms that your readers must refer to it constantly, you probably need to simplify your technical language.

Writing Your Glossary

Your glossary entries should be brief, concise, and clear. Be particularly careful not to use unknown terms in your definitions; this will only produce more confusion. You should define any term you feel would be unfamiliar to some part of your audience; remember to consider the education and technical background of your readers when you analyze your text. For convenience it is sometimes a good idea to make your glossary a detachable page at the end of your discussion; then your readers can place the glossary alongside the text for easy reference as they read.

Occasionally you will also see lists of symbols or lists of formulas at the end of a report. These serve the same function as a glossary, helping readers who may not be familiar with the technical terminology used in that particular discussion.

Appendixes

The appendix section at the end of a formal report contains necessary information that is supplemental to the main body of the discussion. An appendix can include many things: it might present the raw data the report interprets; it might have explanations that are too long for footnotes; it might have documentation for statements made in the report; it might present extended versions of procedures of which brief accounts were included in the report discussion; and it might discuss material of interest only to part of an audience, such as extended calculations for expert readers.

The chief danger in using an appendix is that it may turn into a dumping ground for miscellaneous material. Never put material into an appendix just because you can't find a way to work it into the discussion itself; either the material should be eliminated or your report may need reorganization. Any material you place in an appendix should be relevant to the central point of the report; your main criterion for placing it in the appendix should be your need to avoid cluttering the thrust of your discussion with details.

Writing Your Appendix

When you compose an appendix, your first concern should be length. Try to keep the appendix short (five pages or less). You should have a separate appendix with a title for each item you include. Be sure to mention the existence of each appendix in your text, and direct your reader's attention to it at the appropriate time with a statement such as "for extended calculations see Appendix 2." If you can't find a way to work a reference to the appendix into your text, the chances are that the material in the appendix is irrelevant and should be eliminated.

Finally, your report should be understandable without reference to the appendix; any data or explanations that are so vital that the report is unclear without them should be placed in the discussion where they belong. Appendixes are solely a convenience to those readers who want more information about individual points in your text.

APPEARANCE

We would all like to believe that the content of our reports is more important than their appearance, and it is true that an elegant presentation won't save a report that has nothing to say. But it is a fact of life that readers notice the way reports look as well as the things they discuss. Consciously or unconsciously many people equate a sloppy appearance with sloppy thinking. You cannot overlook presentation if you want your report to have the best chance of being accepted.

Report Requirements

You should begin with the basics. Use a good quality, 20-pound (at least) bond paper. Erasable bond is convenient for a poor typist, but messy for a reader. Use standard type, not script; *never* submit a handwritten report unless you have specific approval in advance. Many firms and almost all journals have standard rules for report formats, including details like margin size. If you're given such "Rules for Authors," be sure to use them: In scholarly journals particularly, correct presentation may mean the difference between acceptance and rejection.

Preparing Your Final Draft

When your report is being typed, make sure to leave margins of 1 inch on top, bottom, and right side of each page; leave 1½ inches on the left side in order to leave room for a binding. When you have a title at the top of the page, leave a top margin of 3 to 4 inches. Double-space your report except for small units such as indented quotations, abstracts, or instructions, which should be single-spaced. Occasionally, for short informal reports or memos, you may single-space throughout; in this case, be sure to double-space between paragraphs.

You should begin counting your pages with the title page, but no number is printed on the title page (if your letter of transmittal is bound after the title page it may also be unnumbered). Your preliminary sections, including your table of contents and abstract, should be numbered with lowercase Roman numerals (i.e., ii, iii, iv, etc.—Roman numeral i is not used because it would be the title page) usually centered at the bottom of the page. Other material, including any appendixes, should be numbered with Arabic numbers (appendix pages are usually numbered as a continuation of the text, although they may be numbered separately with letters, e.g., A-1, A-2, A-3). Placement of the page numbers may differ with your organization, but customarily pages of text are numbered in the upper right corner.

Finally, when you receive your report from your typist (or even if you're typing it yourself), be sure to proofread it. After all the time you've spent putting your report together, don't let yourself be sabotaged by typos.

SUMMARY

1. Formal reports require one or more of these supplemental elements: title, abstract, letter of transmittal, table of contents and list of illustrations, glossary, headings and subheadings, reference list, and appendixes.
2. Titles should be brief but precise, using audience-oriented language.
3. Abstracts are brief summaries of reports; descriptive abstracts are prose tables of contents, and informative abstracts include conclusions and recommendations along with major points.
4. A letter of transmittal is directed to the primary readers of the report, telling them why the report was written and what it contains.
5. A table of contents gives the readers an overview of the report as a whole and allows them to select individual sections for reading; a list of illustrations is a comparable guide to the illustrations included in the report.
6. Headings are brief and informative phrases indicating the subject of the material to follow; they can be made more or less prominent through the use of capitalization, underlining, and placement on the page.
7. Documentation in technical reports frequently uses a reference list: a list of sources cited in the report, arranged

either alphabetically or in order of citation; many possible formats exist for citation entries.

8. A glossary is an alphabetized list defining terms that may be unknown to at least part of the audience.
9. The appendix contains supplemental information that may be of interest to some readers.
10. Reports should be typed with correct margins and pagination; proofreading is essential.

EXERCISES

1. Here are the headings and subheadings from a report, without any indication of degrees of prominence: Summary, Introduction, Need for home blood-glucose monitoring, Need for control of diabetes, Need to prevent complications of diabetes, Types of home monitoring, Ideal monitoring system, Urinalysis, Advantages, Disadvantages, Home blood-glucose monitoring, Reagent strips, Advantages, Disadvantages, Reflectance meter, Advantages, Disadvantages, Conclusions, Recommendations, Reference list.

   Write these out as if they were headings and subheadings of your report, using capitalization, underlining, and placement on the page to indicate which are major headings and which are subheadings.

   Using the same headings, create a table of contents for a report on home monitoring of blood glucose by diabetics.

2. Here are some report references:
   a. An article called Assessment of the psychological factors and responses in self-managed patients written by Andre Dupuis for the third volume of Diabetes Care magazine, issue 1, on pages 117–119. It appeared in the January-February issue of 1980.
   b. A report by George Rudins issued by the Defense Advanced Research Projects Agency in Santa Monica, California, called US and USSR electrode materials development. It was published in December of 1974.
   c. A book called Nondestructive testing handbook that was published by the American Society for Nondestructive Testing Inc. and written by R. C. McMaster in 1959. The place of publication is Columbus, Ohio.
   d. A book by Nicholas Cheremisinoff which was published by the Ann Arbor Science Publishers of Ann Arbor Michigan in 1979. It is called Gasohol for energy production.
   e. An article published in the Winter 1983 issue of the Diabetes Educator (volume 8, issue 4) called Self blood-glucose monitoring: an update by LeAnn McNeil. It was on pages 15–18.

Using the format from the American National Standards Institute on page 152 write these entries as they would appear on a reference list.

Report Supplements Case

THE MYSTERY REPORT

As a technical editor you've had some interesting assignments, but this one is unique. A scientist in the petroleum division has written a report about the general feasibility of ethanol as a fuel; it is directed to some marketing executives who were interested in the subject. The report is written, but the scientist has been unable to complete the supplemental material because of the press of other projects: he hasn't even had time to write a title!

TEXT

Ethanol, that is ethyl alcohol, has been proposed as an alternate automotive fuel. Those who back its use claim that any type of carbon-based material, for example, coal or even municipal solid wastes, can be used to make synthetic gas (carbon monoxide and hydrogen), which can be converted to ethanol using standard available processes. Ethanol therefore could be used as an alternative to petroleum-derived fuels. However, gasoline, like other liquid hydrocarbons, can also be produced from carbon-based material. Thus the question becomes: Is ethanol preferable to gasoline as a motor fuel? This report evaluates the costs of ethanol and suggests the conditions under which ethanol and gasoline are competitive

Ethanol has been used for some time to fuel racing boats and cars; its technical feasibility as a fuel is not in question. Since it is a safer fuel than gasoline in that it is less flammable, it is a preferred fuel for racing. Some racers also assert that engines run cooler with ethanol than with gasoline and that performance at high speeds is better because of increased horsepower. Ethanol has a higher octane number than unleaded gasoline; this permits higher compression ratios. The compression ratio for ethanol is 15:1, rather than the 8:1 of the typical engine designed to use unleaded gasoline.

Ethanol's toxicity and handling properties are roughly equal to gasoline. Ethanol produces less carbon monoxide and fewer nitrogen oxides than gasoline in combustion; it also produces no hydrocarbon vapors. However, ethanol does produce aldehydes and ethanol vapors, unlike gasoline. In addition, engines powered by ethanol are more difficult to start in cold weather because of ethanol's low vapor pressure and higher heat of vaporization. The addition of manifold heat, fuel injection, special carburetion, or the use of a blended-in starting fuel can counter this problem. Ethanol fuel systems must be manufactured from ethanol-resistant materials since ethanol can corrode some metals and cause deterioration in some plastics.

The major difference between gasoline and ethanol is combustion heat. Ethanol contains 50 percent oxygen by weight; thus it is already partially oxidized. For this reason ethanol's heat of combustion is less than one-half that of gasoline. Since twice as much ethanol must be used to deliver equal amounts of energy, the resulting transportation and distribution costs are twice those of gasoline.

Ethanol is capable of mixing with water in all proportions. Therefore ethanol must have special handling during transportation and storage to prevent dilution with water. Minor concentration of water would not affect ethanol's use as a motor fuel; however, appreciable amounts of water will affect the performance of an engine adjusted to run on a given fuel concentration. Environmental water—rainwater, condensation of humidity in tanks, and water added inadvertently in manufacturing procedures—is commonly present in most petroleum products. However, with petroleum-based products, the water layer settles to the bottom of the tank and is not regarded as a hazard. With ethanol this water layer would result in serious dilution of the product; thus the design of the current distribution system would have to be seriously modified to avoid water contamination.

The following section analyzes the impact of distribution costs on retail automotive fuel prices. The figures presented here are for straight fuel ethanol rather than for blends of ethanol and gasoline or ethanol and water.

To repeat, ethanol has higher distribution costs than gasoline. Moreover, because of the lower heat content, more than twice as much ethanol is needed to power an automobile, thus involving increased weight problems as well. For example, a gasoline tank with a capacity of 12.5 gallons (78 pounds) will give a car a range of 250 miles (assuming 20 miles per gallon). An ethanol tank for the same range would have to have a capacity of 28.2 gallons (186 pounds). Automotive fuel economy is inversely proportional to vehicle weight. The average ethanol-fueled automobile would lose 2 to 4 percent in fuel economy (on the basis of miles per million Btu) because of the increased weight of the fuel tank, along with its contents and associated piping and tank supports.

To offset this penalty in fuel economy, the price of ethanol would have to be 2 to 4 percent lower than gasoline at the pump. However, there are several factors which may work against this lowered price.

Biomass conversion for the production of ethanol can be influenced by many things. Regional variations will affect the cost of process equipment. Labor and material costs will likewise vary. Finally, energy prices in different regions can differ markedly.

One major limitation on the use of biomass processing nationwide is the availability of resources. There may be high costs for materials and components for plant construction, or inadequate raw materials for the conversion process. Process assemblies that have been tested only in pilot projects may prove inadequate when installed in large-scale production. There may even be difficulties in finding supplies of raw materials.

Evaluated strictly on the basis of the current prices of gasoline and ethanol at the point of production, ethanol is not competitive with gasoline as an automotive fuel. Either the price of ethanol would have to drop substantially or the price of gasoline would have to rise substantially before ethanol could be considered as a likely alternative.

For this reason I cannot recommend investing in any type of bioconversion plant at this time. The only economically acceptable way to work with ethanol currently would be to invest in a small plant which would process no more than 500 to 600 tons of biomass per day. The market simply will not support any large-scale investment.

If the price of oil should rise sharply in the near future, then the ethanol question can be reevaluated. However, at present any major type of alcohol production is not economically feasible.

ASSIGNMENT

Put this report into finished form. Compose a title, descriptive and informative abstracts, letter of transmittal (to Ms. Harriet Chernoff, Assistant Marketing Director, from Dr. Leo Goralsky, Petroleum Engineering), table of contents, and headings and subheadings (decide how to divide the report into sections).

CHAPTER · 8

Graphics

Introduction

The graphic elements of your reports help you explain relationships that are difficult to express in words; they also provide timely eye relief and allow you to emphasize important points. This chapter discusses how to decide when to use graphics; how to design a graphic to suit your audience and purpose; how to work with a professional artist; how to integrate graphics with your text; and finally how to choose among the common types of graphics by considering the uses of tables, graphs, photographs, line drawings, and organization and flow charts. We will start with the questions you should ask when you think you need a graphic.

Deciding When to Use Graphics

When should you use a graphic? Consider your audience first, by thinking through the items on the audience profile sheet in Chapter 1. Does your audience expect to see a graph of productivity whenever annual profits or output are discussed? Do they expect to see particular types of visual aids? For example, urban planning proposals typically include a map, an aerial photograph of the city (with important areas highlighted) and a time line showing the complete schedule of the work that is to be done by the planning staff. A group of proposal evaluators would probably feel that something was missing if these basic graphics were not included in a city planning proposal. Every audience will be familiar with certain types of graphics and unfamiliar with other kinds. Glance through reports or papers similar to the one you are writing. What did the other writers choose to use? When did they use graphic support? In other words, sometimes you should choose to use a graphic because it is conventional to do so in a specific setting. Once you have a sense of the audience's expectations, you can move on to other criteria for deciding when to use graphics.

Graphics are best used for showing detail and spatial relationships that are difficult to explain in words. The location of head bolts on an engine or the proper location of a surgical incision, for example, are extremely difficult to explain without the use of photographs or line drawings. Graphics also allow you to emphasize important points. A graph showing a rise in sales can underscore the percentages or gross sales figures you report in the text.

Graphics can even help you in your problem-solving process before a report is written. Putting your data in a table can help you see relationships and make comparisons. A graph can give you insight into trends; a diagram might help you to understand a system more thoroughly. By designing simple graphics as you study your data, you not only make your own writing task easier but also anticipate the needs of your readers for graphics.

Graphics can also be used to condense information that takes pages of text to present, and condensing information can make relationships memorable. Finally, sometimes you will decide to use a graphic because it is an easy way to convey information.

DESIGNING GRAPHICS

Graphics should be designed for simplicity of use and understanding. By following a few simple rules, you can design graphics that communicate effectively. Here we will consider simplicity of design and some common problems with the production of poorly designed graphics.

Audience Perception

Once again you should begin with the audience. What types of graphics is your audience expecting? What does the audience normally see in this type of report? What kinds of graphics are familiar to this group?

Next you should consider how graphics are viewed. In America we are used to viewing from left to right and from top to bottom. A graphic that violates those basic conventions will not succeed. Readers also expect to see the most important information in the center of a graphic, not at the sides. When you look at a graph, for example, what do you study first, the lines or the scales printed along the axes? Readers also expect that large print or heavy lines or other large graphic elements will convey information that is more important than anything in smaller print or lighter lines. They also tend to assume that graphic elements that look the same are related. For example, if you saw a pictograph that showed a large person, a small person, a large coin, and a small coin, you would probably begin to assume that there was one relationship between the large person and the large coin and another relationship between the smaller person and the smaller coin. Just as our language use is conditioned by culture and circumstances, our use and understanding of graphics is conditioned by certain conventions.

Simplicity

Make your graphics as simple and easy to understand as possible. Ask yourself what the main point is, and strip away any information or design features that are not necessary to making that point.

Reproduction

Another set of guidelines you must keep in mind as you design graphics concerns the printing, reproduction, and coordination of your graphics. We will consider guidelines in five areas: level of detail, proportions, company guidelines, color, and reproduction and enlarging.

Level of Detail

All graphics of a given type should have the same level of detail. For example, if your first architectural elevation includes people, trees, and details of exterior finish (bricks, wood grain, etc.), all the architectural elevations should have people, trees, and so forth. After seeing the first elevation the audience will expect that level of detail and will think something is wrong if it is not there.

Proportions

Your graphics will look better if they have the same proportions as the length and width of the page on which they are printed. Long skinny graphics are not attractive. Later in this chapter we discuss the special problems of proportion and scale when we discuss graphs.

Company Guidelines

Some corporations, agencies, or even specific projects have guidelines for the production and design of graphics. Ask if there are any that pertain to your report. We once worked on a project where the graphic design was approved by one committee only to find out on the eve of production that an agency policy restricted the use of color in reports to two "official" agency colors. Once again consider your audience and ask questions to learn what the audience expects.

Color

Color graphics do not reproduce well in black and white. A chart that looks great in green and blue can turn into a murky mess when it is reproduced in black and white. You should decide whether the project is going to use color, black and white, or both, and then have the graphics drawn in the appropriate color.

Reduction and Enlargement

Artists often produce oversized graphics. Be certain that your art staff knows what the final dimensions will be. Otherwise, you can face disaster. When fine cross-hatchings and shadings are reduced, they turn into dark grey or black. Percentage signs (%) turn into vague dots, small Greek letters become blobs, and so on. By the same token, a chart that looks fine on an 11-by-14-inch art board can look sloppy and unprofessional as a 2-by-3-foot lecture chart.

WORKING WITH A PROFESSIONAL ARTIST

As a student you will do most of your own graphics, but when you take a professional position, much of your graphic work will be done by professional artists who will work from the "roughs" or notes you provide to them. In this section, you will learn a few rules for working with graphic artists.

Arrange to meet with the artist to discuss the project you are doing and the main ideas you need to get across. Ideally the artist will sit in on some of the staff meetings. Discuss with the artist budgetary and other limitations, such as limits on color, transparencies, total number of graphics, style, and size. If possible provide the artist with samples of previous reports that are similar to what you are creating.

If you have specific graphics in mind, you might provide rough drawings that the artist can use as a basis for his or her own work. Include a paragraph or two explaining (1) the key points that must be made, and (2) the features of equipment that should show. If the equipment is unfamiliar to the artist, include an explanation or photocopies from handbooks or other product literature.

It is particularly important to provide rough sketches and explanations when the artist will not be able to consult with you directly. If the artist can meet with you, ask if a rough sketch is necessary. Some professional artists prefer to come up with the first sketch themselves, because they are the professionals who know how much art can show. In this situation, the artist will probably approach you with one or two sketches and ask for your suggestions or comments.

Integrating Graphics With Your Text

Graphics should always supplement your discussion, not take its place. You should make your points in the text and reinforce them with graphics. On the one hand you can't count on readers to look at every graphic, so the information in the graphic must also be presented in the text. Some readers use charts or photographs as occasions to rest their eyes. If that material is not discussed elsewhere, your point will be lost. On the other hand graphics must be able to stand alone; readers should be able to understand them without reading the text. Sometimes a reader will turn to the graphics only after finishing a chapter of text. Consequently, the graphics should have explanatory titles that remind the reader of their contexts and the points they are intended to make.

To relate your graphic to the text properly, you should both introduce it and explain it. The introduction can be quite simple; a sentence will usually be enough (e.g., "The relationship between the time lapse and the sample failure is shown in Figure 4."). But this introduction must be placed before the graphic in the text; otherwise, your readers may search your text in vain for some indication of the illustration's significance.

The explanation of the graphic can vary from a simple sentence (e.g., "As the graph indicates, production dropped off sharply after the first quarter of 1986.") to a short paragraph, depending on the complexity of the graphic and the sophistication of your audience. As a general rule, however, a graphic that requires an extended explanation may actually need to be redesigned.

Graphics should appear on the same page as the discussion to which they relate. Placing all your graphics in an appendix means that your reader must flip back and forth between text and appendix, which makes it difficult to follow the discussion. Many readers will skip such graphics altogether, and your labor will have been wasted. If it is not possible to put the graphic on the same page as the reference, most readers prefer that the graphic appear on the next page or sometimes on the preceding facing page if the graphic takes up most or all of a page. If a graphic is referred to throughout a text, you might consider the expense of a foldout page or an envelope in which the graphic can be stored when it is not in use. In every case, you should consider whether a graphic is necessary or merely of interest. If a graphic is merely interesting, you should dispose of it or place it in an appendix where it does not interrupt the reader's progress through the report. But if it is necessary, you should place it where your readers can use it easily for reference.

Remember that by the time readers get around to looking at a graphic, they may well have forgotten why it was important. Graphics should be able to stand alone. When you first mention a graphic in your introduction to it, tell the reader where to find it. A parenthetical notation such as this (Fig. 7) will do the job, but you are more likely to get the reader's attention if you write: "See Figure 7 on page 31," or even "Please turn to Figure 7 on page 31." It is even better to explain why the reader should make this effort by pointing out what Figure 7 contains. A full first reference, then, might look like this:

> Turn to Figure 7 on page 31 for a graph comparing the results of Groups A and B.

Every graphic should have a number and a full title that describes its content. Your graphic titles are similar to your report title in that both should be short and specific. A lengthy title such as "A Summary of Production Characteristics, 1985–1986" can be shortened to "Production Characteristics, 1985–1986." Extended discussion of the graphic belongs in the text, not in the title. You can either number all of your graphics consecutively from the beginning of the report, or with each chapter (as in this book). Tables and figures are numbered separately. Technically, a figure is anything not a table, including drawings, graphs, charts, and photographs. Figure numbers and titles generally appear at the bottom of the graphic, table numbers at the top.

Finally remember that every graphic should be clearly related to the discussion. Do not use graphics simply because someone has one around or because at one point in the development of a report someone asked an artist to draft one. Earlier in this chapter, we pointed out that drafting a graphic is a good way to help yourself see relationships and identify important points that you will need to make. But keep in mind that at a later stage of report development, you may need to discard that graphic or replace it with one that is better suited to your audience or organization. We once worked on a project where a map had been included in an earlier edition of a manual. The material in the map referred to a government subsidy program that had long ago lapsed, but the production staff continued to use the map for three reasons, all of them bad: The map had been expensive to print; it was the only chart in color; and the agency still had several thousand copies in storage. As a writer or a group member, you must be on the lookout for graphics like this that have a kind of momentum even though they have no purpose. The remainder of this chapter will review five common kinds of graphics and present their uses, strengths, and weaknesses, as well as some simple rules for constructing each type.

Choosing a Graphic to Fit Your Need

The following sections describe five of the more common types of graphics: tables, graphs, photographs, line drawings, and organization and flow charts. We do not provide exhaustive directions for constructing these graphics. You are probably familiar with the construction of some of them, such as line graphs; others, such as drawings, are beyond the scope of this chapter. However, after reading these sections, you should be able to recognize these graphics and construct simple versions of your own.

Tables

Tables are the best and simplest way to present numerical data. By using a table, you can handle many numbers in a relatively small space. We were once shown a report written for an aircraft company in which a good technical editor had replaced six pages of complicated specifications with a single half-page table. In a long report of twenty or more pages, such reductions can make a major difference.

Tables also make comparisons easier for your reader. Data can become lost in a block of prose; it can be difficult for a reader to trace what is being compared to what. In a table, on the other hand, the data is presented clearly, surrounded by white space. In a well-constructed table, comparisons are merely a matter of checking column entries. Table 8-1 is a typical numerical table.

Tables are also useful for nonnumerical data. Consumer groups, for example, frequently use tables to present the test results of various products and their subsequent ratings. Tables can be used in most situations where you are presenting a comparison. Table 8-2 is a sample nonnumerical table.

Tables do have serious drawbacks. Unlike graphs, tables can only present data; they cannot make any interpretations of that

TABLE 8-1 A Table with Numerical Data. Courtesy Southern Gas Association Pipeline and Compressor Research Council.

Compressor Manifold Mode Analysis

Mode description	*Measured frequency, Hz*	**Predicted frequency, Hz*
Cylinder/Manifold	15.2	15.8
Cylinder Vertical, Cantilever I	26.0	26.4
Cylinder Vertical, Cantilever II	31.0	29.7
Cylinder Resonance I	32.8	32.0
Cylinder Resonance II	36.4	33.4

*The predicted frequencies were obtained from a simulation model using measured values for distance piece, clamp, and flange stiffnesses.

TABLE 8-2 A Table with Nonnumerical Data. Courtesy of Keuffel & Esser Company, Parsippany, New Jersey.

Spectra Processor Trouble Shooting Guide

Problem/Symptom	*Probable cause*	*Solution*
1. DEVELOPER HEATER Push-Switch indicator light does not go on when switch is depressed. Developer reaches proper temperature after 30 minutes. DEVELOPER READY light is lit.	1. Defective indicator light.	1. Consult Technical Representative, switch may need replacement.
2. DEVELOPER HEATER Push-Switch indicator light does not go on when switch is depressed. Developer heater does not appear to be working. DEVELOPER READY light is not lit. DEVELOPER LEVEL LOW light is lit.	1. Solution is low in developer tank.	1. Fill Developer Tank with diluted 1:4.5 Developer Solution until solution touches bottom of fill pipe.
3. DEVELOPER HEATER Push-Switch indicator light does not go on when switch is depressed. DEVELOPER READY LIGHT does not light after 30 minutes.	1. Power to machine has been disconnected. 2. Fuse has blown. 3. DEVELOPER HEATER Push-Switch is defective.	1. Check incoming electrical. 2. Check fuse. 3. Check solution temperature and level.

DEVELOPER LEVEL LOW light is not lit.	4. The over-temperature control unit in electrical box has shut off heater.	4. Push reset button on Cromalox over-temperature control unit located in electrical box. 5. Consult Technical Representative. Switch or light may need replacement.
4. Drive and Pumps do not operate when DRIVE AND PUMP Push-Switch is depressed. Indicator light is not lit.	1. Power to machine has been disconnected. 2. Fuse has blown. 3. DRIVE AND PUMP Push-Switch is defective.	1. Check incoming electrical. 2. Check fuse. 3. Consult Technical Representative. Switch may need replacement.
5. Right developer solution pump is not working. DRIVE AND PUMP Push-Switch is illuminated.	1. Fuse has blown.	1. Check fuse.
6. Left developer solution pump and rinse solution pump is not working. DRIVE AND PUMP Push-Switch is illuminated.	1. Fuse has blown.	1. Check fuse.

data. Thus when you use a table you must include interpretation in your text; otherwise you run the risk that your readers may interpret your data differently than you do. Tables are also less suitable for showing continuity among data points. Graphs are far preferable for showing trends.

For an example of what a table can do, consider the following paragraph of data.

> During the last 10 years the cost of shooting six 30-second commercials for Taxon has risen as follows: expenses for the production company cost $75,000 in 1974, editing cost $12,000, while talent, music, and casting cost $11,000. Ten years ago the agency commission was $17,000. We also spent $4,000 for travel for locations and $6,000 for miscellaneous expenses. Total expenses were $125,000. In 1984 the production company expenses were $339,000 and editing cost $34,000. Talent, music, and casting cost $17,000 while the agency commission rose to $69,000. Travel for locations cost us $11,000 and miscellaneous expenses were $9,000. The total expenses for 1984 were $479,000.

How many times would you need to read this paragraph to get all the comparisons? At least twice, probably, and perhaps more than that. Consider the same data placed on a table as shown in Table 8-3.

In this case, you could probably make all the comparisons in one reading, and the total is even more dramatic. Now that you have seen the advantages and disadvantages of tables, you are ready to learn some of the details of table design.

There are two types of tables: formal and informal. The informal type is simple with minimal data; it has no title or number and frequently no lines. The informal table is really a convenience for the reader; it provides eye relief and it simplifies data. An informal table using some of the data in the previous example would look like this:

Production costs in 1974 were as follows:	
Production company	$75,000
Editing	12,000
Talent, music, and casting	11,000
Agency commission	17,000
Travel	4,000

Formal tables have specific parts and format. The parts of a formal table look like this:

Table Number ***Table Title***

Stub Heading	*Column Heading*	*Column Heading*
Line heading	Data	Data
Line heading	Data	Data

All tables have a title and a number, like any graphic. But your tables should be numbered separately from your figures. Place the title of the table at the top, unlike figure titles which are placed at the bottom of the figure. The headings at the top of the table are the stub and column headings; they have vertical reference, that is, the stub heading identifies the line headings while the column headings identify the items in the columns along with any units of measurement necessary.

The title of the table and the stub/column headings are separated by a line. Then the stub/column headings themselves are separated from the line headings and data by another line.

If footnotes are necessary to give any additional information about the table, they are placed at the bottom of the table

TABLE 8-3 *Costs of Six 30-second Commercials for Taxon, 1974 and 1984.*

Expenses	*1974*	*1984*
Production company	$75,000	$339,000
Editing	12,000	34,000
Talent, music, and casting	11,000	17,000
Agency commission	17,000	69,000
Travel for location shoot	4,000	11,000
Miscellaneous	6,000	9,000
Total	$125,000	$479,000

rather than at the bottom of the page to avoid confusion with footnotes cited in the text. Use lower case letters (a,b,c) or nonnumerical symbols (*,#,@) to indicate the notes in order to avoid confusion with data on your table (see the footnote on Table 8-1, for example). The independent (unchanging) variables in your data go in the far left column as line headings; the dependent (changeable) variables go in the columns to the right.

Always identify the source of your data on the table itself. Place the source directly below the title (as in Tables 8-1 and 8-2) or in a footnote at the bottom of the table. If your entire table is taken from another source, you should acknowledge this as well; place the acknowledgment either at the top of the table under the title or at the bottom of the table in a footnote.

Here are some final points to remember:

1. Align the numbers in your columns either by the farthest right numeral or by the decimal point if one is present.
2. Surround your tables with white space, but be sure not to extend the table beyond the margins of the page.
3. Try to fit the table on one page if at all possible. If it must be continued on a second page, write *continued* on the bottom of the first page and repeat the title and stub/column headings on the second page.
4. Double-space your data and triple-space between the title and the stub/column headings.
5. Avoid cluttering your tables with lines.
6. Arrange your line headings and column headings using some form of logic, such as alphabetical order, greatest to least quantity, most recent to most distant data. This will help your reader to make the necessary comparisons.
7. Make sure all the units of measurement in one column are the same.

FIGURE 8-1a.
Line graph of x = y, where x scale = 1.6 y scale.

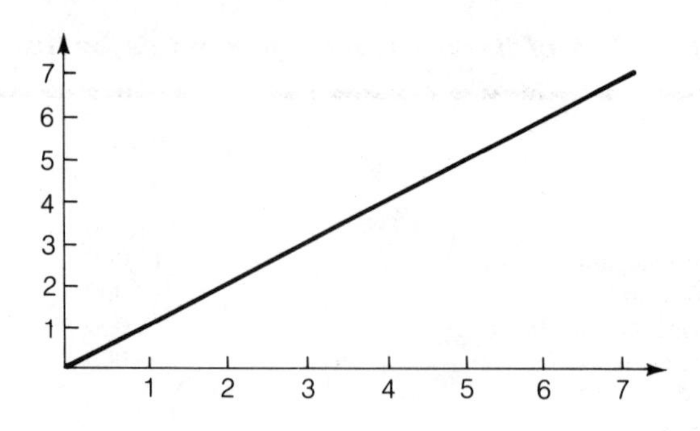

8. Use standard abbreviations, such as *temp* for *temperature*, in order to reduce the width of your column headings.
9. Do not use periods with abbreviations except for abbreviations that could be mistaken for words (e.g., *no.* for *number*).
10. Always capitalize the first word of your column headings, even if that first word is an abbreviation. Generally you will capitalize all major words in stub/column heads and only the first word (and proper nouns) in line headings or entries.

Graphs

Graphs come in many forms, including frequency polygons, bar and line graphs, and pie or circle graphs. We will consider the design of each of these varieties after we present five basic considerations: when to use a graph, how to label a graph, how to design accurate scales, how to place the zero, and when to limit the complexity of graphs.

Graphs are better than tables for showing continuity and trends, but graphs do not emphasize single points effectively, especially points that do not fit a general trend. And graphs are not a good way to show the relationships among several trends. Thus while tables emphasize discrete data points and make general comparisons difficult, graphs have the opposite strengths and weaknesses: Graphs make general comparisons easy and obscure the importance of discrete data points.

In line and bar graphs, it is conventional to plot the independent (unchanging) variable along the *x* or horizontal axis and the dependent (changing) variable on the *y* or vertical axis.

The scales of your graph should accurately and aesthetically present the information you are presenting. First, the relationship of the overall height and width of a graph should approxi-

FIGURE 8-1b.
Line graph of x = y, where x scale = ½ y scale.

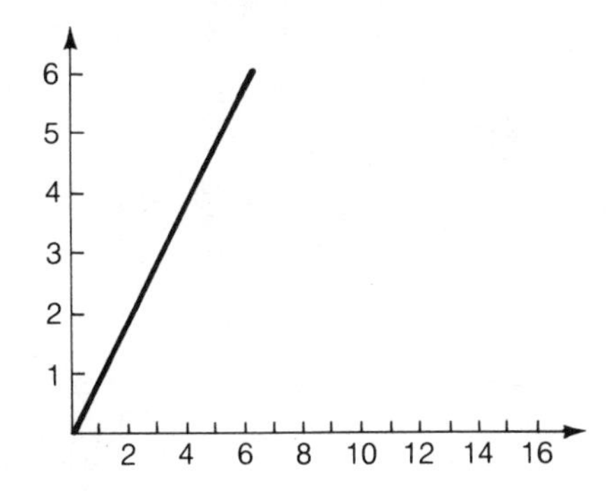

mate the "golden mean." In approximate terms, this means that the x axis should be 1.6 times as long at the y axis, and the scale should be designed according to the same proportions. In Figure 8-1a, the x and y axes are correctly proportioned. In each of the two panels, we have plotted the equation $x = y$. In the first panel of the figure, the units on the x scale are 1.6 times the units on the y scale. Notice that the line has a moderate slope and fills the entire graph. The effect is visually pleasing. In Figure 8-1b we exaggerate the scale. The units of measurement on the y scale are twice those on the x scale. A scale such as this might misrepresent whatever is represented by the y variable: progress, corporate profits, etc. By adjusting the scales, you can make a boom year look like a slump and vice versa.

Zero placement is related to the choice of scale. Usually, the x and y scales begin with zero. But suppose that the value you are plotting begins with 8 rather than 1, or 8 million rather than 1 million? You may suppress the zero value and start your scale at a higher value, as in Figure 8-2. But you must leave the axes separated or use a broken line at the bottom of the axes, so that your reader will notice that the zero value is not included. If you do not include the zero value, you can easily misrepresent the facts. If, for example, a company made $9 million one year and $10 million the next, a bar graph that suppressed the zero and started with $8 million would make the second year's profits look twice as large as the first year's profits. Such a graph would show one unit of $1 million for year one and two units of $1 million for the second year. The context would be lost.

Finally, graphs should not attempt to show too much. Circle graphs lose their value if the pie is cut into more than five or six sections, because it becomes too difficult to notice the differences in size among the slices. Line graphs become confusing if there are more than two or three lines. Always consider

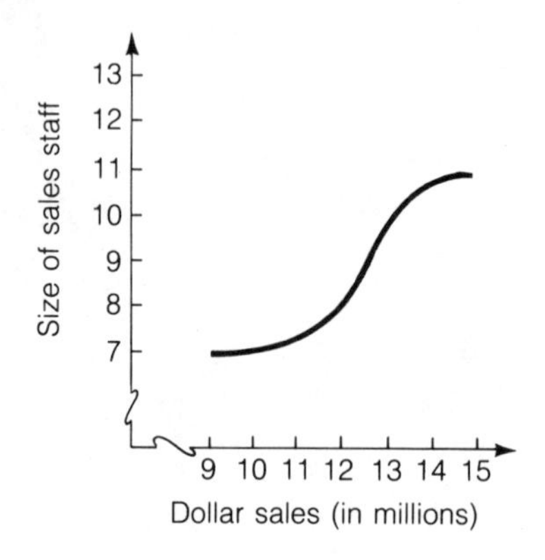

Figure 8-2
Graph with suppressed zero.

whether some level of detail can be eliminated without changing your point. With these basic considerations in mind, you are ready to examine several of the common types of graphs used in technical reports.

Frequency Polygons

Frequency polygons are used to show at a glance the relationship among the frequencies of scores in a statistical distribution. As with other forms of graphs, the lines between points do not necessarily represent a continuous sequence of data points; in fact, in a frequency polygon the lines do not represent data points at all, but merely the general shape of a hypothetical distribution curve. Figure 8-3 is a frequency polygon. Frequency polygons are related to other curves drawn to approximate a statistical distribution. If you present frequency polygons or any of these other graphs to a general audience, it may be important to point out to the audience that it is not correct to interpret every point on these graphs as actual data derived from samples or experiments. These lines show trends by connecting only a few data points.

Bar Graphs

Bar graphs are used for noncontinuous variables. Consequently, the true bar graph has space between the bars to indicate the nature of the variables. For example, in the hypothetical graph in Figure 8-4, it makes no sense to ask what might fit into the space between the bar for writing and the bar for ethics.

If a bar graph is used to represent a continuous variable, then it is called a histogram and there is *no* space between the bars. Histograms are useful if for some reason you want to emphasize changes in the slope of a distribution curve. A continuous curved line on a line graph might not reveal a point of change as clearly as a series of bars standing side by side. To

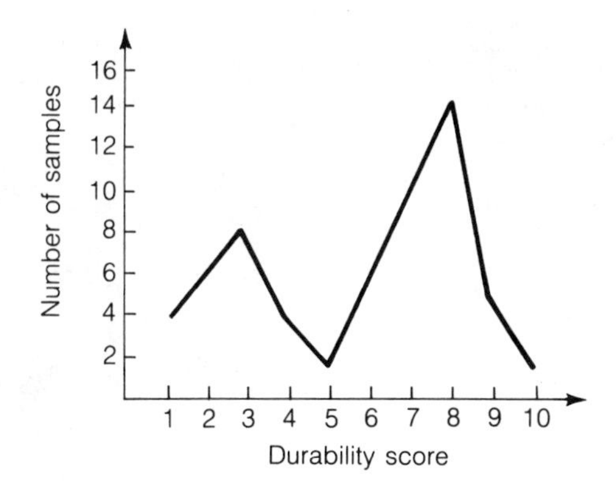

FIGURE 8-3
Frequency polygon. Note that the lines are not continuous.

make a histogram, you must decide how the scale along the x axis is going to be divided into bars of equal width. Then total the values in the area represented by a single bar; this total becomes the height of the bar. If you try this out, you'll find that depending on how you slice up the x axis, representing the data in this way may skew your results either toward the zero end of the scale or toward the high end.

A final type of bar graph is the divided (or layered) bar graph, which uses layering to show the distribution of the variables. When you layer your variables, the largest should always be at the bottom and the smallest at the top, with the other variables distributed according to their size. Figure 8-5 is a divided bar graph.

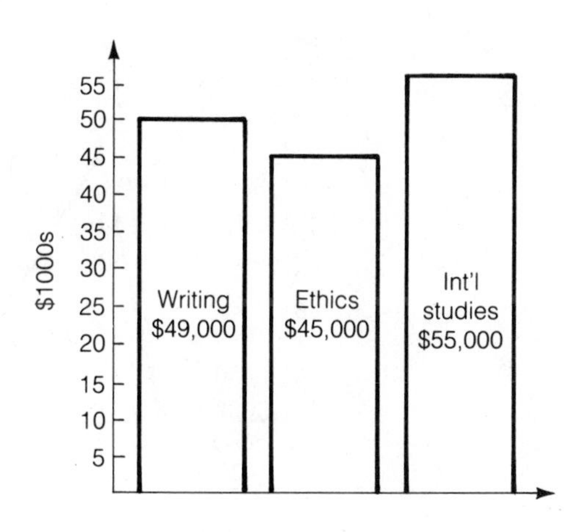

FIGURE 8-4
Bar graph of total grant funding: 1986–1988.

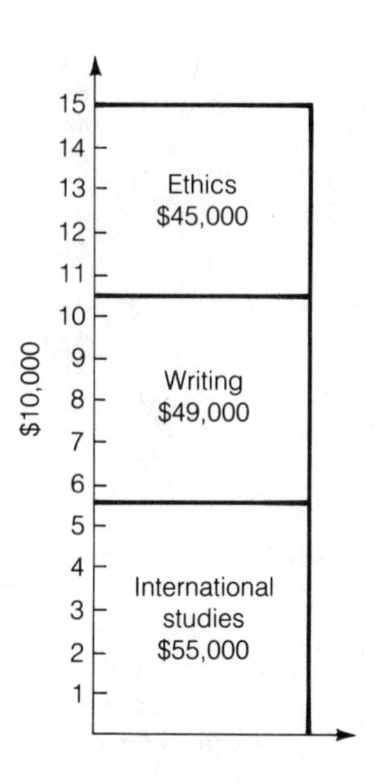

FIGURE 8-5
A divided bar graph of total grant funds: 1986–1988.

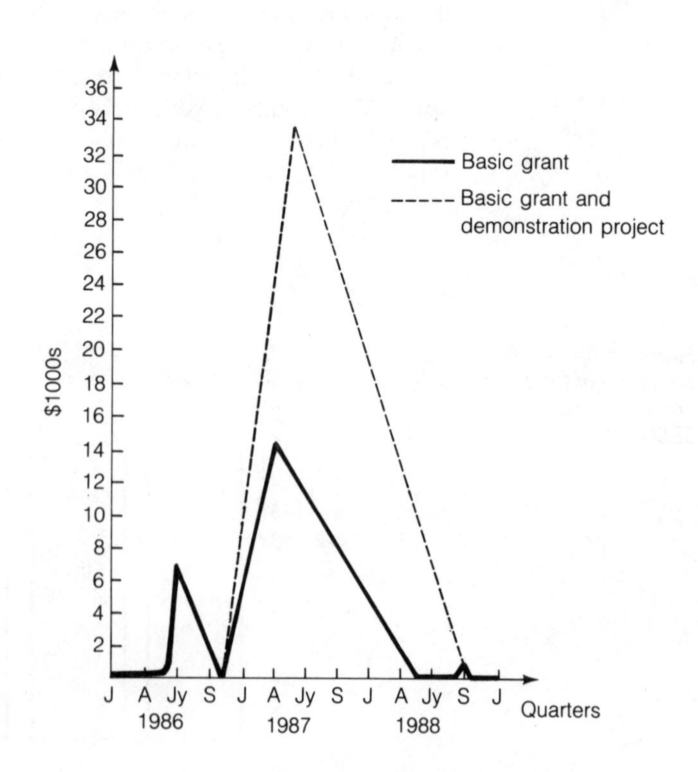

FIGURE 8-6
Line graph of grant funding for cross-campus writing project: 1986–1988.

FIGURE 8-7
Pie graph of projected college enrollment by age group: year 2000.

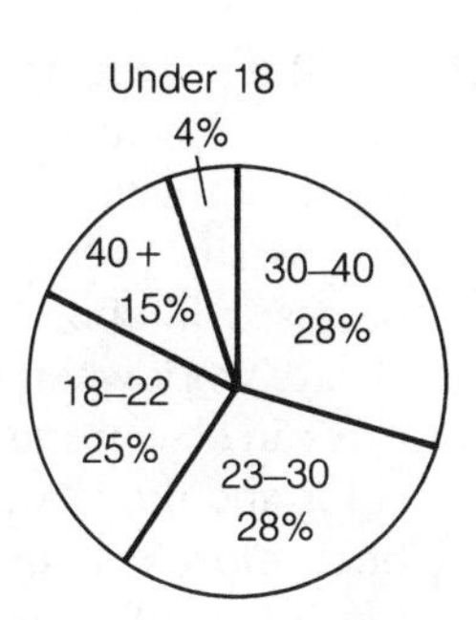

Line Graphs

Line graphs are perhaps the most familiar form of graph. When you label a line graph, it is important to let the reader know whether the curve represents a continuous series of points (as in the plot of an equation like $y = 2x$) or an approximation drawn from a few points. Line graphs with more than three lines are difficult for most audiences to read. Line graphs show trends well, but they are not ideal for showing relationships between trends, especially if several lines on the graph intersect. Figure 8-6, a simple line graph, reveals a final weakness of line graphs: They do not show absolute values well.

Pie Graphs

Pie or circle graphs are useful when all the values being represented constitute a whole. Pie graphs are often used to represent budgetary information. The circle represents the whole budget, and the slices represent the segments dedicated to specific items. Pie graphs are useful only when limited to five or six sections. Beyond six sections, it is difficult to compare segments. Pie graphs can clearly show either rough equality of segments or large differences in size between segments. On a pie graph, it is difficult to see the difference between values that are nearly identical. Finally it is hard to gauge the total size of the pie: A multibillion dollar budget looks the same as a budget of $100 when both are small circles on a page. Figure 8-7 is a pie graph.

Photographs

Photographs are ideal for showing objects under realistic conditions. They can be used to supplement a description, and they are unequaled as evidence when you are trying to convince an audience. A reader is far more likely to believe a photo of a smashed piece of equipment after a failure than a line drawing of the same object. In some fields, such as microbiology and astronomy, photographs also provide a higher degree of precision than any other graphic.

But photographs have several drawbacks. Photographs can provide too much extraneous background when you wish to point out a few important details. Photographs are also not necessarily precise and, unless you can tear the subject to pieces, will show only external detail. Photographs will not always work when you want to show fine detail because shadows can obscure small parts and other information, such as true color and the characteristics of finish. Finally, photographs are often more expensive to reproduce than other graphics. With careful handling, however, a photograph can be effective.

The most economical way to use a photograph is to reproduce it through offset lithography. The first version of the photograph is likely to be an 8-by-10-inch glossy print, which is best for reproducing highlights and details. This print is then turned into a "screened print," which reproduces the solid section of the negative as dots, ranging from 65 to 120 per inch. The screened print is turned into a printing plate, and the plate is inked and printed on a rubber roller, which then prints the image on the page.

From this review of the printing method, you can understand the problems that photographs pose both in terms of preserving fine detail and expense of reproduction. In each of the steps from original object to negative, enlarged print, screened print, printing plate, rubber roller, and page, some sharpness and detail is lost. The end result will be much less detailed than the original object. And although printing many copies of a photo in this way is relatively inexpensive, a report that requires only a few copies would cost far more using the offset method.

You can reduce the loss of detail in photographs by using great care in your composition. Work with your film processor to make your negative as useful as possible. Cropping the negative will assure that the picture focuses on the most important features of the subject. Retouching can remove unnecessary details and background. You can even go to the extent of having an artist airbrush your photograph to highlight details. Certain papers are better for reproducing photographs; you should consult with your printer about choosing the best paper for your job.

The expense of reproducing photographs is hard to reduce since photography itself requires initial expenses. Sophisticated equipment is mandatory; the average home camera will not reproduce your material in enough detail to be useful. This expense can be mitigated somewhat if you work for an organization that has its own photographer and darkroom. The ex-

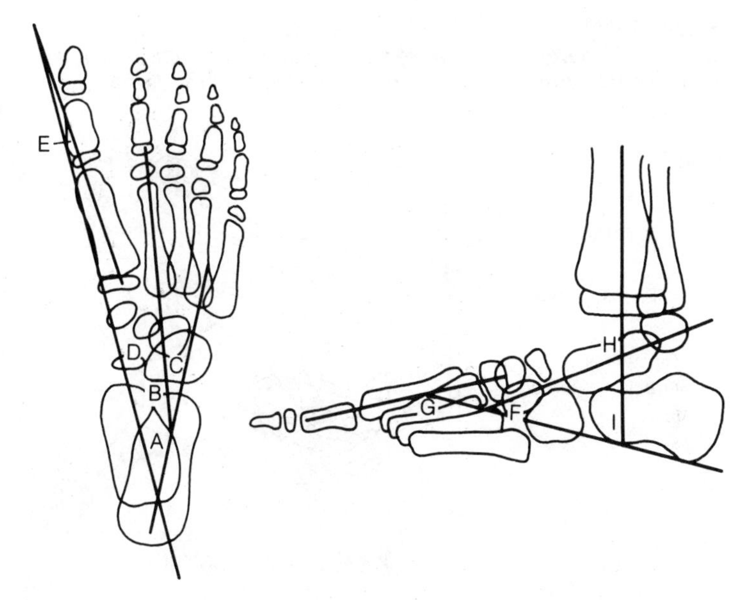

***FIGURE* 8-8** *Angles used to diagnose club foot and evaluate surgical results. Courtesy of Dr. George W. Simons.*

pense of reproduction for a small number of reports can be limited by having several prints of each photo made and inserting them in each report by hand. In this case, you should mount each photo on paper or art board with rubber cement, surrounded by sufficient white space. Number and title each photo at the bottom, as you would any other figure.

Keep in mind, finally, that there are situations in which photographs simply cannot be used. If your organization has no facilities for reproducing photographs and is unwilling to go to the expense of using an outside professional, you should probably use another type of graphic. Many journals will not accept photographs, and others charge the authors anywhere from $100 to $1500 per photograph to reproduce them. Consider your options, but if a photograph is your best graphic, and if your organization is willing to use it, do the job well. Take enough photographs to ensure a good result.

Line Drawings

Line drawings range from simple outline drawings of circuits or parts to sophisticated shaded drawings that suggest depth and lighting. Compared to photographs, line drawings have two advantages. First, the amount of detail can be controlled, so that the reader's attention is directed toward important features or parts. Second, it is possible to draw views that would be difficult, if not impossible, to photograph, even if a mechanism were cut into parts to reveal inner workings. In Figure 8-8, the diagnostic

***Figure* 8-9**
Drawing of a developer and rinse flow system for a technical manual.
Courtesy of Keuffel & Esser Company, Parsippany, New Jersey.

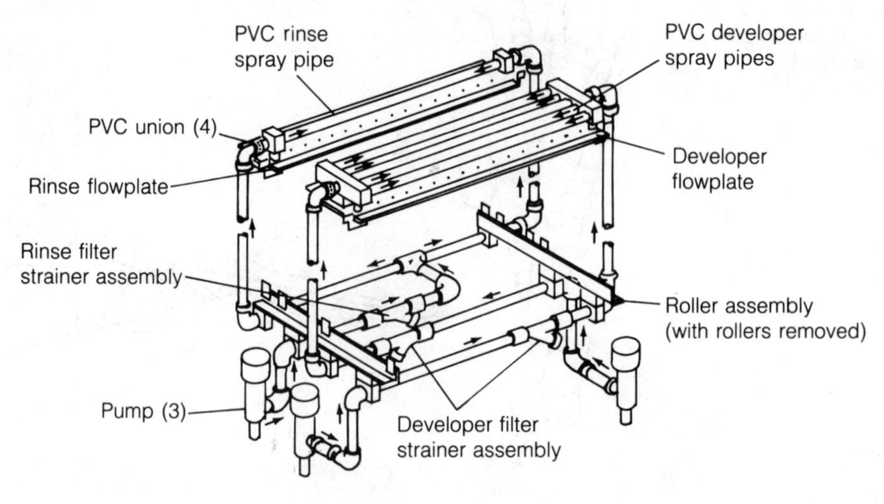

Developer and rinse flow system

relationships between the bones are indicated in a way that would be impossible to show in dissected specimens and difficult to show on radiographs. In this case, the drawing has an additional virtue. It can show general characteristics or ideal bone shapes, while actual radiographs would always be limited to the specific features of particular specimens. Figure 8-9 is a drawing from a technical manual. Again the drawing shows features and relationships that could not be shown easily on a photograph.

Organization Charts and Flow Charts

Organization charts reveal the levels of authority in organizations; flow charts illustrate processes or logic paths.

Organization Charts

Organization charts show how responsibility is divided in companies, agencies, universities, or other organizations with more than one level of officers or departments. Organization charts are common in proposals, where they are used to document management proposals and clarify relationships among major departments, researchers, or contractors.

Individuals or departments with the most authority are placed at the top of the chart with lines showing who reports to whom, what subdivisions are part of what major divisions, and so on. Organization charts may diagram either the relationships

FIGURE 8-10
Organization chart. Courtesy of Society for Technical Communication.

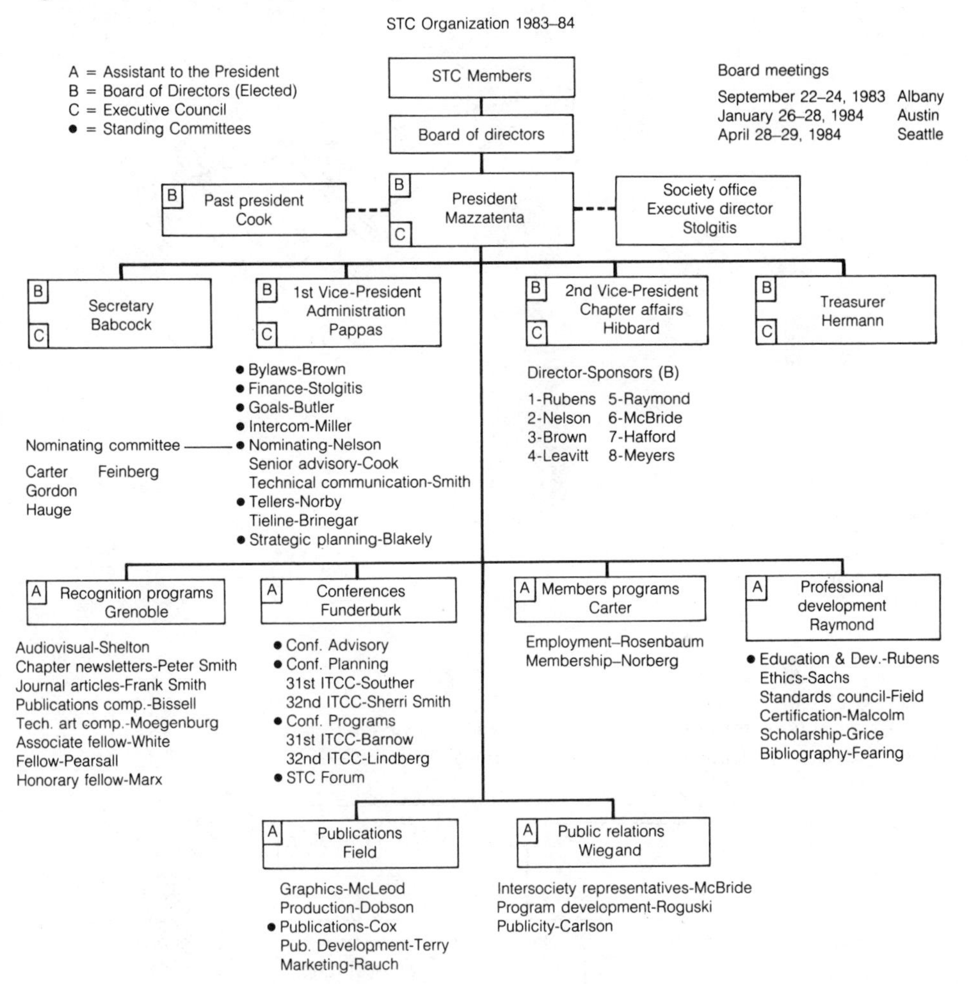

of individual officers or the relationships of entire corporate divisions. A particularly complex organization may require more than one organization chart. As you put such a chart together, however, remember that organization charts can show things only the way they are, not the way they should be. It is possible to come up with a poorly organized chart simply because the organization itself is shaky. A chart in this case may lead to efforts to reorganize the agency itself, but you as a writer are responsible for charting the organization only as it currently exists. Figure 8-10 shows a typical organization chart for a simple organizational hierarchy.

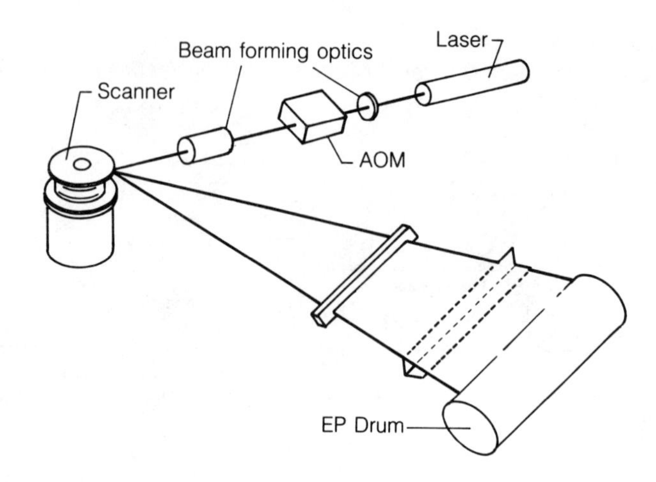

FIGURE 8-11
Process flow chart: Scanning the drum with the 9660 Laser Printer. Courtesy of Datapoint Corporation.

FIGURE 8-12
Process flow chart: Information flow—revisions to service manuals. Courtesy of Keuffel & Esser Company, Parsippany, New Jersey.

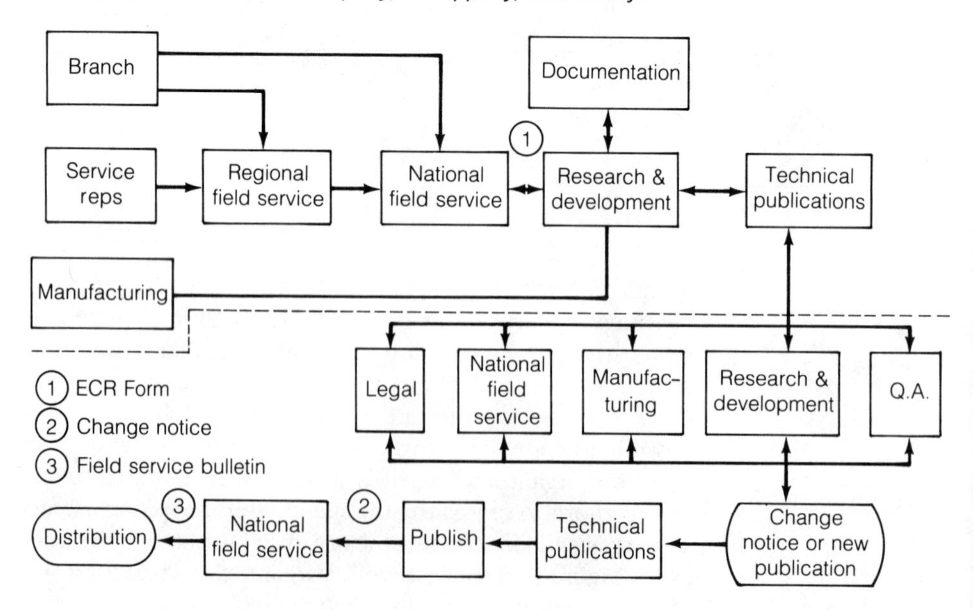

FIGURE 8-13
Symbols for a programming flow chart. Courtesy of Datapoint Corporation.

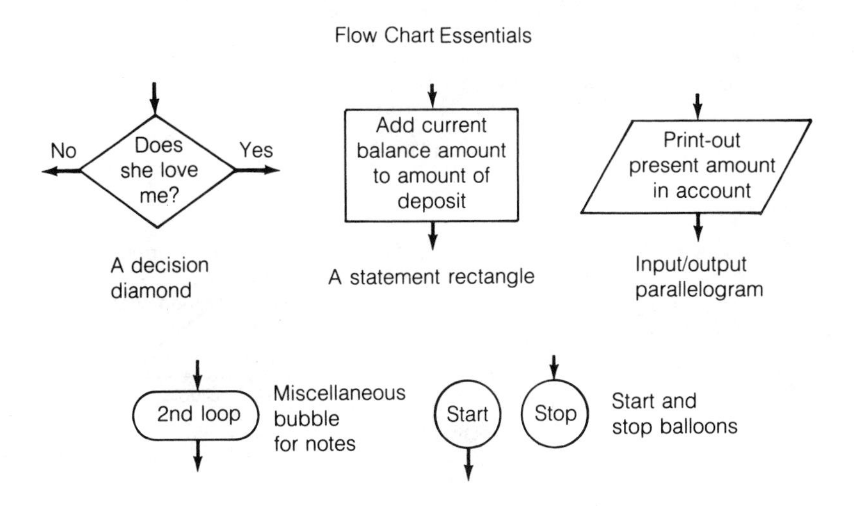

This kind of chart is usually composed of rectangles or circles containing names and connected to one another by lines. Divisions or officers at the highest level are placed on top; those who report to them directly are placed immediately underneath. It is conventional to place divisions or persons who have the same level of authority at the same level on the chart.

Lines showing authority should be heavier than the lines outlining the rectangles or circles. Sometimes broken lines can be used to show consulting or liaison relationships between blocks, as in Figure 8-10.

The blocks can contain either titles alone or titles with individual names. Blocks with titles alone will have less chance of becoming obsolete when individuals change jobs. Some organizations, however, prefer the more personal effect of listing individual names. All organization charts should show an effective date near the title to show how recently the chart was prepared.

Flow Charts

There are two types of flow charts: those used to diagram processes and those used to diagram logic paths. The former are used to outline manufacturing or personnel procedures; the latter are used in computer programming.

Process flow charts diagram manufacturing and personnel processes, using symbols for each step. Customarily the diagram moves left to right (occasionally right to left). Arrows

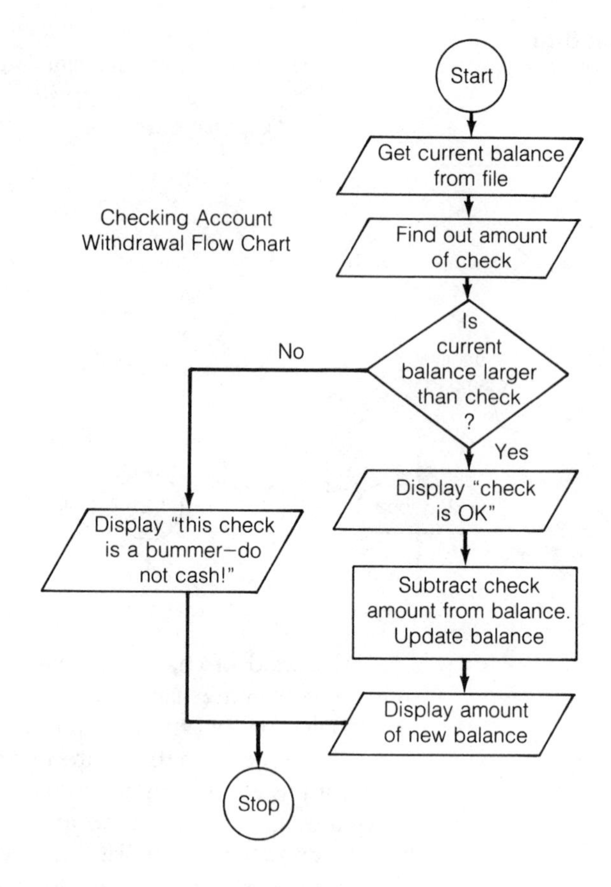

Figure 8-14 *A typical programming flow chart: Checking account withdrawal. Courtesy of Datapoint Corporation.*

indicate how each step follows another. The symbols used to indicate each step in the process are sometimes standardized; the American National Standards Institute has a list of symbols that are used for particular industrial components. Figure 8-11 is a typical process flow chart outlining the steps in a laser printer; Figure 8-12 shows a personnel process, in this case the steps a manual must go through during revision.

Programming flow charts diagram logic paths. This is a standard part of computer programming, but it can be used to illustrate any thought process. This type of flow chart customarily moves from top to bottom. It also uses standardized symbols (the standard symbols for flowcharting are shown in Figure 8-13). Figure 8-14 is a typical logic flow chart showing the process a check goes through before being approved by a bank.

SUMMARY

1. Technical writing usually requires a graphic component; graphics are best used to show detail and relationships that are difficult to explain in words.
2. Always integrate graphics with your text by using both an introduction and an explanation.
3. Place graphics as near as possible to their description in the text.
4. Give every graphic a title and a number; number figures and tables separately.
5. Keep graphics simple and consider in advance any possible problems with reproduction.
6. When working with an artist, be sure to provide him or her with enough information so that the graphic will meet your requirements.
7. Tables are effective for showing numerical data and allowing the reader to make comparisons; they do not show trends well.
8. Graphs are effective for showing numerical data; they show trends well but do not emphasize single data points as well as tables. Common types of graphs include frequency polygons, bar and line graphs, and circle or pie graphs.
9. Photographs are effective for showing objects under realistic conditions, but they are expensive and do not reproduce fine detail.
10. Line drawings are effective for reproducing detail, seeing inside objects, and directing the reader's attention to specific features of an object.
11. Organization charts diagram lines of authority in organizations; flow charts diagram processes or logic paths. Organization charts are common in proposals; logic paths are common in computer programming.

EXERCISES

DIRECTIONS: Convert each of the following paragraphs into some kind of graphic. Consider the best possible format (more than one may be possible) and the best possible design. When you have created an effective graphic, write both an introduction and an explanation for each one (see instructions on p. 167) as if you were inserting the graphic into a report.

1. According to George Freire radon emitted by decaying uranium in the soil reacts with nitrogen and oxygen in the air, creating negative and positive ions. Negative ions are absorbed by mist particles, making clouds slightly negatively charged. Polonium ions also emitted by the radon are also absorbed by the mist particles, creating positive particles. The positive particles attract the negative particles, forming a raindrop. When enough negative particles are attracted, the mist particles are neutralized; the raindrop is then large enough to fall through the cloud, colliding and combining with other particles.
2. The production costs for steel are as follows. The cost of labor at a typical U.S. integrated steel mill was approximately $150/metric ton of steel produced; for a U.S. minimill the cost was approximately $65/metric ton. Costs for iron ore and coke were $65/metric ton and $50/metric ton respectively for the integrated mill; minimills do not purchase either iron ore or coke. On the other hand, minimills expended approximately $1000/metric ton on purchased scrap, while integrated mills spent only $20/metric ton. Other costs were approximately the same: $50/metric ton. Other energy cost both mills approximately $75/metric ton.
3. Young Penelope wanted to know something about her family, so she asked her mother, a genealogy buff. This is what she was told:

    My grandmother, Kathryn Worthington Blimp married Swafford Grimsby in 1892. They had five children: Agatha, Terrence, Worthington, Ermentrude, and my father, Philpott. Agatha married Coleman Porter; their children were Matilda, Corwin, and Coleman, Jr. Terrence married Letitia Hartshorn and they had Agatha and Toby. Worthington ran off with that awful Emma Snipes and was never seen again; no children there, or so we all hope! Ermentrude married Harmon Kildare, but they had no children. Philpott married my mother, Honeychile Leverett, and they had me—Kathryn. I married your father, Corbett McGonigle, and we had you.
4. The optimum pulse rate for aerobic exercise should be between 70 percent and 85 percent of the maximal heartrate. For someone aged twenty to twenty-five, the minimum heart rate would be 138 and the maximum would be 167 (at twenty-six to thirty the minimum is 134 and the maximum 163). But if you were between sixty-six and seventy, your maximum would be 129 and the minimum would be 106 (for ages sixty-one to sixty-five the minimum is 110 and the maximum is 133). Someone thirty-one to thirty-five would have a minimum of 131 and a maximum of 159; for thirty-six to forty the minimum is 127 and the maximum is 155. At forty-one to forty-five the minimum drops to 124 and the maximum to 150; for

ages forty-six to fifty it is 120 and 146. And for ages fifty-one to fifty-five the minimum pulse rate is 117 and the maximum is 142 (for fifty-six to sixty it's 113 and 138).

5. Pregnant women require at least 2200 calories daily. 400 calories (700 grams) should come from protein, 700 calories (around 80 grams) should be fat, and 1100 calories (275 grams) should come from carbohydrates.

6. In 1981 real estate was a tax shelter for $1,600,000,000; in 1983 the figure was $4,477,000,000. Oil and gas accounted for $2,884,000,000 in 1981 and $2,995,000,000 in 1983. In 1981 equipment leasing sheltered $200,000,000; in 1983 $338,000,000. Research and development accounted for $25,000,000 in 1981, $237,000,000 in 1983. The 1983 total for film—$141,000,000—almost doubled the 1981 total of $80,000,000. Cable TV did double: from $35,000,000 in 1981 to $71,000,000 in 1983. Agriculture increased from $21,000,000 in 1981 to $44,000,000 in 1983. However, transportation dropped from $39,000,000 in 1981 to 0 in 1983.

7. The Arctic Ocean has ice year round; however, ice breakers operate only from June through October. On the Great Lakes ice breakers operate from November through April—there is no ice during the remaining months; this same schedule applies in the Gulf of Bothnia and the Baltic Sea. In the Western Arctic Ocean of the USSR, ice is present year round, but ice breakers operate only from July through November.

8. The Caribou herd population in Newfoundland was measured at 5000 in 1957 and 1959. It rose to 7000 in 1961 and 1964, and to 8000 in 1966. By 1971 the population had increased to 13,000 and it peaked at 15,000 in 1973.

9. Nuclear waste materials are enclosed first in a shell of reinforced concrete. This is in turn encased in a fairly thin stainless steel cylinder, which is surrounded by a layer of asphalt. This entire enclosure is then placed in a cast iron receptacle and placed in a rock chamber with a layer of concrete backfill between rock and iron.

10. In 1950 there were fewer than 10 social security beneficiaries for every 100 active workers; by 1960 the number had risen to 20 per 100; by 1970 it was 28 per 100; by 1980 it was 32 per 100. For 1990 the proportion should be roughly the same, rising to 34 per 100 in the year 2000. By 2010 the number should rise to 35 per 100; then 40 per 100 in 2020 and 50 per 100 in 2030.

11. Ten years after burial a cannister of nuclear waste will emit a temperature of 392°F. Sediment within 1 foot of the cannister will have a temperature of 302°F; at 1 yard the temperature will be 212°F; at 1½ yards, 167°F; at 2 yards, 122°F; at 3 yards, 77°F. Below 212°F no significant chemical changes are predicted.

12. A decibel (dB) represents the intensity of a sound. Zero dB is the lowest sound an average human can hear. Sixty dB is the level of normal conversation. Seventy dB is the noise of a vacuum cleaner. Eighty dB is the noise of an alarm clock. Eighty-five dB represents potential ear damage. A child screaming for eight hours would register 90dB; a four-hour convertible ride on a freeway would register 95dB. Two hours' worth of jackhammer noise registers 100dB, while thirty minutes with a live rock band registers 115dB. An auto horn blasting for 7½ minutes registers 120dB, while 3¾ minutes of an air raid siren registers 130dB. The threshold of pain is reached at 135dB and a jet engine at 140dB reaches the level at which hearing is damaged.

13. Hair consists of a cortex surrounding the hollow medulla, all of which is sheathed by a scaly cuticle layer. Harsh shampoos can ruffle the cuticle, causing tangles; bleaching, curling, and blow-drying can damage the cuticle, exposing the cortex. Conditioners make the cuticle lie flat.

14. Of the mediterranean fruitfly population 4 percent are adults; 25.5 percent are eggs within the fruit; 60.5 percent are larva, which bore into the fruit and later into the soil; 10 percent are pupae, which rest near the soil surface. Sprayed pesticides affect only the exposed adult flies.

15. Black holes begin when the hydrogen at the core of a large star fuses to helium and then to carbon. Further fusions produce layers of silicon, oxygen, neon, carbon, helium, and hydrogen. Gravitational compression raises the temperature of the core until it becomes unstable and collapses, blowing off the outer shell in a supernova explosion. The remnants of the star are blown off into space and the core collapses to form the center of a black hole.

P · A · R · T

Applying the Standard Formats

CHAPTER • 9

Short Reports

INCIDENT REPORTS, INVESTIGATIVE REPORTS, AND PROGRESS REPORTS

INTRODUCTION

You will probably write more short reports than any other kind of document, with the exception of letters. And as you will see in the business correspondence chapter (Chapter 13), many of the standard letter formats, such as the bad news letter, are in fact short report formats modified to meet the needs of specific situations. Most of the formats introduced in this chapter, unlike the correspondence formats of Chapter 13, are designed for in-house use, for reports you will write to other workers in your own company or to contractors. In other words, your audience will usually be known to you personally. The purposes served by short reports, however, tend to be multiple and complex. One of our goals in this chapter is to reinforce an important lesson about purposes: the audience seldom just wants the facts. In addition to information, most audiences expect formal or informal recommendations and proposals. Your job as a report writer, then, is twofold: to *report* the relevant facts and to *interpret* those facts to your readers.

This chapter includes five main sections: the basic principles of short reports; a basic outline that can be used for most short reports; and guidelines and examples of incident reports, investigative reports, and progress reports. The basic principle that underlies all of these formats is: Don't just report, show how you interpret the findings and facts.

EXHIBIT 9-1
Report on a Defective Fan System

On January 26, I inspected tractor #1708, which was reported to factory services because of a whining fan system. After investigating the problem, I instructed the shop to install a new engine.

Inspection showed that heavy gear oil was sprayed over the engine housing in the vicinity of the fan. Fan belt condition was good. A high whining noise emanated from the clutch housing but was traced to the main bearings. Fan blades are molded plastic. Unit has new traction tires. Engine began to grind and whine at about 400 RPM. Operator has been squirting gear oil on fan shaft.

Shop replaced the engine with a new diesel unit.

BASIC PRINCIPLES: RECORDING AND INTERPRETING THE FACTS

At a bare minimum, your short reports will record the facts of incidents, investigations, or progress on a project. Your two fundamental purposes in a short report, in other words, are to record and present information and, on the basis of your interpretation, to formulate recommendations and proposals, coordinate your work with other projects, and even initiate new actions.

Present the Relevant Facts

If you want your readers to reach the same conclusions you do and to trust the validity of your conclusions, then you must present all the relevant information honestly and completely. But what is relevant? One rule is to present all the information that you need to develop your conclusions as well as the key information that might be used to support responsible alternative interpretations.

In the report in Exhibit 9-1 there are two irrelevant facts: the design of the fan blades and the new tires on the tractor. On the other hand, many important facts are missing, including information about the engine's age and general condition and the importance of keeping the tractor in service. This report would not meet the basic requirement of providing the audience with the relevant facts about the fan system. The writer has also failed to interpret the facts.

Interpret the Facts

You are almost always expected to interpret the facts you discover: your employer has hired you for your analytical skills and training as well as for your ability to notice details. In the

EXHIBIT 9-2

Report on a Defective Fan System

ENGINE REPLACEMENT AND FUEL EFFICIENCY UPGRADE: TRACTOR 1708

A new fuel-efficient diesel engine was installed in Tractor 1708 after the main bearings failed on its 20-year-old gasoline engine.

On January 26th, I inspected Tractor 1708, which was reported by the yard crew on the 24th with a whining noise in the engine. The engine (A-63-7794310) had logged 8000 service hours and was overhauled four years ago. The engine was 20 years old with low fuel efficiency. I traced the noise to the main bearings. Parts are no longer available. In view of these facts, I instructed the shop to install a new fuel-efficient diesel, since #1708 is otherwise in serviceable condition. The defective engine can be used as a parts farm for the remaining tractors of this type.

Number 1708 has been returned to service and placed on the annual maintenance schedule. Fuel consumption figures on this tractor should show improvement.

simple case in Exhibit 9-1, the writer needed to explain the significance of the oil sprayed over the engine housing and other wear data, as well as the reasons for the choice to replace the engine rather than tear it down for repairs. Those reasons might have included the time the tractor would be out of service, the availability of parts, the service age of the engine, or the extent of the damage. The revised report might look like Exhibit 9-2.

While this report is extremely brief, it demonstrates the basic outline of a short report.

THE BASIC SHORT REPORT OUTLINE

Short reports follow this basic three-part format:

1. Provide an overview.
2. Discuss the data, your interpretation of the data, and your conclusions.
3. Summarize your main points. This outline is presented in more detail in Figure 9-1.

The Overview

In your overview provide a clear indication of your purpose, the subject matter, and the organization of your report. The

FIGURE 9-1
Basic short report outline

1. Overview
   Subject
   Purpose
   Forecast of organization (optional)
2. Body
   Context (who, what, when, where, why, how)
   Data
   Interpretation of data
   Recommendations, conclusions, proposed solutions, or action taken
3. Summary
   Restatement of main facts and interpretations
   Conclusions, action, or further action to be taken

report reproduced in Exhibit 9-2 is so short that organization forecasting is not necessary, but the purpose and subject matter are clearly introduced in the title and first sentence. For a longer report, you may need forecasting.

The Body

The body of your report should provide the context of your investigation, the data, your interpretation of the data, and your recommendations, conclusions, proposed solutions, or the action you have taken. The context section should answer all the basic journalistic questions: who, what, when, where, why, and how. In Exhibit 9-2, the second sentence names the problem (what and why), who reported it and where (the yard staff), when the report was made (January 24), and when action was taken (January 26).

The Summary

A final section should repeat the main points of your report, including your conclusions and the action taken or recommended. An important feature of the short report is the way the summary is often used to indicate a future course of action. The report is viewed as part of a continuing process rather than the final work on a subject or investigation. In Exhibit 9-2 the last sentences look forward to the next scheduled maintenance and fuel economy check.

All of the short report formats presented below follow this basic outline. Short reports can be written as memos or letters or in more formal layouts. You should use headings to help your readers skim the report contents. We will discuss the incident report first.

EXHIBIT 9-3
Incident Report

INCIDENT REPORT: MORTON GRAVEL COMPANY,
BROKEN WINDSHIELD.
On June 23d, 1985, the yard cleaning staff discovered and reported a broken windshield on the plant utility van, an '84 Ford, license AMT 694. Reports were filed with maintenance, the police, and the insurance carrier.

At 7:30 AM, June 23d, Frank Connors, yard chief, found the windshield smashed on the '84 Ford utility van. Mr. Connors reported the incident immediately to the insurance and security offices. Several large stones were found near the van. A late load of river stone was shipped out by truck from the adjoining yard at 11 PM Friday night. Children have also been seen in the adjoining stone yard over the weekend, and the stones rarely fly off the delivery trucks. A vandalism report was filed with the county police, who assigned the case number 85-16987. Officer Robert Cizewski investigated. A telephone claim was filed with the insurance carrier on the 23d at 4 PM. The maintenance crew drove the truck to Cedar City Ford on the 25th.

Reports were filed with the police, insurance carrier, and maintenance office. The accident review committee will receive a copy of this report for recommendation and action.

Bob Smith (signed)
Morton Gravel Company Insurance Office

INCIDENT REPORTS

Incident reports are used to report information that can later be fully analyzed and interpreted. In this case you need not interpret the facts you present. Accidents, equipment failures, and preliminary sales contracts can be reported in this format. An incident report is sometimes written on the assumption that additional information will need to be gathered before final conclusions can be drawn. For example, in Exhibit 9-3 the broken windshield is reported, but the cause is not determined and preventive actions, such as changes in security policies or rock-loading procedures, are not discussed.

A related report form is the conference report memorandum discussed in Chapter 13. That format is used to report on meetings and telephone conversations so that all participants have a copy of what was decided. This type of incident report, like the accident report in Exhibit 9-3, allows all participants to see what facts have been recorded and make corrections before a final report is written.

INVESTIGATIVE REPORTS

Investigative reports differ from incident reports in their emphasis on analysis. The formats are the same. When you are asked to write an investigative report remember your responsibility to analyze and interpret the data you present. We will review three types of investigative reports: (1) trip reports, (2) inspection reports, and (3) troubleshooting reports.

Trip Reports

Engineers and other professionals are often expected to write reports when they return from field trips. We have placed the trip report under the more general heading of Investigative Reports to emphasize that any trip, no matter how routine, should be considered an opportunity to carry out an investigation, to be an active observer. The trip report in Exhibit 9-4 records initial observations about new computer graphics equipment an engineer saw at a convention. In this case the need to write a trip report led the engineer to think through the limitations of the company's graphics programs. Given a chance, report-writing time can often turn into a time to reorganize information you already know and, in some cases, achieve new insights about your work.

The short trip report in Exhibit 9-4 presented few organizational difficulties. Inspection reports are usually more complicated and require that you take good notes and organize them carefully before drafting your report.

Inspection Reports

Inspection reports are written with specific purposes in mind. You might be asked to inspect a factory for fire safety or for the capacity of its electrical power distribution system. Or you might be sent out to test effluents for the presence of a specific contaminant. While you should keep your eyes open for other interesting or relevant data, your inspection reports should always state the main purpose clearly and use the purpose as the organizing principle of the report. Exhibit 9-5 is a fire insurance inspection report. Note the use of underlining, which helps the reader follow the general organization of the report.

EXHIBIT 9-4
Trip Report

TO: Bill Wilcox DATE: May 8, 1985
FROM: Nancy Press
SUBJECT: CDC Conference Trip Report

The CDC Conference I attended April 27-28 in St. Louis provided useful information and several contacts that may prove of value to the Engineering Department. Because this was a worldwide meeting attended by more users of Control Data equipment than have ever assembled at one place before, CDC took the opportunity to announce several new products. Among these were a new series of computers, the new VIking terminals, and the latest operating system.

An 825 mainframe was operating at the conference site, supporting live demonstrations of CDC products. Of these, I found the Remote Micro Facility (RMF) particularly interesting. RMF enables several types of personal computers to interface effectively with a Cyber mainframe in the local processor, interactive terminal, or file transfer mode.

Members of the European ECODU graphics committee were concerned with American acceptance of the Graphics Kernel System standard, which may soon be adopted by the International Standards Organization. The American National Standards Institute has shown a preference for the Graphics Planning Standards Institute (GPSI) Core System, which has been designed into several systems marketed in the United States. As we consider which graphics software to implement in our system, we will need to watch the International Standard Organization and consider the implications of its actions.

The most exciting technical session at the meeting was the report on CDC-funded graphics research with Romulus and MOVIE.BYU software. CDC may soon offer graphics capabilities at the level of its number-crunching capacity.

The proceedings of the meeting will be published and mailed within the next month. At that time, I will make copies available.

Troubleshooting Reports

Troubleshooting reports present analyses of problems and recommend possible solutions. As with other short reports, you should be careful to present all the relevant data, clearly explain

EXHIBIT 9-5
Inspection Report

United Insurance
Field Inspection: Agri-Spray, Inc.
35 North Wilson Blvd., Bellingham, WA

Date: May 3, 1986
Asst. Underwriter: H. Smith
Policy # CD-567948-82

On May 3, I made an initial visit to the shop and storage yard of Agri-Spray, Inc., to determine the level of risk and compliance with applicable fire-safety standards. We are writing a new fire policy only. In this report, I will first present what I found, then assess the risks.

Agri-Spray is a commercial insect control company which does contract spraying throughout Washington State. The company maintains a fleet of four trucks, two helicopters, and one fixed-wing aircraft. Tanks of insecticide are stored in a separate, fenced area at the north end of the property; vehicles are kept at the south end. A fuel storage tank is buried in the truck yard adjacent to the single corrugated steel garage and shop building. Agri-Spray maintains a dirt landing strip on adjoining property. Fire extinguishers are mounted in all vehicles; these are routinely inspected under contract with AA Fire Supply and meet all FAA requirements, as well as NFPA industrial fire codes. The fuel pump has an automatic fire-control system, which meets Washington State Bureau of Licensing standards.

The garage is divided into two halves, a general vehicle service area and an engine repair shop. Spraying equipment is stored in the service area. The building is all steel with a concrete floor. Two fire extinguishers are kept in each area.

In the vehicle service area, both oxygen/acetylene and electric arc welding are performed for about two hours each month for cutting and loosening parts, such as spray heads and mufflers. The welding practices meet all NFPA requirements. Combustibles are more than 35 feet from the welding area.

In the engine repair shop, both oxygen/acetylene and electric arc welding are performed on a large steel table. Welding is done about ten hours each month. The main activity is fabricating small steel steps for the helicopters and supports for the spraying units; however, per FAA requirements, the insured is not allowed to perform major welding on the helicopters themselves. As a result, large welding jobs are performed by an outside shop. Welding practices deviate from NFPA standards in that there is less than a 35-foot clearance to a combustible cabinet that contains aircraft paint. However, I consider the hazard limited and well controlled because the cabinet is 25 feet from the welding table, the area is kept clean, and the two extinguishers are mounted on the wall near the welding table.

Summary. At this time Agri-Spray meets all fire and safety standards, with the exception that a paint cabinet is 25 feet rather than 35 feet from the welding table in the engine repair shop. I suggested that the cabinet be moved or that the paint be stored at the far end of the vehicle service shop, where there is less welding done. I will check on this at my next scheduled visit in December.

how you interpret the data, and then specify and support the solution you propose. Exhibit 9-6 is taken from the correspondence between two specialists. A surgeon has asked another professional to comment on the results of a research project and point out how the various research results ought to be presented to an expert audience. Notice how the reviewer identifies the key issues or problems, offers an interpretation of what the study shows, and then proposes several ways that the research report can more persuasively present the results of the study.

Progress Reports

Progress reports are used to inform other people and companies working on a project about the work you have completed, the problems and delays you have encountered, and the work you intend to do next. Progress reports are also used to help coordinate the activities of different work groups and to recommend or propose changes in schedule, equipment, staff, or design. Proposals often include a specified number of progress reports for the duration of a project. Progress report organization differs slightly from the organization of other short reports because the progress report is arranged chronologically.

There are two types of progress reports, the occasional report and the periodic report. Occasional reports are filed whenever they are needed. In large projects, working groups are often asked to file weekly, monthly, or quarterly reports using a uniform format. When these reports are bound or filed together, they form a running record of everything that happened in the course of the project. Both forms use the same general outline.

The Progress Report Outline

The progress report outline (see Figure 9-2) differs slightly from the standard short report format because data must be presented in chronological order. Also, at the beginning of the report, reference is made to the last report, while at the end of the report, the date of the next anticipated report is given.

EXHIBIT 9-6
Troubleshooting Report

Dr. Alan B. Sithe January 26, 1986
Chief of Surgery
Mercy Hospital
Brandywine, Pennsylvania

Dear Alan:

In this letter and the enclosed article (where I have marked changes) I have provided what I think you asked for: (1) a letter suggesting major changes and (2) answers to the specific questions you asked in your letter. As I see it, your primary purpose is to demonstrate that your Group III patients, who received the new surgery, have much better results than your Group II patients, who were treated several years ago with the most widely used procedure. You have also studied a smaller group of patients who presented much less complex surgical problems, and you enumerate various special cases and complications, most of which were caused by earlier treatments. Consequently, these complications have little bearing on the comparison between your two major surgical groups.

Many of the changes I recommend here are based on three conclusions. I'd like to discuss these conclusions first in order to emphasize them as strongly as I can. If you think I am dead wrong, fine. First, I think you should delete Group I (minor procedures) from the study and include those seven cases among the others you did not choose for comparison. If you do this, you will focus the paper on the comparison of Groups II and III, which are well-balanced with twenty-five and twenty-six cases respectively. Throughout the paper, the constant comparisons to Group I are of limited relevance.

Second, you should briefly summarize pages 32-41. Such a long discussion of a few minor complications steals fire from your conclusions. A single page should do the job.

Third, I think you need tables of statistics that show the effect of postoperative changes. You mention postoperative losses of correction, but you do not explain how these differ from one group to the other. I recall that the last version of this paper contained a few paragraphs that showed that most of the postoperative changes had no effect on function. If you make these changes your tables will be much simpler, you can eliminate several tables, and you will reduce the length of the report by about 13 pages, which will make the paper about the right length for a journal article.

[The letter continues with two pages of specific comments on statistics, the design of tables, and the logic of the research. The concluding paragraph repeats the main point.]

Your research results are strong, and the comparison of the two major procedures is convincing. The only problem I see here is that the journal's review panel could easily be sidetracked by your references to a control group that was not focal to your study. I hope you can make these changes without recalculating many statistics.

Exhibit 9-7 is an occasional report in letter form. In it a contractor informs a subcontractor about the winding down of two projects and his latest estimate of the beginning of the next project. Notice that the general outline for progress reports has been followed in most details.

Exhibit 9-8 is a periodic progress report that is much more technical in content. In this example, the writer uses headings effectively.

FIGURE 9-2
Progress report outline

1. Overview
   Subjects and content
   Date of last report
   Main points
2. Body
   Work completed in last period
   Work scheduled in next period
   Problems encountered, solutions developed, and adjustments made or requested
   Adjusted schedule (if required)
3. Summary
   Summary of main points
   Date of next report

EXHIBIT 9-7
Occasional Progress Report in Letter Form

R. W. Wilson
1100 Woods Creek Road
Carnation, WA 98014

January 21, 1984

Thomas N. Redding
3389 N.E. 35th Street
Seattle, WA 98112

Dear Tom:

This is the latest payment on the slowing sale of the Masterson book—this is for the period August 1984 to the end of December. Sales are not good, and unless we do a good deal more advertising, I don't see any reason for them to start up again. I do have more copies available, in case you need a hundred or so.

I talked with Matthew Stevens the other day, and he was very pleased with your work on the word-processing book he is doing. I don't know if I told you I had recommended you to him.

The good thing about Matt is that he is a hard worker, has a contract for the book, and is serious about getting good talent to help him. I'm glad you could work with him and hope it is proving to be a good thing for you as well.

My computer company is inching closer and closer to starting. I have taken on another man to be president of the company. We are rewriting the business plan and should be finished with that about this time next month. Then we will start getting some money (we do have some lined up already). After that, we will begin the process of getting the program here, writing the screens, and drafting the user's manual. That's where you come in. When we get closer, I will call you to set up some time for us to meet, for you to see the specifications and screens, and for us to talk about the contract.

Sincerely,

Bob

EXHIBIT 9-8
Periodic Progress Report in Memo Form

DATE: August 2, 1985
TO: James Easton
FROM: Nancy Press
SUBJECT: Bimonthly Progress Report, June-July 1985

MODEL DEVELOPMENT

Potential Equation Revised

Lois Morantz and Gerald Lewis have reformulated the potential calculation in the rotor analysis program in order to eliminate the solution's dependence on velocity distributions. The pressure closure has been modified to reflect the new potential. Program editing is currently in progress.

Computer Code for Airfoil Design

I obtained a flow program, which was made operational on the mainframe system. Lois Morantz wrote a supplementary program to generate plots of the grid network and surface pressure contours.

Aerothermal Modeling Program Nearing Completion

A number of "bench" tests have been run to determine the accuracy of the physical submodels. The stress model developed during this period showed significant improvements over the previous model for the entire group of flows under analysis. Further improvements will be required to improve the accuracy of the predictions.

For combustion tests with high inlet temperatures, the old model provides better predictions. For cold inlet conditions, the new model provides more accurate predictions.

Additional computations and model refinements are scheduled to be completed by September 1, and the final report will be due on November 15.

Summary

Programming and modeling research are currently on schedule. The next progress report will be submitted on October 2.

SUMMARY

1. Short reports follow a three-part format: overview and forecasting, body, and recapitulation.
2. Short reports must provide their contexts: who, what, when, where, how, and why.
3. Short reports usually inform and record, but they should also recommend, analyze, propose, coordinate, or initiate action.
4. Your chief responsibility in a short report is to interpret the data you present (except in incident reports).
5. Your second responsibility is to present all the relevant data.

EXERCISES

1. Write an incident report using either an incident in your own life this coming week, or the data provided in the cases following the exercises.
2. Write a progress report for your term paper in this class or another course in which you have a major project. Use the report as an occasion to analyze problems you face, develop solutions, and adjust your schedule.
3. Take a trip downtown or somewhere else for three hours and write a trip report focusing on the events you observed that have a bearing on your career plans.
4. Write an investigative report comparing one of the following:
   a. stereo equipment
   b. two cameras
   c. two or three software packages for mathematics, statistics, or word processing

See Chapter 10 for advice about the development of criteria.

Short Report Case 1

THE STUDENT LOUNGE

Six months ago your Student Government Association (SGA) came up with what seemed like a brilliant idea: a student lounge in the basement of the library. They had in mind a place where students could take a break from studying and perhaps get a cup of coffee or a snack. Since you thought the idea was a good one, you were chosen as chairman of the committee to carry the idea out. It's been a real education in more ways than one!

First, you had to get the approval of both the school administration and the library staff, no easy task as it turned out. The administration was worried about costs and the librarians hated to give up their basement storage space. But a compromise was worked out. Only half of the basement will be used, and the SGA will foot the bill for the renovations; in return the SGA will receive the revenues from the vending machines that will be installed.

You've met with the campus architect, and he's submitted three plans for approval. One of them was accepted by a committee vote last week and will be presented to the SGA at the end of the month. You'll be talking to the construction chief next week to get exact figures on the costs for the project.

You still have to investigate costs for both vending machines and furniture. That will be done after the plan is approved so that you'll know exactly how many machines and how much furniture you need. After the plans are approved, construction could start during the next summer term.

The SGA is pleased with what you've been telling them, but they've decided they want a formal progress report for the record. Guess who gets to write it?

ASSIGNMENT

Write the progress report for the SGA on the library lounge project.

Short Report Case 2

The Nursing Home Problem

As the district manager for Carling Nursing Homes, you have come to know and like the supervisors for many of these institutions. One home in particular, Shady Rest in Blanco County, has always struck you as pleasant and well administered. On your last visit, however, you found what seemed to be a real problem.

One of the residents, Mr. Clarence Biemer, had complained that several of his personal possessions were missing and intimated that one of the nurse's aides had stolen them. On investigation, the nursing home staff located two of the items, a gold watch and a picture in a silver frame, in Mr. Biemer's room where they had apparently been mislaid. But the other items, a locket belonging to Mr. Biemer's wife and a music box, could not be found.

The supervisor of Shady Rest, Carol Mahaffey, had questioned the accused aide and searched both Mr. Biemer's room and the adjoining rooms, but no trace of the items was found. Ms. Mahaffey had decided that Mr. Biemer had perhaps mislaid these items as he had the others, and had taken the question no further. But Mr. Biemer continued to complain; the nurse's aide in question finally resigned in protest, saying that she felt she was being harassed.

Ms. Mahaffey has been left with an unhappy resident—Mr. Biemer—and some unhappy staff members who feel that Mr. Biemer has been responsible for forcing out a first-rate aide. Both sides—Mr. Biemer and the staff members—make their feelings known to you on your next visit.

ASSIGNMENT 1

Write an incident report to your supervisor describing what has happened. Remember, with this type of report you are passing on information to be analyzed later; use only the details provided in the case.

ASSIGNMENT 2

Write a troubleshooting report based on the same incident. Here you will analyze the problem and present possible solutions. How would you deal with this situation?

CHAPTER · 10

Recommendation Reports

INTRODUCTION

In Chapter 2 we stated that one purpose in technical writing was to recommend. When your purpose is to recommend, you are suggesting a course of action. Usually you present alternatives, evaluate them, and direct the reader toward one that you think is preferable. In the process, you present your readers with criteria—measurable standards of judgment—on which your evaluation is based.

Many kinds of reports are recommendation reports. Feasibility reports, for example, study the practicality of a proposed course of action and make recommendations about its adoption. Investigative reports discuss existing programs or products and make recommendations about their continuance. Even proposals are a specialized form of recommendation report since they make recommendations about undertaking a particular project. In fact almost any report can include a recommendation along with its other objectives. In this chapter, however, we will discuss reports whose sole purpose is to make a recommendation, reports that exist only to get that recommendation across.

Recommendation reports are sometimes the result of months of work. The investigations they are based on may be the work of many people, and you may not be involved in every aspect of them. Indeed you may simply receive the data concerning a study and be expected to review it and arrive at a recommendation. Whether or not you are the principal investigator, the process of preparing a recommendation can be broken down into five steps:

1. Formulating a purpose statement
2. Formulating criteria
3. Formulating alternatives
4. Evaluating the alternatives according to the criteria
5. Formulating conclusions and recommendations

In this chapter we will discuss this recommendation process. You will learn ways of developing criteria and alternatives, as well as the process of arriving at a recommendation. Next we will consider the problem of audience for a recommendation report. Audience analysis is always an important factor in any report, but it is vital for recommendations. In this type of report your audience's reaction will determine whether your objective is achieved, whether your recommendation will be accepted or rejected. Finally, you will learn how to organize your report. Recommendations depend on logic; the reader must be made to see the way in which you derive your recommendations from your data. Your reader should never wonder how you got *that* conclusion based on *those* facts.

THE RECOMMENDATION PROCESS

The process of arriving at a recommendation is based on the five steps we mentioned in the previous section. We will examine each step in turn.

Formulating Your Audience, Purpose, and Scope

In recommendation reports as in most other reports, your first step must be to define and analyze your audience. Once you have the audience in mind, you can move on to your purpose. Suppose you're studying word processors for your office. You might have several purposes: to help the clerical workers, to improve accuracy in typed documents, to improve record keeping, to reduce the number of clerical personnel, to extend computer capabilities, even to help you correct your spelling errors. Deciding which purpose or purposes you want to achieve will clearly affect the kind of research that you do. If you're not really interested in reducing the number of clerical workers, you wouldn't need to consider how processors could consolidate the existing clerical tasks.

If you have been told to write a report, the process of formulating your purpose may be easy. You can simply ask whoever gave you the assignment what purpose they had in mind and make sure you understand exactly what their objectives are. If you are the originator of this report, however, you'll have to spend some time defining precisely what your objective is to be. Consider what decisions you will be making and what recommendations you need to come up with.

Scope is closely associated with purpose. Scope in this case refers to any limitations you impose on your study and thus on your purpose. Cost might be a limiting factor, for example. Will you consider costs or will they be immaterial? You might also consider the area you intend to study. In the word processor example you might decide not to consider any processors that would be hooked into a mainframe computer; this would reduce the number of processors you would need to investigate. The scope of your investigation will vary with each project you undertake, but seldom will you undertake a project with no limitations whatsoever. Whatever limitations you impose on your study will affect your purpose and, more directly, your criteria.

Formulating Criteria

Criteria are the standards you devise to evaluate the alternative solutions to your research problem. Formulating criteria is one of the crucial steps in the recommendation process; your readers will judge your recommendation on the basis of how well it meets your criteria. Make sure those criteria are precise, measurable, and based on an accurate assessment of the needs of your audience.

First of all, your criteria should be clear and concrete. Your readers should be able to measure how well an alternative meets a criterion. Nebulous or abstract criteria are difficult to assess. If you were recommending the purchase of a car, for example, gas mileage would be a measurable criterion; however, a requirement that the car "look sexy" would be less so. Your reader might reasonably ask, Look sexy to whom? Me? My best friend? My mother?

You can base your criteria on several definable factors. First, you can consider the goals of your organization: What values are important to them? Then you can look at the budget: What costs are acceptable? Finally, you can consider your needs and the needs of your readers: What standards must an alternative meet to be successful from your point of view?

You should be realistic in forming your criteria, but don't take anything for granted. You need not automatically assume that company policy must always be followed. Just because the company has always purchased limousines, don't assume that they always intend to, particularly if your purpose is to reduce transportation costs. In this case, consider the objective behind the company's limousines: Are they used for prestige or comfort? Is there any way that you can meet that criterion and still accomplish your purpose?

Once you've decided on your criteria, try to make them as precise as you can. Write the criteria out as complete statements. In the preceding example "comfort" would be a very imprecise criterion; try writing it out as "The car selected must have adequate head- and legroom for four average adults." As you make your criteria precise, you'll be further defining and limiting them. Then when you come to evaluate your data, your standards should be quite clear.

Finally consider your criteria in terms of importance. Some criteria will be absolute, and they will eliminate a range of alternatives. If you have an absolute budget limit of $10,000 per car, for example, you have effectively ruled out many cars on the basis of that criterion. But some criteria are negotiable; that is, they may depend on other criteria or they may be subject to change under certain conditions. Thus, for example, you might have a limit of $10,000 per car *unless* the car comes complete with service contract, in which case you could spend up to $15,000.

Once you have decided which criteria are absolute and which are negotiable you can list them in order of importance. Put your absolute criteria first since they are the ones that must be met without change. Then list your negotiable criteria in order of their weight in your decision. If you have more than about seven criteria, you'll need to group them for the convenience of your reader (see Chapter 5 for more information on grouping). Consider whether some of these criteria concern the same general standard; try to combine them into a more generalized statement that covers the same material.

Formulating Alternatives

Having developed your criteria, you're now ready to begin studying alternatives. At this point you must begin looking at the data you have collected. Developing alternatives means bridging the gap between that data and the criteria you have just established; you are answering the question: Given these facts and these necessities, what can I do to accomplish my purpose? There are no particular rules for devising alternatives, but you should allow yourself plenty of possibilities. This is the time for creativity, taking into consideration the limits within which you must work. Devise as many alternatives as you can. Don't limit yourself to the obvious; consider alternatives that might, at first, seem absurd. (In the Environmental Impact Statement process discussed in Chapter 16, for example, the federal government mandates that one alternative *must* be no action at all.)

Evaluating Alternatives by Criteria

The final step in this process is to weigh your alternatives against your criteria. It is best to do this systematically. List each alternative and consider it in relation to each criterion you have established. You should try to be honest and objective. It's natural to favor some alternatives, but don't cheat on your criteria. Your readers may catch you out and reject your recommendation if they don't think you've been rigorous in your process of elimination. It is best to get this entire process down on paper; it will help you make the comparison if you can see things in black and white. It will also make it easier for you to write your final report, since many of your preliminary steps will already be written out.

Let's look at an example of how this process works, establishing purpose, criteria, alternatives, and, finally, weighing criteria and alternatives to come to a decision. It's the first night of finals week; it's dinner time; and you're hungry. Your roommate comes in and informs you that dinner in the cafeteria tonight is meatloaf. Clearly you have a potential problem here. Your purpose is to get a dinner that will satisfy your hunger in a way that you like. There are some limitations, however. You have only $5.75 on hand because you forgot to cash a check today, and you can't take more than an hour for dinner because you have to study for your physics exam tomorrow. Your criteria are:

> ***Absolutes***
> Dinner must cost less than $5.75, including tax.
> Dinner must take less than one hour.
>
> ***Negotiable***
> Dinner must be something you like to eat.

You have the following data to work with:

> You don't like the cafeteria's meatloaf.
> You like Tico's Tacos.
>> Tico's Tacos is a 20-minute walk from the dorm.
>> Tico's tacos cost $1.50 each and take five minutes to prepare.
>
> You like Burger Delight.
>> Burger Delight is a 25-minute walk from the dorm.
>> Burger Delight burgers cost $3.00 and take ten minutes to prepare.
>
> You like Gino's Pizza.

Gino's Pizza is a 15-minute walk from the dorm.
Gino's Pizza costs $5.85 with tax and takes 20 minutes to prepare.
Gino's Pizza delivers.
Your roommate has not eaten.
Your roommate has money.
Your roommate loves pizza.

Using your data you can develop the following alternatives (among others):

You could eat in the dorm cafeteria.
You could go to Tico's Tacos.
You could go to Burger Delight.
You could go to Gino's Pizza.
You could suggest to your roommate that the two of you order a Gino's pizza delivered and split the bill.

If you analyze your alternatives using your criteria you would arrive at the following conclusions:

	Dorm	*Tico's Tacos*	*Burger Delight*	*Gino's*	*Gino's Delivered*
Money	Yes	Yes	Yes	No	Yes
Time	Yes	Yes	No	Yes	Yes
Taste	No	Yes	Yes	Yes	Yes

You now have two alternatives that meet all your criteria: Tico's Tacos and a Gino's Pizza delivered to the dorm and split between you and your roommate. Each alternative has advantages and disadvantages: Tico's wouldn't require your roommate's help, but would take more time; Gino's would be faster, but must have your roommate's approval. Your recommendation in this case will probably depend on how heavily you weigh the advantages and disadvantages of each and perhaps on some immeasurable factors such as whether you're more in the mood for tacos or pizza. Most decisions involve a certain amount of personal feeling; the best you can do is try to be objective and make sure your decision isn't unreasonably weighted with a bias.

Formulating Conclusions and Recommendations

Your final step in the recommendation process is to formulate your conclusions and recommendations based on your evaluation of your alternatives. Conclusions are the convictions you have arrived at based on your analysis of your purpose, criteria, data, and alternatives. Conclusions are statements of facts that have arisen from your research. A conclusion in the previous example would be that eating at Burger Delight would require 70 minutes. Recommendations, in turn, are action statements based on your conclusions. A conclusion, in effect, summarizes a fact or series of facts. A recommendation then tells your readers what they should do because of that fact or series of facts. In the case of Burger Delight your recommendation would probably be not to eat at Burger Delight tonight because of the limit on your time. To cite another example, if a marketing report concluded that 35 percent of the customers for Gino's Pizza are under the age of twenty-five, one possible recommendation might be for Gino's to increase its advertising in the college paper to try to attract a larger share of the eighteen to twenty-five-year-old market. The conclusion states a fact; the recommendation calls for action based on that fact.

Conclusions should always be based on facts rather than opinions. If your data is so inconclusive that your conclusion is partly interpretive, be sure to explain this when you state the conclusion. You should also be careful not to overstate your case; as with most scientific work, your conclusions should follow logically from your data. Any attempt to inflate your conclusions will be painfully obvious.

Your recommendations should grow logically out of your conclusions. It should be easy for your readers to see why you are recommending this action, given the conclusions you have just stated.

PERSUADING THE AUDIENCE

You have done all your preliminary work: decided on your purpose, collected (or reviewed) your data, developed criteria and alternatives, and decided on conclusions and recommendations. It may seem that your major work is over, but now you must return to the element you first considered when you analyzed your audience: the people who will read and act on your recommendation. You should not base your conclusions and recommendations on your audience's preferences, any more than you base them on your own preferences. Your recommendations should always be based on your analysis of your data, considering your purpose and criteria. But having arrived at your recommendations, you must now decide how to present them to your audience. Although you can't change your conclusions, you can look for a way to present them so that they have the best chance of succeeding, even with a hostile audience. Along with the audience analysis steps we presented in Chapter 1, consider the following suggestions.

Considering Reactions

Your first step in considering your audience should be to weigh the consequences of your recommended course of action. Who is likely to be affected by your recommendation? Only a few individuals or entire departments? What changes will most probably come about if your recommendation is accepted and how will those changes be received? Are you suggesting an adjustment in an unpopular program, or are you tampering with something that people like?

Most probably you see your recommendation in a positive light; after all, it is *your* recommendation. But try to look at it from the other side now. What objections could you raise to this idea? Who might be liable to raise those objections, that is, who is most likely to be hostile to your idea? If you can see possible objections, you can consider counterarguments before the objections are raised.

Developing Strategies

If you anticipate resistance or even hostility to your recommendation among your readers, you should consider strategies for counteracting it. Your best defense, of course, is to be able to demonstrate conclusively the correctness of your argument. Logic will help you here, and we will discuss logic in the next section of this chapter, but there are other ways to counter negative reactions.

First of all you should be careful about your tone when you fear your audience will be hostile. Be straightforward but tactful: Bluntness, even if it's intended to be honest, may be seen as an attack. Rather than saying "The problem in this department is extreme inefficiency," for example (even though that may, in fact, be the case), you could say "Other departments have found this system more efficient than the one we are now using." Try to find a way to phrase your points that will avoid hurt feelings.

You can also alter your organization if you suspect that your audience may favor another alternative than the one you are recommending. Begin your discussion by analyzing the alternative your audience may favor; be thorough and fair, and try not to slant your discussion. Close your analysis with a study of the disadvantages of that alternative, leading directly to the discussion of the alternative you favor. This not only prepares for your discussion, it demonstrates your fairness and impartiality.

Finally, be sure to supply ample support for your conclusions and recommendations in the form of hard evidence: statistics, laboratory results, test cases, and so forth. If you can show that you have studied this data and arrived at a logical conclusion based on it, you can counter a great deal of criticism.

ORGANIZATION

As you can see, once you have developed your purpose, criteria, alternatives, conclusions and recommendations, you will have generated most of the information needed for your report. Your organization then becomes simply a process of putting that information in the most effective order. But don't underestimate the importance of this ordering process; the success of your report will depend on whether your reader can follow a clear, logical progression in your argument. Your reader should have no doubt about the way your purpose leads to your criteria, which in turn lead to your alternatives, your conclusions, and your recommendations. Each step should lead inevitably to the next, so that by the end of your report your reader is convinced that you have reached the best possible recommendation given the circumstances.

It may be best to begin the process of putting your report together by checking your logic, making sure that each step does, in fact, lead to the next. You can begin with a form like that in Figure 10-1 to give you the skeleton of your argument. Now consider these questions:

1. Is my purpose correct?
2. Do my criteria serve as appropriate standards for my purpose?
3. Have I weighed all possible alternatives?
4. Are my conclusions correct given my purpose, criteria, and alternatives?
5. Are my recommendations reasonable given my conclusions?

If you can answer yes to these five questions, your report should be based on an effective logical progression. You can now proceed to organize your material into introduction, discussion, conclusions, and recommendations.

The Introduction

As usual you should begin your report with an introductory section. In this introduction you should give your readers all the information they need to understand the circumstances of your report. First of all you must explain your purpose. This may involve a statement of your research problem or it may simply be a brief statement of your objective. You should include necessary background information, a discussion of the project involved, for example, or an explanation of how you came to work on the idea. Since your purpose is basic to the logical development of your recommendation, it should be your first concern.

Figure 10-1
Recommendation outline

1. My purpose in this report is

2. The criteria I have developed are

   a. Will these criteria need justification? The justification is

3. The alternatives I have studied are

4. The evaluation of the alternatives in terms of the criteria is

5. My conclusions are

6. My recommendations are

Next you can explain your criteria and their derivation. You may need to go into whatever scope limitations you faced and explain why these criteria were chosen and, if necessary, why others were rejected. At this point you should also describe your methods of research. Tell your readers how you obtained the data you are using: Did you conduct interviews, use laboratory data, or rely on studies done by other researchers in other areas? You can include a review of the literature about your topic at this point if it's necessary to describe the work done previously by others.

Finally, you should describe the alternatives you have developed. At this point a detailed description will probably be unnecessary; you'll do that in your discussion. But you should include a basic outline in your introduction so that your readers know from the beginning what possibilities you considered. Again, if you rejected some alternatives that your readers might expect, you should explain why here.

The Discussion

You have a choice when it comes to ordering the body of your recommendation report: You can either use your criteria or your alternatives. However, in most cases it is best to order your report around your criteria. Both of the sample reports included at the end of this chapter are ordered in this way, although the short report does not state the criteria initially. If you order by criteria it will be much easier for your readers to see the comparisons you are making between your alternatives. Say, for example, you're comparing four alternative cars to be purchased for your company fleet on the basis of price, maintenance costs, mileage, and size. If you arrange by alternatives, your discussion will look like this:

Car A costs $15,000, has maintenance costs of $200 per year, gets twenty-five miles per gallon, and has room for five people. Car B costs $16,500, has maintenance costs of $150 per year, gets twenty-two miles per gallon and has room for four people. Car C costs $14,500, has maintenance costs of $230 per year, gets thirty miles per gallon, and has room for five people. Car D costs $18,000, has maintenance costs of $200 per year, gets twenty-four miles per gallon, and has room for six people.

Consider what your readers will have to do to make comparisons between those four cars. They will have to skip back and forth between four sections, trying to keep track of each figure as it applies to each car as they do so. Even with all the figures on one page this is time-consuming; imagine what it would be like with more data, strung out over ten or twenty pages.

If you set your comparison up by criteria, the reader could see this:

Car A costs $15,000, Car B $16,500, Car C $14,500, and Car D $18,000. Car A has maintenance costs of $200 per year, Car B $150 per year, Car C $230 per year, and Car D $200 per year. Car A gets 25 miles per gallon, Car B gets 22 miles per gallon, Car C gets 30 miles per gallon, and Car D gets 24 miles per gallon. Car A has room for five people, Car B for four people, Car C for five people, and Car D for six people.

As you can see, this comparison is simply much easier for a reader to make if you organize it by criteria (for further discussion of criteria arrangement, see Chapter 16).

The only exception to this rule comes when you are comparing only two alternatives. In that case you can sometimes make an effective comparison based on those two, particularly if you're concentrating on the advantages and disadvantages of each.

As you order your sections you can use a parallel structure for each one. You can begin with a definition of the criterion, justifying it if you think the reader may not understand its importance. Then you can present and interpret the data for each of your alternatives in order. If this data is extensive (enough so that a full-scale presentation will bog down your discussion), you can place it in an appendix at the end of your report.

In deciding which criterion to discuss first, and which second, third, and so forth, consider the needs of your readers. It is usually best to use order of importance here, beginning with the most important absolute criteria and working down to the negotiable ones. Some readers, particularly management ones, prefer the most important data at the beginning of the report so that they can skip over the less crucial material. On the other hand, if you feel that some of your criteria are likely to be controversial, you can begin with your least controversial ones and place those more likely to be disputed at the end of the report. In this way you can build up a foundation of agreement, and, perhaps, credibility before you challenge your readers' assumptions. Likewise if your criteria are particularly complex you might begin with the simplest and work up to the more difficult.

The Conclusions and Recommendations

Your conclusions and recommendations are usually placed in a separate section with a separate heading (or headings) since they represent an important separate element of the report growing out of the discussion. A major concern about this section is its placement.

Many organizations prefer that you place your conclusions and recommendations at the beginning of your report, before your discussion. This is, after all, the section of most interest to your readers; they want to know what you've discovered and what you recommend more than anything else. It is also typical of the deductive order common to scientific work: you begin with your conclusion and then present data to support it.

Some writers, however, prefer to put the conclusions and recommendations at the end of the report so that the readers will see the entire argument and become convinced before they reach the conclusions. They argue that placing the conclusions at the beginning of the report breaks up the logical chain you have produced, with each link dependent on the preceding one. If you decide to place your recommendations and conclusions at the end of your report, however, be aware that many readers will skip to them directly anyway rather than waiting to read your entire discussion. As with the placement of your criteria, the placement of your conclusions and recommendations will be influenced by the needs of your readers.

It is possible to compromise between these two views and put your conclusions and recommendations both at the beginning and at the end. At the beginning you place them in a formal section complete with heading, using the format we will discuss below; at the end of the report you provide an informal recapitulation of conclusions and recommendations to remind your readers without forcing them to turn back to the beginning of the report.

In terms of format it is sometimes easiest to present your conclusions as a list of numbered statements. In the car example your conclusions might look something like this:

1. Car C has the lowest purchase price.
2. Car B has the lowest maintenance costs.
3. Car C has the best gas mileage.
4. Car D has the greatest interior size.

Remember, as we stated earlier, your conclusions should be statements of facts, summaries of the data you have presented. Be careful not to inject opinions or subjective judgments at this point; your conclusions should be easily verifiable by referring to your data. You should order your conclusions from general to specific, beginning with broad statements and ending with the specific conclusion that will lead to your first, and most important, recommendation.

Your recommendations can also be presented as a list, or they can be written up in a short paragraph following the conclusions. Your recommendations should be action statements that tell your readers what should be done based on the conclusions you have just given them. Your first recommendation should fulfill your purpose; that is, it should be the recommendation your report was designed to provide. If the purpose of the car report were to recommend the most economical car to be purchased for the company's fleet, your first recommendation might be:

1. **We should purchase Car C for our company fleet because of its cost effectiveness.**

The other recommendations that follow your principal recommendation can then be implementing recommendations, statements that explain how the first recommendation should be carried out. In the car example other recommendations might include

2. **We should discuss fleet purchase options with Dealer C.**
3. **We should invest in yearly maintenance contracts.**
4. **We should investigate future leasing options.**

You may have only one recommendation or you may have several, based on the amount of data you have amassed and the type of report you are writing. You should always have at least one recommendation, however, even if it is a recommendation that *no* action be taken.

No matter where you put your conclusions and recommendations and how you write them out (either as a numbered list or as a short paragraph), you should not fail to include them. You may sometimes feel that your data, criteria, and alternatives are so clear-cut that no statement of conclusion or recommendation is necessary: Your readers can simply figure it out for themselves. But it is a mistake to let your readers draw their own conclusions. Although the matter may seem obvious to you, there is always the risk that it may seem less so to your readers; they may well draw quite different conclusions from the ones you expected. You can simplify their reading and also maintain control of your material by spelling out exactly the conclusions you want them to make, as well as the action you recommend.

EXHIBIT 10-1
A Professional Recommendation Report

Refer to:
FB: JRF: 023851
Date: June 23, 1982

TO: David Gonzalez
FROM: Jane Feldman
SUBJECT: IFTRAN

I strongly recommend that Engineering Sciences take steps to keep the IFTRAN preprocessor even as we move to FTN5 and the native mode operating system.

All current research in software engineering recognizes that structured, readable code improves program quality and programmer productivity, particularly in the maintenance phase of the software life cycle. IFTRAN encourages generation of good code by providing control structures not included in Fortran. Using these, even nonprogrammers such as engineers are likely to produce structured code without extra effort.

An added benefit, possibly of greater importance than the extra constructs, is the systematically indented IFTRAN program listing. Systematic indentation stresses the relationships between code segments and gives the reader a quick grasp of the global as well as the local structure of the code.

Admittedly the Fortran code generated by IFTRAN is even less readable than most examples of Fortran code generated by people. Yet if we lose IFTRAN, maintenance costs for the 50 or so IFTRAN programs now in the library could easily double. AXCAPS, ATTILA, and NANCY are among the 50 IFTRAN programs.

The Software Workshop of General Research Corporation claims to have a version of IFTRAN which is compatible with Fortran 77, though I believe it may require access to some non-ANSI FTN5 constructs such as Hollerith constants. The cost of this IFTRAN is $1875. If this is indeed a product which will run on the CDC 800 series machines in native mode, it is a very good buy. Converting our current source versions of IFTRAN would require about one month. I recommend that we purchase the new GRC IFTRAN preprocessor.

EXHIBIT 10-2
A Student's Recommendation Report. Courtesy of Ralph Voss, Jr.

THE FEASIBILITY OF A SOLAR SPACE-HEATING SYSTEM FOR A SAN ANTONIO HOME

INTRODUCTION

The purpose of this report is to present the results of a cost-benefit study that evaluated the feasibility of using solar energy to space heat a home in the San Antonio area. To prove feasible, the solar installation should provide enough savings to pay for itself within a maximum time period of seven years. This report discusses the various types of solar systems which are available today. The main criteria are cost, complexity, ease of installation, and savings produced by the solar system. Before specific solar systems are analyzed, the San Antonio climate and the general role of a solar system must be considered.

Climate of the San Antonio Area

The continental United States can be divided into four predominant climatic areas: a cool region, a temperate region, a hot-humid region, and a hot-arid region. The San Antonio area is included in the hot-humid region.

Many climatic factors differentiate these four areas. When space heating is the primary concern, the most important of these factors is the annual number of heating degree-days of the area (Total Environmental Action 1975, p. 26). Heating degree-days is the sum, over a specified period of time, of the number of degrees the average daily temperature is below 65°F (18°C). In other words, the annual number of heating degree-days is a direct measure of the amount of heat that will have to be supplied to space heat a home during the course of the year. Therefore, it is obvious that this one factor alone will play a major role in dictating the required size of the solar space-heating system. Table 1 gives an example of a city in each of the four climatic regions and its annual number of heating degree-days.

Table I Heating Degree-Days of Various Regions

REGION	CITY	STATE	HEATING DEGREE-DAYS
Cool	Minneapolis	Minnesota	8159
Temperate	Boston	Massachusetts	5634
Hot-humid	San Antonio	Texas	1546
Hot-arid	Phoenix	Arizona	1552

Although San Antonio's heating requirements are not as high as some cities', they still require an efficient system.

Auxiliary System

A solar space-heating system designed to provide 100 percent of a home's heating needs would be very expensive and oversized. Most solar heating systems are designed to provide from 50 percent to 80 percent of a home's yearly heating needs (Bendt and Soto 1980, p. 9). This is known as the "solar fraction" of the system. Therefore, if a solar space-heating system has a solar fraction of .5, you can expect the system to provide 50 percent of the total heating requirement.

Since in most cases solar energy is unable to supply all the heat needed to space-heat a home during the course of the year, an auxiliary or backup system is required. In most cases, the auxiliary system is a conventional electric or natural gas system (U.S. Department of Energy 1981, p. 26). Often when a solar heating system is installed in an existing house, the heating system that was in use prior to solar installation is kept and used as the auxiliary system.

SOLAR SPACE-HEATING SYSTEMS

There are basically two methods used for space-heating a home with solar energy. One method uses a passive system; the other, an active system.

Passive Systems

A passively designed system is one that uses the design and mass of the building as a heating system, without the use of mechanical devices for moving and distributing heat (Frank 1981, p. 205). For example, some of the most efficient passively designed houses have large areas of glass on the south-facing side so that the sun's heat can be used directly to heat the house as it shines through the glass. Because this kind of system depends greatly on the initial design of the house, a passive system would not be practical for the majority of San Antonio residents who are confined to neighborhood-style housing.

Active Systems

Active systems can be defined as systems that "employ a distinct collection device and use a mechanically propelled heat transfer fluid to transport heat from collector to storage" (Szokolay 1980, pp. 23–24). The most practical, and therefore most common, active space-heating systems use flat-plate collectors mounted toward the south at an angle that will capture as much of the sun's direct radiation as possible during the year. To account for the difference in the winter position of the sun as opposed to its higher summer position, some collector arrays are mounted on brackets that can be adjusted up and down. Usually, the collectors are installed on a south-facing roof if possible. However, the collectors can be mounted almost anywhere, including the ground; just as long as they are unshaded and face the sun directly.

The storage system can also be located at any convenient and accessible place, garages and basements being the most commonly chosen spots. Furthermore, it is preferable to place the storage terminal as close as possible to the collecting panels in order to minimize heat losses through the distribution system (Texas Energy and National Resources Advisory Council 1982, p. 16).

There are generally two kinds of active systems. The basic difference between these two systems is the type of heat transfer medium used.

Water Systems. A majority of the solar-heating systems today use water as the transfer and storage medium. The typical water system consists of a collector; a store; a system of piping, pumps, and controls for circulating water from the store through the collector; and another system for transferring stored heat to the dwelling.

Advantages of water systems:

- They have repeatedly been proven to work well.
- Water is a cheap and efficient heat transfer and storage medium.
- Piping, unlike ductwork, uses little floor space, is easily connected, and can be routed easily to remote places and around corners.
- The circulation of water uses less energy than the circulation of air with corresponding heat content.

Disadvantages of water systems:

- Steps must be taken to prevent the occurrence of corrosion, scale, or freezing.
- The domestic hot water supply may be contaminated if a leaking pipe allows treated water from the store to enter the domestic water system (Architects and Planners Inc. 1976, p. 6).

Air Systems. Air systems use basically the same principles as water systems, except the former uses air as the transfer medium between collector and store. Because of this difference, the storage site in air systems is most commonly made of stones 2 to 4 inches in diameter. Ductwork in air systems has the same function as the pipe system of water systems.

Advantages of air systems:

- There is no corrosion, rust, clogging, or freezing problem.
- Any leakage that occurs will not cause damage to the building fabric and will not saturate the insulation.
- Rock-bin storage produces better temperature stratification.
- Because no heat exchanges are used, there is no loss of temperature (Szokolay 1980, p. 220).

Disadvantages of air systems:

- Ductwork is more expensive to install and harder to accommodate than a pipework system.
- Air has a lower thermal capacity than water and thus requires more energy to transfer a given amount of heat.
- Leakage is hard to detect and can significantly lower system efficiency.

- Air systems are much more complex and thus more difficult to design.

The overall efficiency of a well-built water system is about the same as that of a well-built air system (Szokolay 1980, p. 220). However, because air systems are more complex and more expensive than water systems and take up considerably more space, they are less frequently used than water systems. Until the efficiency of air systems is improved enough to justify the greater cost and trouble involved, the water system will remain the most popular and commonly used solar space-heating system.

FEDERAL TAX CREDIT

The federal solar energy tax credit was increased in 1980 by the U.S. Congress. Federal taxpayers can now subtract up to 40 percent of the cost of eligible solar, wind, or geothermal energy systems from their federal income taxes. Credits can be claimed for the first $10,000 expended, for a maximum allowable federal credit of $4,000.

The new federal credit became effective January 1, 1980, and runs through January 1, 1986. To claim the credit, the solar system must be installed in or on the individual's principal residence, which can be owned or rented.

COST-BENEFIT STUDY

The purpose of this cost-benefit study is to provide information that will prove useful in determining the feasibility of installing a specific system in a specific type of house located in a specific area.

Assumptions Made During Study

- The house is located in the San Antonio area.
- The house is reasonably, but not exceptionally well insulated and has a floor area of $150m^2$.
- The house already has a conventional electric system, which will be used as the auxiliary system.
- The solar-heating system that will be installed is a typical active water system and will supply 60 percent of the house's total heating needs.

Calculations

First, the total cost of the solar space-heating system must be established:

	$3150.00	flat-plate collector panels
	900.00	storage tank
	300.00	pumps
	275.00	blower and ductwork
	150.00	pipes and fittings
	200.00	labor
Total	$4975.00	

Note: the size and price of the system were established by conferring with local installers.

Next, the specific heat loss rate (q) of the house must be established. For a reasonably well insulated home, the specific heat loss rate can be assumed to be 2.5 W/deg C per m^2 of the house (Szokolay 1980, p. 221). Therefore,

$$q = (2.5 \text{ W/deg C m}^2)(150\text{m}^2) = 375 \text{ W/deg C}$$

The annual number of degree-days for San Antonio is 1546 (Montgomery 1979, p. A-21). This figure is converted to degree hours:

$$(1546)(24) = 37104 \text{ deg Ch}$$

This value multiplied by the specific heat loss value (q) gives the annual heating load:

$$\text{Annual heating load} = (37104 \text{ deg C})(375 \text{ W/deg C}) = 13{,}914 \text{ kWh}$$

The annual solar contribution is found by multiplying the solar fraction by the annual heating load:

$$\text{Annual solar contribution} = (.6)(13{,}914 \text{ kWh}) = 8{,}348.4 \text{ kWh}$$

The value of this contribution depends on the cost of heating by alternative conventional means. For the conventional electric system, the national average cost of heat was 6.06¢/kWh (U.S. Department of Energy 1981, p. 43). The annual benefit of the solar installation is found by multiplying the annual solar contribution by the heat cost:

$$\text{Annual benefit: } (8348.4 \text{ kWh})(\$.0606) = \$505.91$$

The crude payback period is found by dividing the annual benefit into the total cost of the system:

$$4975/505.91 = 9.8 \text{ years}$$

It should be noted that the above figure of 9.8 years was calculated without considering the 40 percent federal tax credit. If the homeowner takes advantage of the tax credit, the payback period of his solar installation is greatly reduced:

$$\$4975 - (.4)(4975)/\$505.91 = 5.9 \text{ years}$$

CONCLUSIONS

1. A passive solar system would not be practical for the majority of San Antonio residents.
2. A water system is a more practical investment than an air system when considering active solar-heating systems.
3. When the federal tax credit is taken advantage of, a solar water system installed in a San Antonio home for the purpose of space-heating is feasible.

RECOMMENDATIONS

The San Antonio homeowner should:

1. Install a solar-heating system using water as the transfer medium.
2. Take advantage of federal tax credits to offset the cost of the system and shorten the payback period.

SUMMARY

1. The recommendation process begins when you analyze your audience, formulate your purpose, scope, criteria, alternatives, conclusions, and recommendations.
2. Criteria are standards of judgment that are precise, measurable, and based on an accurate assessment of audience needs.
3. Alternatives are formulated by analyzing data in terms of purpose and criteria.
4. Conclusions are formulated by evaluating alternatives in terms of purpose and criteria.
5. Conclusions are statements of fact; recommendations are statements of action to be taken because of those facts.
6. The audience for a recommendation report should be considered in terms of possible reactions to the recommendations and possible strategies for countering negative reactions.
7. Introductions to recommendation reports include a purpose statement, an explanation of criteria, and a description of alternatives.
8. Discussions in recommendation reports are usually organized around criteria, which are ordered by importance.
9. Conclusions and recommendations in recommendation reports may be written as numbered lists and are usually placed before the discussion.

EXERCISES

1. Analyze the recommendation reports in Exhibits 10-1 and 10-2. Identify the purpose, criteria, alternatives, conclusions, and recommendation of each.
2. Consider a purchase you would like to make: a car, a typewriter, a home computer, a bicycle, a stereo, a radio-cassette player, or some

other product. Decide on criteria for your purpose; select three or four alternative types and analyze them; then formulate conclusions and recommendations.

3. Using the same product you studied in Exercise 2, rewrite your analysis for another audience: your parents, your boss, your best friend, etc. Assume that they want to make the purchase and you are doing the analysis for them.

4. A student working on a similar exercise came up with the following criteria for a home computer:

   It should have a keyboard that is easy to use.
   Software.
   It should have an expandable memory.
   External disk drive is handy.
   Games.
   It should look high tech.
   It should come with a dot matrix printer.
   I don't want to pay more than $2500.
   A mouse.
   Is a modem a good idea?

   Obviously these criteria need work. Group them to reduce the number to seven or less. Make them precise, parallel, and measurable.

5. Define a problem that you see on your campus: too few study carrels in the library, no change machines around the vending machines, poor visibility in a lecture room, and so on. Choose some relatively minor problem that is nonetheless an irritant. Develop criteria for a solution. Consider possible alternatives. Write a short report to the university administration recommending your solution to the problem.

Recommendation Case 1

The Underground House

Since you completed your architecture degree you've been looking forward to this moment: your first commission. In many ways it's an ideal situation: The customer is your friend Nora Friedrich who knows and admires your ideas. But when Nora begins to outline what she's looking for, some of the glow begins to wear off.

Nora is committed to energy conservation and alternate energy sources. She wants a home that reflects her beliefs, and she feels she's come up with the ideal solution: earth-sheltered housing. She wants you to design a house that is at least partially underground! She feels that such a house would represent a significant savings in energy costs and could be both attractive and unique.

Nora is enthusiastic about the idea, but she doesn't know many specifics; she expects you to know more. Unfortunately that isn't the case. You know almost nothing about earth-sheltered housing and whether it can be built in your area. After discussing the matter with Nora, you decide your first step must be a feasibility study to determine whether such a house could be built in your city and whether the savings in energy costs would be sizable enough to offset construction costs.

After a lot of research, these are your notes:

Earth-sheltered house = house with earth cover over roof and walls. Like other below-ground things: basements, storm cellars, etc. Not for places with heavy flooding, but takes tornadoes, light earthquakes, other disasters better than conventional. Needs ventilation and dehumidifying. Claustrophia? Skylights, windows on noncovered walls give light, space illusion. Earth good soundproofer; can build near noise pollutions (e.g., airports, highways, etc.)

Construction costs. Somewhere from 25 percent to 10 percent more than conventional. Depends on design. Materials cost more—concrete. But less labor intensive. To get comparison, compare cost of labor with cost of materials.

Maintenance costs, up to 30 percent savings. Less repainting; roof covered with soil, less roof repairs. Less expansion and contraction with weather. Almost rot and termite proof: fewer perishable materials, better protected.

Energy: better insulated. Consistent temp. held by earth shelter. Passive solar; no mechanical equip. needed. Heat from household stuff: dryer vented to heat exchanger, lights, appliances, wood stoves, warm waste water, exhaust air. Heating costs 50 to 95 percent less. Air conditioning? Less uncontrolled air leaking in through cracks, holes, etc. Ventilation air controlled (heat loss/gain depends on ventilation load for heating or cooling air from outside, heat transmission through building to outside). Less temperature fluctuation in earth; below .2 meters, almost none. Earth temperature about 55° in winter; outside air 31°. Easier to heat to room temperature (75°?). Thermal mass of earth-sheltered greater, can store more energy; temp. raised slowly during day, drops slow at night. Also less cooling needed: earth's coldest temperature lags two months behind air's; walls of earth sheltered coolest in April, warmest in Oct. Rain water held on roof and walls by sod; slow evaporation cools. Earth sheltered in warm climate 40 percent (average) less cooling costs than conventional.

Technical: Clay unsuitable (too heavy, can plug drainage); other soils o.k. South and west of city, heavy clay concentrations. Northside, limestone under topsoil. Light enough to excavate; no settling problems. Bedded limestone can support approx. 40 tons/sq. ft. Northside water table o.k., not saturated soil. Building codes o.k. Insurance available (First Federal offers premium reductions for earth-sheltered).

ASSIGNMENT

Write the feasibility study for Nora. Remember, she wants to know whether the house can be built in your city and whether the savings she might get from such a house would equal her outlay. Formulate purpose, criteria, conclusions, and make a recommendation.

Recommendation Case 2

The Parking Problem

That's it! You've had it! You just spent thirty minutes cruising the parking lots looking for a space and *not one* was available. The same thing happened yesterday; you ended up parking in a reserved space so you could make it to your 10 o'clock class, and when you got back you had a parking ticket. Now it looks like you'll have to do the same thing again or miss your biology lecture. Is this what you get for your $15 parking sticker?

Discussing the situation with your friends, you confirm what you already suspected: The parking situation on campus is a mess. The university has an enrollment of 13,500 this year with a projected increase of 3 percent per year for the next five years. There are currently 5230 parking spaces available to students in the campus lots. That might be enough if the students came and went at different times. There are always parking spaces before 8 A.M. and after 1 P.M. But during the peak class periods from 9 to 12 the lots fill up fast and few leave until after lunch. The campus police report that parking violations are up 8 percent this year, and they expect the situation to get worse.

After researching the problem you decide that the campus needs a minimum of 2000 new parking spaces within the next two years. These spaces would relieve the current problem and help take care of the continuing growth of the student body. Brainstorming with your friends you come up with some possible programs: building new lots; establishing carpool lots; establishing lots for small cars, which would have more parking spaces; taking out the landscaping medians in the existing parking lots to create more spaces, and doing away with campus parking altogether (shuttle buses from various locations around the city would be substituted).

To gain 2000 new spaces the university would have to build twenty to forty new lots, at an average cost of $20,000/lot. Removing medians would gain 500 spaces for around $10,000. Using carpool lots the university could gain the equivalent of 2000 spaces (figuring four potential drivers to a car) with five to ten lots at $20,000/lot, assuming that the students would use carpools. Using small car lots, the university could gain 2000 spaces in 17 to 30 new lots at $20,000/lot (small car lots have an average of 20 more spaces per lot). Banning parking altogether would require no construction of lots, but would require building at least ten shuttle bus stops at $1000/stop. It would also deprive the university of approximately $150,000 per year in parking fees and require the university to negotiate a contract with the city to provide buses and routes.

ASSIGNMENT

Decide on your purpose. Develop criteria to meet it. You may use the alternatives provided or develop alternatives of your own (either new ones or combinations of those listed). When you have evaluated your alternatives by your purpose and your criteria, develop conclusions and recommendations. Then write a report to the president of the university outlining your recommendation for dealing with the parking problem.

CHAPTER · 11

Proposal Writing

INTRODUCTION

No matter what your field of work or study, you will write proposals: sales proposals, program proposals, and proposals to do studies or win grants. Proposals differ from other reports in one important respect: They require more negotiating between the writer and the audience. Because of the importance of the negotiating and research stages of proposal writing—because proposal writing, in fact, is just as much a political process as it is a thinking, organizing, and writing task—this chapter will cover both proposal formats and proposal politics.

In this chapter we will describe the process that most proposals undergo on their way to acceptance or rejection. More than other types of writing, proposals are responses to specific audience criteria; thus it is important to be aware of the procedures writers go through in order to discover and fulfill those criteria. We will also introduce the basic parts of formal and informal proposals and explain why many proposals are accepted or rejected. Since proposals are often rejected because of breakdowns in preproposal negotiations, we focus on how you can organize a proposal development effort that will satisfy both your needs and the objectives of your audience, the funding agency. (When we say *agency* in this chapter, we mean any organization, firm, governmental body, or research institute—anything that acts.)

PROPOSALS DEFINED

A written proposal can be thought of as a sales document that is designed to answer four questions:

1. What is the service (or product), and why is it needed?
2. How and when will the service be performed?
3. How much will it cost? and
4. How will the results be evaluated?

Consider this simple example. While riding your bicycle, you crash and bend the front wheel. You take the bike to a nearby shop, where the bicycle mechanic says she can straighten the wheel by adjusting the tension on the spokes. She can do this in half an hour, and the charge will be $10. When you come back, the mechanic promises to demonstrate that the wheel is true by placing it in a vise and using a set of calipers to measure any variance from perfect alignment. Here, the service is trueing the wheel; the procedure is adjusting the spokes (as opposed, say, to hitting the wheel with a hammer or jumping on it); the price is $10; the time half an hour; and the method of evaluation a test performed with the wheel off the bike.

In many proposals these four questions are addressed in separate sections called the technical, management, cost, and evaluation proposals. Sometimes these individual sections are prepared by separate groups of specialists. Thinking about proposals strictly as documents, however, can lead to misconceptions about the ways they are presented and assessed. Sometimes a written proposal is not even necessary, and often the written proposal is prepared after an agreement is reached between the contracting parties.

Proposals can also be classified in terms of the type of service to be provided, such as sales, research, and design or development of demonstration or model programs. Whatever the service, the basic formats and strategies remain the same.

Solicited and Unsolicited Proposals

Proposals can be usefully classified according to how they are initiated. If someone asks you to submit a proposal, it is a **solicited** proposal. On the other hand, if you approach an agency with an idea, the proposal is **unsolicited**. Many agencies use an additional formal screening process in which they issue a brief request for qualifications (RFQ) that asks only for company credentials and experience. After the RFQs are screened, a group of three or more finalists are asked to prepare full-scale proposals, a task that may cost each bidder thousands of dollars. A Request for Proposals (RFP) is just what the name implies: a request that an agency (or agencies) submit a proposal for a particular project. Some RFPs ask for specific products or services, while others ask for the development of solutions to problems without specifying the form the solution is to take. Medical foundations often advertise the availability of funds for studies in particular branches of medicine, for example, such as cardiology or pediatrics. The foundation's aim is to fund a certain number of well-conceived projects, whatever their focus within the field.

You can submit an unsolicited proposal to stimulate a funding agency to solicit that type of proposal at a later date. Your company might discover, for example, that only one foundation has an interest in the kind of research you would like to do. Unfortunately, this year that foundation has earmarked its money for some other area of study. In spite of that, you initiate a discussion with the officers of the foundation and submit a draft proposal, hoping that next year the foundation will set up a program in your area of interest. By this means, an unsolicited proposal can be turned into a solicited proposal. Excerpts from an actual RFQ and RFP are reproduced in Exhibit 11-1.

An extended example of a proposal is provided in Exhibit 11-2. This proposal was written by a group of students, and it contains a slightly different set of sections than the bicycle repair proposal mentioned earlier. After a cover letter addressed to the audience the students have already contacted, the proposal itself includes a statement of problems and needs, a solution, a cost analysis, a discussion of a possible management plan, and finally a summary. One important feature of this proposal is the way it is tailored to meet the concerns of the university administrators to whom it is addressed. The students saw the pleasures of a new outdoor eating area as most important, but they recognized that the administration would see student retention, student recruitment, and the financial security of the student union cafeteria as more important. The appendix of the proposal, which provides working drawings for the construction, is not included.

Notice that the authors of the proposal in Exhibit 11-2 took the time to discover which faculty and administration committees would have the power and responsibility to act on such a proposal and contacted at least the Vice President for Operations before submitting their proposal. In other words, the student group engaged in a process of investigation, negotiation, and lobbying. Moreover, since the proposal they submitted is still open-ended in many areas, it is possible that one of the faculty committees would ask for revisions or further discussion

EXHIBIT 11-1
RFQ and RFP from an Urban Transportation Project

Statement of Qualifications [RFQ]
Firms shall include the following information, limited to about 20 pages, and submit four copies:

1. Introductory remarks should include at a minimum the history of the firm, size, financial capability, ability to meet schedules and projected workload, and familiarity with Urban Mass Transit Administration projects on this scale.
2. Brief outline of its approach to the project and special problems identified by the firm.
3. Past experience of the firm and its subconsultants, if any. Note any experience with Urban Mass Transit Administration projects in detail.
4. Expertise of the project team, including consultants. Information for each should include (a) discipline, (b) years of experience, (c) years with the firm, (d) education, (e) role of team member, (f) expertise with related projects, (g) experience with other projects.
5. Brief discussion of general project management philosophy.
6. Description of how you will meet minority and women's business participation goals.

Consultant Scope of Work [RFP]
[The full RFP includes four tasks: urban design, air quality and noise studies, transit mall alternatives, and terminal alternatives. Only the first is reproduced here.]

Consultant Subtask 1: Urban Design Objectives
OBJECTIVE: To incorporate into the project from the beginning a high priority for good urban design.
METHOD: Develop some urban design objectives that will be used in the development of alternatives. Identify opportunities and constraints particular to downtown Seattle and the efficient operation of the transit system. Develop a sensitivity to the quality of life in downtown Seattle, so that the transit alternatives developed in the project blend with existing conditions. Participate in local agency, public, and Urban Mass Transit Administration reviews.
INPUT: Metro has done a considerable amount of work on this subject. Documentation will be available from Metro.
OUTPUT: Written urban design objectives and graphics to be used to evaluate the alternatives and for review by the local agencies, the public, and the Urban Mass Transit Administration.

EXHIBIT 11-2
Sample Proposal with Letter of Transmittal

George Hooper
Room 216
Hill Hall
Pacific University
May 3, 1986

Dr. Richard Martin
V.P. Operations
Pacific University

Dear Dr. Martin:

We are proposing the development of an outdoor eating area on the unused space northeast of the security office. We feel that such a development could greatly enhance the university campus, attract new students, and meet off-campus restaurant competition. It is also our feeling that the benefits, both aesthetic and economic, can well justify the project's modest initial cost. In the proposal that follows, we have outlined what we feel are the problems and needs that the project will help to meet. We also provide an overview of the proposed outdoor eating area and its benefits. The appendix contains detailed construction plans and a materials budget.

Thank you for taking the time to discuss this plan with our group. We look forward to the response of the committee on plant and operations and the faculty student affairs committee.

Sincerely,

George Hooper
Students for a Planned Environment

PROPOSAL

Statement of Problem
As students at Pacific University, we have observed several growing problems that have a common, if partial, solution. In these tough economic times, everything that is feasible must be done to maximize the financial potential of this institution. Declining enrollment and an increasing number of student transfers are becoming critical problems.

Last spring, Paradise Sam's restaurant opened across the street from the student union. Sam's business has increased steadily since that opening, and a large part of that business is coming from people who used to eat at the student union. According to the director of the union cafeteria (SUB), the union is losing between 4 and 7 percent of its share of the market. This, of course, means a loss of revenue to the school. During peak hours, the SUB is overcrowded and has difficulty providing enough space to serve those people who do want to eat and study there.

Study space is another critical commodity on campus. Increasing numbers of students are going off campus to find a place to study. Other university libraries in the city receive an increasing number of Pacific University students who are searching for quality study areas. This seems unnecessary when one walks around our campus and sees the large amount of undeveloped outdoor area.

Solutions

It is in this light that we propose the development of an outdoor eating and study area on the unused space northeast of the security office. The development of this small area (approximately 700 square feet) will include the addition of tables and minor landscaping. Five tables will give seating for twenty people. The tables will be weatherized so that they can be left outside year-round. The landscaping will consist primarily of the addition of two small privacy hedges: a row of Siberian spruce trees along both the east and south edges of the area. Other landscaping will involve leveling and sodding the area. Finally, camellias and trelliswork will be used to fill the gap in the shrubbery along the west perimeter. Construction details and an illustration of the plan are given in the appendix.

The benefits of such an area would be many. By enhancing the appearance and atmosphere of the campus, the new area will be a "drawing card," both attracting prospective students and keeping current students. It will also increase the capacity and desirability of the SUB as an eating facility, since this area is directly across a major pathway from the SUB. This area will also augment available study space on campus, because small groups or classes may use this area as a meeting place.

Security should be no problem at all. The tables will be permanently anchored to the concrete platform so that theft will not be a concern. Furthermore, the area is directly adjacent to the security office. That makes it one of the safest places on campus since the security office is open twenty-four hours a day. The spruce hedge will be only four feet high.

Cost

Only minimal additional costs will be incurred for maintenance. Since the tables will be weatherized, there will be no reason to move them inside. They are designed to give many years of trouble-free service. Maintenance and care of the lawn and shrubs should also cause only minimal additional expenditure. The present hedge, shrubs, and lawn are already being maintained. The spruce trees will not need trimming for the first three to four years, and the camellias will need only an annual pruning.

The total cost of materials for the project is approximately eleven hundred dollars. This price was arrived at by using the current retail prices and does not include any commercial discounts that the university may have with local businesses. A detailed breakdown of material costs is listed at the end of the appendix.

Management

Labor costs have been excluded from the cost estimate because volunteer labor will be used. Candidate groups include the volunteers from the Associated Student Body or a dormitory or service group. Last year such a group provided landscaping services to the community surrounding the campus. If a volunteer group is used, students might oversee the project as an independent study in personnel management. Alternatively, the project could be turned over to one of the design classes (possibly Art 2203, 3212, 3213, or 4222). There are many ways to incorporate this project into the curriculum or at least to take advantage of volunteer labor.

Summary

It is our conclusion that the development of this eating area will

1. Help this institution attract new students and maintain enrollment
2. Help revitalize the Student Union and meet off-campus competition
3. Enhance the appearance and atmosphere of Pacific University's campus
4. Provide students with a much-needed outdoor study and meeting place

We also feel that the new area will provide these benefits with a minimal initial investment and that the eating area will not require a large commitment of future resources for maintenance, upkeep, and security.

before approving the plans and budget and making money available. As one executive remarked of a seventy-five-page business proposal—the product of eighteen months of intensive research—"Sometimes a proposal is like a resume; it just gets you in the door." The process of negotiation is common in proposal writing, and it is the focus of the next section of this chapter.

FIGURE 11-1
Communication in the proposal-writing process

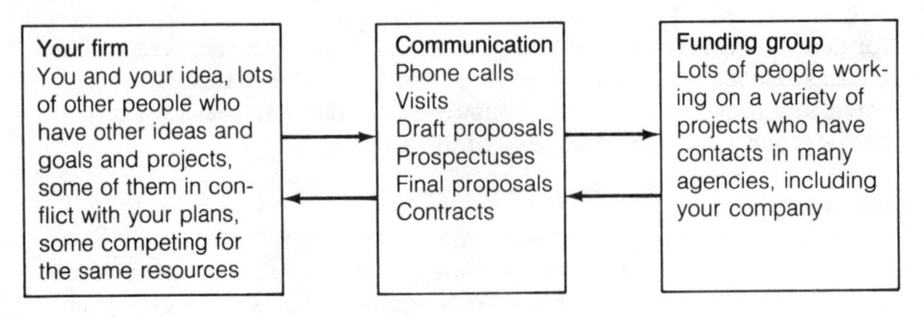

THE POLITICAL PROCESS

Proposals start in one of two ways, with ideas or opportunities. You can develop a new idea, or you can find out that money is available and respond by developing a project to fit the opportunity. Before you begin to approach a customer, however, you must assess your proposal's suitability and timeliness in terms of your own goals and those of your employer. Unless you are a completely independent researcher, your proposal will become your employer's proposal. For this reason, it is useful to consider the proposal process as having two main stages. In the first stage, your employer (or department, agency, or division) adopts some form of your idea, while in the second stage, your employer enters into negotiations with the funding agency. During the later part of those negotiations, your firm will write a formal proposal and, with luck, you will win a contract.

Some view the proposal-writing process in a different light: as one person or group of persons writing a proposal to an anonymous panel of impartial reviewers or customers in the target agency. We find it more realistic to look at the process as it is drawn in Figure 11-1.

Successful proposal writing requires extensive communication between you, your employer, and the funding group, starting with an abstract.

Phase 1: Writing the Abstract

Once you have a rough idea of what you plan to propose, you should draft a one- or two-page document explaining what you want to do and why it is important. If you wish you may also estimate costs, explore management options, and counter possible objections. This document is called a proposal abstract, and its purpose is to generate open and frank discussions. For that reason, it's all right for an abstract to look rough; it should leave room for other people to make suggestions and identify themselves with the project. If an abstract is too polished, it can be perceived as an attempt to railroad an idea.

Before you develop the abstract, you might start even more generally by making a short oral presentation to a committee or staff meeting. You can distribute copies of your revised abstract after the group has had a chance to respond to your presentation. With this groundwork in place, you are ready to consider how you can get the support of your colleagues and immediate supervisors. Exhibit 11-3 is a typical draft abstract. Notice how general it is.

Phase 2: From Writing the Abstract to Organizing a Proposal Team

After the preparation of your abstract your next concern will be to make your idea acceptable to your employer. You must consider whether your proposal will fit in with the long-term goals of your employer and your particular work group. Try to anticipate possible conflicts between your idea and the ideas and goals of others in your division. If you determine that such conflicts exist, you must find a way to deal with them, perhaps by appealing to the special skills, talents, and interests of your associates.

At this stage, a proposal group will be formed to carry the work through to completion. The students who wrote the sample proposal in Exhibit 11-2 formed their own group, yet they too needed to assess their skills and interests to determine which group members had talents for writing, drafting, information gathering, and interviewing university personnel. Because these students were taking a class together, they also had to determine whether their proposal was in conflict or competition with designs being developed by other class groups. (See Chapter 15 for further ideas about writing or working in a group.)

Once you gain the approval of the group and the company, your employer will begin to take a more active role in managing and supporting the project. Three basic systems of proposal management are described in the next section.

Systems for In-House Organization of the Proposal Process

Your employer will probably have a system for managing proposal development, because every proposal, however small, commits company resources of time, building space, support staff, accounting services, and so on. In some cases, individual researchers are allowed considerable freedom to make proposals on their own authority, although they may be expected to request the assistance of a proposal manager and notify the firm of any possible commitments of facilities. This model is most

Exhibit 11-3
A Typical Draft Abstract

COMPUTER SIMULATION OF ORTHOPEDIC SURGERY

The Problem
Surgeons who correct deformities of the foot and ankle work with a limited knowledge of the dynamics of the human foot. Anatomical dissection provides only a partial view of the movement of the living foot, because when the specimen is cut open, some of the critical alignments are altered. Contrast studies (using radiopaque dyes injected into the spaces between bones) are not fully satisfactory because the dye often leaks or obstructs the view. Moreover, in the infant foot, many of the surgically important features are not visible on X rays.

Recent work in the field of pediatric orthopedics suggests that more detailed information about the dynamics of the foot is needed before surgical procedures can be perfected. In order to gather this information (such as the location of the axes of rotation of the bones of the ankle) a new method of modeling or analyzing the structure of the foot must be found.

A Possible Approach
In the years since 1972, when a human hand and face were first modeled by computer scientists at the University of Utah, computer simulation of the human anatomy has progressed rapidly. Computer simulation, however, has tended to focus on modeling the surfaces of structures. At the same time, the medical applications of this technology have been limited to systems that can distinguish between tumors and normal soft tissue or that can model parts of the cardiovascular system. While these systems, including CAT scanners and the Mayo Clinic's Dynamic Spatial Reconstructor, are revolutionizing some areas of medical practice, they are not designed to analyze bony tissue or to distinguish among bone, partially ossified bone, and cartilage. It is possible, however, that a system of this type could be developed.

At this time, it would not be necessary to develop a full-scale diagnostic system that could be used to study individual patients. Surgical research would be assisted by development of a computerized model of the dynamics of the human foot, both adult and infant. This model would require simulation of the movement and elasticity of the skin, tendons, bones, and other tissues. Once such a model were designed and programmed, it would then be possible to experimentally simulate various types of abnormal feet by varying the values assigned to bone length, bone position, tissue elasticity, and so on. This procedure of varying parameters could also be used to estimate the effect of specific surgical procedures.

Specific Proposal
Development of a computerized model of the foot would probably require the coordination of several specialties, including anatomy, bioengineering, photogrammetry, and computer simulation. This study would assess current technology and would also include a search of the literature to discover how much of the required anatomical data is now available. A special focus of the pilot study will be placed on methods for estimating the elasticity of the tissue types found in the adult and infant foot.

common in university departments where research projects can be carried out by small groups, such as a lead professor and several graduate students.

Two more common organization models are the **centralized** system and the **federated** system. Under centralized proposal management, the decision to proceed, client-lobbying activities, and document preparation are handled by proposal-writing specialists who are called proposal managers, senior proposal engineers, and similar titles. When you have an idea, you might be asked to write an abstract and participate in planning and authorship, but many of the key decisions, such as those concerning the choice of a target agency, the method of approach, and the membership of the research team will be made by the proposal office. The sample abstract in Exhibit 11-3 was prepared for a centralized proposal office at a large university, where the proposal officer planned to take control, gather a work team, and initiate discussions with a potential sponsor. The federated system of organization, by contrast, allows individual research units to develop their own proposal contacts, while a central office provides coordination and document preparation services.

Whatever the system, there must be some method for assuring that proposal teams do not approach the same client at the same time or trip over each other in other ways. Once you are tied in to your management and support system, you and your team are ready to look for a funding agency.

Selecting a Funding Agency

The process we have described so far will take place in some form whether your company is responding to a request for a proposal (RFP) or generating an unsolicited proposal. If the proposal is unsolicited, you and your employer must now select a likely sponsor for the project: a buyer. Final selection of the target agency will be made with two sets of information in mind: (1) what you know about the decision-making processes of possible funding agencies, and (2) what you know about your company's previous relationships with these buyers, markets, or funding sources.

Market analysis is beyond the scope of this chapter; we can, however, list a few of the information sources used by proposal developers who are looking for funding. The U.S. Government Printing Office publishes notices of federal government RFPs in the *Commerce Business Daily*; local government jurisdictions print similar newspapers that announce local contract and grant opportunities. *The Annual Register of Grant Support* is a good general index of government and foundation programs. These references provide information about types of funding, annual levels of expenditure, program descriptions, program limitations, and deadlines. Professional journals and newsletters are an excellent source of current information about awards and funding opportunities.

As you narrow your focus to a few possibilities, you can begin to dig deeper for personal contacts who can provide market information. Even if you have received a specific RFP from a granting agency, you will still want to learn as much as possible about the agency and their preferences before writing the actual proposal. Sometimes you can find a recent grant recipient who will share information about how to build a relationship with an agency and how to write the proposal. Some contractors are even willing to pass on copies of their proposals. In a highly competitive market, even small scraps of information can be useful. The title of a winning proposal, for example, can reveal the preference of the funding agency for particular jargon or particular approaches to problems.

While you gather outside information about the client you plan to approach, you will also want to spread the word through your own company that you are collecting information. Your firm may have present or past contacts with the target agency; former employees of your firm may now work for the funding group; or members of your staff may have come from the target company.

However you proceed, your object is to select a *single* funding group with which you can cultivate a strong working relationship. As a general rule, it is a poor idea to try to sell a project to two or more buyers at the same time. If your clients find out, they may drop you cold. Besides, it is a better use of your energy to cultivate one strong connection than to develop two or more weak ones.

The students who developed the outside eating area proposal knew that the university was likely to be the source of funding for their project, although they could have sought outside funding (from an alumnus, for example) and then presented a completely funded project to the university for approval. Still, the student group had to identify the university administrators and faculty committees that had the power to approve such a project and authorize funding. Once they identified these groups, the students were ready for the next step: making preliminary contact with the potential sponsor.

Phase 3: Making the Approach—First Contacts, Bidders' Conferences, and Prospectuses

First Contacts

When you have received an RFP or selected a target for your proposal and have gathered as much preliminary information as you can, you are ready to make the first contact. Depending on your employer's proposal development system and past relationships with the target agency, the first contact can take one of several forms, including a direct phone call to a proposal evaluator, a personal visit, an abstract, a prospectus, or even a final proposal. By all of these means you are trying to get feedback so that your final proposal represents a negotiated compromise between what you want to do and what the funding agency is willing to pay for.

Bidders' Conference

When a funding agency issues an RFP, it anticipates that bidders will make various legitimate approaches to gather information, clarify the project description, and renew old contacts. To simplify the process, many agencies invite probable bidders to a conference so that common questions can be answered. The bidders' conference does not eliminate the need for other contacts; it simply provides an opportunity for the sponsor to avoid answering the same questions repeatedly, as might easily happen if the RFP runs to dozens of pages of specifications.

The Prospectus or Short Proposal

A prospectus proposes a project for further investigation. It can be a long abstract or a short proposal, and in fact there is often little difference between a well-developed prospectus and a short proposal. We have reserved the term *abstract* for the two- to three-page sketch of the technical content of your proposal. Abstracts can include some information about management and costs, but often the abstract is no more than a discussion of a problem or need and a suggested solution. A prospectus, on the other hand, includes more of the information that will appear in the final proposal, such as a literature review (for a research project), a statement of needs, an enumeration of specific objectives and procedures, a description of key personnel and facilities, or an estimate of the total cost. Prospectuses can be as short as three pages or as long as ten. Their object is to allow the funding agency to comment on all the elements of the proposal. In essence, a prospectus is a summarized proposal that gives the agency an opportunity to comment on the major points before the detailed version of the proposal is put together. It offers the proposal writer one more opportunity to construct a proposal that will meet the agency's specifications. A prospectus may be submitted during any of the personal contacts described in the earlier sections of this chapter.

When you have the sponsor's comments in hand, you are ready to write the final proposal. Not every proposal will require all of these steps, but the more research and groundwork you do, the more likely you will be to develop a good relationship with your client and sign a contract. The whole strategy of proposal development is to cultivate a limited number of customers who are likely to need your services or support your projects. Think of the final contract as a marriage license. You don't propose to six people, you lovingly cultivate one relationship. And you don't marry someone you've never met. Proposals work the same way; pick a likely partner and communicate. The example in Exhibit 11-4 could be called either a short proposal or a prospectus. The aim is to get the city council to discuss the project. A full-scale proposal would include additional pages of information about the proposing company and its experience and staff, as well as a more detailed budget.

Proposal Formats

The example in Exhibit 11-4 combines two popular proposal formats: the letter proposal and the short proposal. Proposals generally take three principal forms: (1) the long, multipart proposal, (2) the narrative proposal, and (3) the short, letter proposal.

As you have seen, all proposals address the same questions: What will be done and why? Who will do the work and according to what schedule? How much will it cost? How will the

EXHIBIT 11-4
Example of a Prospectus or Short Proposal

Design Associates
Street Address
City, Washington 90000

City Council
City of Warrenton
City Hall
Warrenton, Washington 90000

Subject: State Route Corridor

Dear Council Members:

As most of you are aware, completion of the new state road is near. Once in service, traffic on the old state road will decrease dramatically. This will have a tremendous impact on the section of town near the old road. Initially, it may be a welcome change from the roar of Trident-bound traffic. But the real impact will be felt when the local merchants realize that this traffic was their lifeblood. Over the years they have been gradually shifting from a locally based economy to a highway services oriented market in response to user demands. This is true of most merchants along this corridor. Unless they begin to reorient their market to a local base, most of the businesses on the old state road will cease to exist.

Now is the time for the city to take action to prevent this area from becoming a ghost town. Once the new state road is open, the state will deed back to the city the old state road, which currently has a 100-foot right-of-way. This will take on the appearance of a desolate expanse of asphalt once the traffic levels are reduced.

The city can help by laying plans now for the redevelopment of this area. Reduce the right of way to a realistic width and develop off-street parking areas that will promote shopping. Extend Warrenton's already developed pedestrian system around the end of Freedom Bay to connect it with this part of town. Create the sense of township that your city is known for in this area. Help the merchants help themselves before it is too late.

My firm, Design Associates, has been involved with the planning and design work for Freedom Bay Square. Mr. Richard Spring, manager of this project, is concerned about the life in this part of town, and for just reason. Freedom Bay Square is a $4 million shopping center designed for use by a local market. If the old state highway is not redeveloped, there is a very real chance that the local commercial area will shift toward the opposite end of town. This would render Freedom Bay Square an economic disaster. With this as our motive, we approached your planning director, Ellen Smith, with this concern. She was most interested to hear from us, as merchants have already approached her about the same matter.

A grassroots committee made up of local merchants has already been formed to discuss the challenges before them. My purpose for writing is twofold; first to express our concerns as development consultants on Freedom Bay Square, and second to present a proposal to the council involving my firm, Design Associates, as consultants to the city for redevelopment of this area.

In anticipation of the many questions you may have concerning the right approach to take and process to follow, I have attached a preliminary outline of activities I recommend following, along with a draft fee proposal and office resume [not included here].

For this part of town to come alive with the spirit that makes Main Street successful, many people will have to take part. It will be a community office involving merchants, planners, engineers, and townsfolk. The effort, however, starts with you. Give this issue your attention. Vote to consider my proposal, and let's start working.

Sincerely,

Hank Scott, President
Design Associates

DESIGN ASSOCIATES' PROPOSAL TO THE CITY OF WARRENTON, WASHINGTON

Dear Council Members,

Because it will take a community effort to clearly identify all of the issues and design considerations that need to be addressed, this proposal does not name definite numbers or dates, allowing for flexibility in all areas.

My proposal consists of four parts: work procedure, proposed schedule, financing alternatives, and fee proposal. These parts are further defined below.

PART 1. WORK PROCEDURES

Our role in this project will be as lead consultant to the city planning department. In this capacity we will perform the following tasks.

ASSIST IN COMMITTEE ORGANIZATION
We will identify and recommend appointment of merchants, citizens, city staff, council members, and other outside consultants to a design committee.

LEAD DISCUSSIONS
We will organize and prepare topics for discussion at each meeting so that these meetings are orderly and effective.

HELP IDENTIFY THE ISSUES
We will introduce important urban design considerations to stimulate discussion. From this we will narrow the issues and identify the approach.

RECOMMEND APPROACH
Once the issues are established, we will tackle the approach to translating them to a graphic design. We will analyze the issues as selected by the committee and recommend the proper design approach.

ILLUSTRATE IDEAS
Following the approved approach, we will illustrate committee-generated ideas and develop an overall schematic master plan for this area. At this point we will bring the master plan to the council for preliminary review and approval.

PREPARE FINAL DESIGN PLANS
Based on council-approved preliminary plans, we will prepare final design drawings illustrated in plan, section, elevation, and perspective.

PART 2. PROPOSED SCHEDULE
The urgency of this project cannot be overstressed. Funding sources available to the city at this time may not exist in six months to a year. We have prepared the following schedule with this time limit in mind. The worst case would require 120 days to complete the design drawings and prepare construction drawings.

Task	Time
Committee organization	2 weeks
Issue identification	3 weeks
Approach recommendation	1 week
Schematic design	4 weeks
Council presentation	1 week
Final design preparation	2 weeks
Council presentation	1 week
Council approval	2 weeks
TOTAL TIME	16 weeks

PART 3. FINANCING ALTERNATIVES
The realization of this project rests with the ability to finance it. We feel that several approaches should be reviewed. Our firm will assist the city in identifying and contacting these sources. At this time, we recommend consideration of the following sources:

1. Establish the project as a Local Improvement District in which benefitting parties would tax themselves for these improvements.
2. Issue city bonds.
3. Apply for Federal Trident Nuclear Submarine Base Impact funds.
4. Apply for community block grants.
5. Request financial assistance from the State Department of Transportation.
6. A combination of the above.

We feel that because this redevelopment is in part due to the impact of the Trident Submarine Base, requesting federal funds is appropriate and timely. In addition, the State Department of Transportation is not returning this area to the city in the same state as when it took over for highway construction and widening. As far as the city is concerned, the right of way is in far worse condition. It is important that state funds be requested to help with redevelopment before the city accepts the return of the right-of-way.

PART 4: PRELIMINARY FEE PROPOSAL
Because it is difficult to estimate the amount of work that will be required, I propose that compensation be based on an hourly rate with a "fee not to exceed" maximum. We have established ranges of fees below for your review.

Task	Fee
Committee organization	$250–$300
Issue identification	$1000–$1500
Design approach	$250–$300
Schematic design	$3000–$4000
Council presentation	$250–$300
Final design preparation	$1000–$1500
Council presentation	$250–$350
Outside consultant expenses	
Traffic engineer	$500–$1000
Landscape architect	$500–$1000
TOTAL RANGE OF FEES	$7000–$10,250

Please give this proposal your thoughtful consideration. I will attend your next council session on Monday, July 31, to answer any questions you may have. Please call me at my office if you have any questions earlier. My number is 765-7890. Upon approval, I will prepare a contract for your review and approval.

Sincerely,

Hank Scott, President
Design Associates

results be evaluated? Variations in the format depend on how the answers to these questions are grouped. We will first present the most formal and elaborate organizational pattern and then introduce and explain the variations. The form you use will depend on two factors: what you have to say and what your audience expects to see.

The Multipart Proposal

Long, multipart proposals are commonly written in large corporations that bid on government research and development contracts. This proposal format is a good one to learn because it makes explicit all the parts that are used in less-formal proposals. A general introduction summarizes the content of the entire proposal and gives an overview of the organization of the subproposals (technical, cost, management, and evaluation). The body then consists of the various subproposals, which are often written by teams of specialists and directed to different audiences of proposal evaluators. For example, the people who evaluate your technical solution might know very little about how to evaluate your firm's cost-estimating procedures. The technical evaluation team will want to read the introduction to understand the range of prices, the proposal schedule, and so forth, but they will probably leave the details of those sections to evaluators who are more qualified to assess them. Figure 11-2 presents an outline of the multipart proposal.

The student proposal in Exhibit 11-2 includes many of these parts: a transmittal letter that serves as an introduction, a technical proposal with a problem statement, a cost proposal, and a preliminary management proposal that calls for the students themselves to do most of the work. Even in the case of such a short proposal, you can imagine how the university administrator might call on several different individuals for proposal evaluation. The plant services manager might be asked to judge the accuracy of the cost estimate; a student affairs administrator might be asked to comment on the feasibility of having students do construction work; the school's legal staff would be asked to comment on possible labor and insurance complications; and of course the architect, planning office, and food service staffs would be asked to assess the accuracy of the problem statement and the value of the proposed solution. Notice that the multipart organization of the students' proposal facilitates this commenting and evaluation process. Now we will discuss each of these proposal parts in detail.

FIGURE 11-2
Sections of a multipart proposal

Part 1: Front Matter and Introduction
- Cover
- Transmittal Letter
- Introduction
- Tables of Contents and Illustrations

Part 2: The Subproposals

Technical Proposal
- Introduction
- Statement of Problem
- Proposed Solution
- Task Statement and Time Line
- Statement of Objectives
- Summary and Appendices

Management Proposal
- Introduction
- Project Management
- Company History
- Administrative Information
- Related Past Experience
- Summary and Appendices

Cost Proposal
- Introduction and Pricing Summary
- Supporting Schedules
- Terms and Conditions
- Cost-estimating Techniques
- Certifications
- Summary and Appendices

Evaluation Proposal

Part 3: General Appendices and Supporting Materials

Front Matter and Introduction

Cover

The cover of a proposal provides an initial summary, introducing the names of the client and the proposing agency, the title of the proposal, and sometimes the total bid unless the bid is to be sent in a sealed envelope.

Transmittal Letter

The transmittal letter points out important features of the proposal. Exhibit 11-4, the city planner's proposal, includes a cover letter that explains most of the proposal that follows.

Introduction

The introduction summarizes the subproposals (technical, cost, management, and evaluation) in language that should be understandable to a generalist.

Tables of Contents and Illustrations

Many readers skip everything else and use the tables of contents and illustrations as introductions.

The Subproposals

In the last few years, agencies have become increasingly sensitive to proposals that drag on. The National Science Foundation recently asked that proposals be limited to fifteen pages, excluding appendices and supporting documents. Other experts use a similar rule of thumb: three to five pages for each of the major subproposals, with supporting schedules submitted as appendices. Up to four subproposals may be included (technical, cost, management, and evaluation); evaluation proposals are most common in education and social service proposals.

The Technical Proposal

This is the section that you, as a technical specialist, will most often help write. It has six common subsections: introduction, statement of problem, proposed solution, task statement and schedule, statement of objectives, and summary and appendices.

Introduction. The introduction, as always, should forecast both content and organization.

Statement of the Problem. Depending on the subject of the proposal, the statement of the problem can define a problem or present a need. You may also include a brief discussion of the background of the problem, but be careful to include only as much background information as is persuasive and absolutely necessary.

Proposed Solution. This section is the meat of your proposal. Write as if you already have the funding; say what you *will* do, not what you would do. You might need to anticipate and overcome objections. This section of the technical proposal can include subsections on methods, procedures, technical problems, or even a study of the feasibility of the proposed solution. This is also the place to introduce and explain any exceptions:

substitutions or changes you have made in the specifications published in the RFP, if you are responding to one. However, exceptions should not be taken lightly, because the people who write the RFP have given careful consideration to what they want, and an exception might be seen as a challenge to their judgment or knowledge. Every exception should be explained and defended.

Task Statement and Schedule. In the task statement and schedule you present a task-by-task breakdown of the project, making sure that the tasks are grouped so that your presentation will be easy to skim. Many evaluators appreciate a graphic presentation of the schedule—a time line—that helps them assess whether a schedule is realistic or calls for conflicting tasks to be performed at the same time (to cite one common problem). Your schedule should provide time for periods of evaluation, report writing, and unavoidable delays.

Statement of Objectives. The statement of objectives is an alternative to a task statement that is appropriate for projects that cannot easily be broken into discrete tasks. Design projects lend themselves to task statements. For example, if you are developing a new microprocessor component, the task of circuit design would probably precede the task of developing a model assembly system. Some projects, however, are less amenable to analysis into discrete tasks or chronological steps. While the development of a new educational or social service program can be perceived as a series of necessary steps—various tasks and anticipated outcomes—it may be extremely difficult to judge how long each step will take, or how various steps will interact. In such cases, a statement of objectives may be used in place of a task statement.

Use a task statement when you can develop a step-by-step plan with a firm schedule for the steps. Use a list of objectives when the steps are harder to define, as when any political activities, lobbying, or sales are required.

Summary and Appendices. Your technical proposal can close with a brief summary, followed in turn by appendices, which are the proper place for lengthy documentation, such as calculations, printouts, sets of drawings and photographs, and additional figures.

The Management Proposal

The management proposal is usually written by a management team and has two goals: (1) to demonstrate that you have the necessary staff and facilities to finish the job and (2) to give evidence that your agency has done similar (or identical) work in the past. It typically has two sections: project management and summary and appendices.

Project Management. The project management section includes a brief introduction and an overview of organization and staffing. It names key personnel and gives their qualifications to direct the project. This data is followed by a series of sections that summarize your employer's history and experience, including information about similar projects that your employer has completed in the past. If your employer lacks the necessary experts to complete a project, you might use outside consultants or subcontractors, whose names will also be included in the management proposal. Notice how Exhibit 11-4 uses the names of past and present associates and contractors.

Summary and Appendices. Some employers submit standard organization "boilerplate" that includes staff members' resumes, bibliographies of in-house publications, and photographs of company facilities. This collection of data may be placed in an appendix to the management proposal.

The Cost Proposal

A full-scale cost proposal is usually prepared by an accounting division and includes not only cost estimates, but also supporting documents that show how estimates were made. Budget development is covered in Chapter 17; in this section we will simply describe the three subsections of the cost proposal.

Introduction and Pricing Summary. The pricing summary is usually a simple one-page table that names major categories and their associated costs.

Supporting Schedules and Cost-estimating Techniques. If prices for specific materials or types of labor are likely to be controversial, the cost proposal will include explanations of how those prices were determined. For example, if the cost of living is higher in your region than the national average, the cost proposal may refer to this fact to explain higher salary schedules. Similarly the cost proposal can include a section describing the techniques used to estimate probable costs in terms of the man-years required to do a job and probable price fluctuations. Should there be cost overruns, these estimating techniques may be a point of contention and are thus included for possible future reference.

Certifications and Summary. A certification is a statement that the offer you make is valid at a stated price for a given period of time. It protects both you and your customer. Your customer knows that the offer is valid, say, for sixty days, and you know that you can adjust the price after sixty days have passed. A summary closes the full cost proposal.

The Evaluation Proposal

Evaluation proposals are common in government, social service, and basic research proposals. In straightforward technical projects, the final test will be whether the product meets the stated specifications. In the case of development projects or projects intended to address a public need, the evaluation can be based on the attainment of measurable objectives or on an expert opinion of the manner in which the program activities are being carried out.

The Narrative Proposal

Narrative proposals exist because many problems cannot be effectively addressed by the four-part format described above: a form best suited to engineering projects. Often it is simpler to combine elements of the management, cost, and technical proposals into a single running discussion labeled Narrative. This approach is justified when you do not have enough data to fill a full-scale cost or management proposal. Narrative proposals can have subparts, however. Here is one typical system of organization.

Introduction
Statement of problem or need
Goals and objectives of the project
Project description
Methods and task statement
Evaluation system
Budget
Future budget
Staff and facilities

We have discussed all of these parts except the future budget section. Sponsors of program development grants often want to know what will happen after the seed money is spent. Will the program just wither away? Or does the agency have a plan for keeping the project going? If so, how much will that cost, and where will the money come from? These are the subjects of a future budget presentation. Exhibit 11-5 is an excerpt from a narrative proposal for the California Farmlands Project, which calls for development of regional conferences

EXHIBIT 11-5

Excerpt from a Narrative Proposal. Courtesy of the California Institute of Public Affairs.

THE PROBLEM

Conversion of prime agricultural land is one of the most important and difficult resource issues facing California. Over the last decade alone, 1.5 million acres of California agricultural land were converted to urban and other nonfarm uses; of these, 500,000 acres were prime farmland. University of California studies have predicted that, unless something is done about it, prime agricultural lands may continue to be lost to other uses at the rate of about 65,000 acres a year. Until now, most of this loss has taken place in Southern California and in the San Francisco Bay Area; in the future, it is likely that much of it will occur in the San Joaquin Valley, which is acre for acre the most productive farmland on earth.

Protection of prime agricultural land has been an issue in California since at least the midfifties. People give different reasons for their concern depending on their perspective. A shrinking supply of land presents farmers with problems not only of space itself, but also of increasing costs in rent and taxes. Environmentalists often view the issue in terms of loss of open space and rural landscapes. Others see the problem in terms of increases in food prices or loss of capacity to feed hungry people abroad.

Nearly everyone agrees, however, about the importance of preserving agriculture as the keystone of the economy of California, the country's number one farm state. Although agriculture makes up less than three percent of the gross state product, farm production is a vital link in an economic chain involving hundreds of thousands of Californians. For some commodities, like grapes, canning tomatoes, almonds, olives, and lemons, this state accounts for all or almost all of U.S. production.

Even though people give different reasons for their concern, there is broad agreement that existing measures designed to limit conversion of agricultural lands, such as local planning and zoning and the California Land Conservation Act of 1965 (the Williamson Act), have been ineffective and that much stronger measures are needed. But regulatory approaches such as those used to limit development in the California coastal zone have become politically unpopular and have little chance of being enacted, except for a few critical areas. A key question is how to compensate property owners for the "wipeout" of property values that occurs when land in the path of urbanization is limited to producing food and fiber. Other important questions include deciding exactly what kinds of lands should be protected; the kinds of exemptions that should be allowed; the respective roles of state and local governments; where new development should be directed if conversion of farmland is restricted; and who pays for farmland preservation.

There is renewed interest in the problem both nationally and in California. The findings of the National Agricultural Lands Study, issued early in 1981 by the President's Council on Environmental Quality and the U.S. Department of Agriculture (a million acres of important farmland are irreversibly lost to agriculture every year in the United States), have received a good deal of attention in the media. National conservation, planning, and agricultural associations have taken up the cause for the first time in a serious way, and new groups have been formed in Washington to focus specifically on preserving agricultural lands.

In California, the State Department of Conservation has launched an inventory of farmlands statewide, and the governor signed a bill early in 1982 requiring records to be kept on how much farmland is lost each year to urban development. People for Open Space, a regional planning group in the Bay Area, recently compiled a major farmlands conservation study. The League of Women Voters of California has adopted agriculture, including farmland conversion, as a priority issue for study and action. The Sierra Club has reactivated its California Agricultural Lands Committee. Local groups, often led by farmers, have organized to work on the problems in such areas as Butte, Sacramento, and Stanislaus counties.

We think the time is ripe to take a fresh look at what can be done about the problem at the state level in California. The role of our California Farmlands Project is to try to pull together these separate efforts, find common ground among all of the major groups concerned, and arrive at specific findings, conclusions, and recommendations.

on the problem of disappearing agricultural lands. The full proposal includes a problem statement, a section on organization and personnel, and a section on objectives and methods. The appendix lists the names of thirty cooperating consultants, in this case a group of legislators, activists, and public organizations. As you read, notice how the narrative format combines introduction, problem statement, and goals in one running discussion.

Letter Proposals

Short proposals are often written in letter form. Letters are less intimidating than long proposals, even though a letter proposal can have the same sections as a long proposal. Exhibits 11-2 and 11-4 provide examples of letter proposals, and the section of this chapter on the prospectus or short proposal (p ____) discusses the general differences in content between short and full-length proposals. Letter proposals also use the strategies and formats of persuasive letters, which are presented in Chapter 13.

Variant Formats

Some agencies provide forms or suggest a proposal format and others expect you to use a familiar, standard form. There might be occasions in your professional life, however, when a traditional proposal format will not have enough impact. In those instances you can be as creative as you like. One successful proposal writer assessed his audience's resistance to an idea and responded with a proposal that confronted each objection in turn. His section headings read Outcomes, Seed Money, Worst Case, Best Case, Fears, and Alternatives.

Every sales letter you get in the mail is a proposal, and any proposal you write can make use of the argumentative and graphic strategies you find in effective advertising.

PROPOSAL EVALUATION: REASONS FOR REJECTION

Proposals fail because of poor communication. Over the years, many agencies have compiled lists of criteria for evaluating proposals and still longer lists of common reasons why proposals are turned down. We reproduce below a typical list of reasons for rejection. As you read it consider how each problem could be eliminated by means of improved communication.

1. The proposal is incomprehensible, written in language only the author can understand.
2. The problem is not within the sponsor's scope, is not appropriate, is addressed to the wrong audience.
3. The bidder proposes to do too much at one time; the program is too ambitious, suggesting that the bidder is naive or underprepared.
4. The procedures or methods are too vague.
5. The evaluation system is inadequate.
6. The relationship between the project and its objectives is not clear or unspecified.
7. A pilot study should be attempted first.
8. The time schedule is inappropriate, too optimistic, unrealistic, or poorly designed.
9. The staff does not have the necessary training or experience.
10. Consultants are needed.
11. The lead personnel are not specified.
12. Specific responsibilities are not assigned.
13. The firm does not have the necessary equipment or experience.

14. The budget is too high, or too low.
15. The bidder has asked for operational or support money (indirect costs—explained in Chapter 17).

A few of these objections cannot be overcome by discussion, such as the lack of experience. Once you are aware that your firm does not possess enough experience, however, you are spared the time and expense of writing a proposal that cannot possibly be accepted. A phone call will establish whether a particular agency is the appropriate audience for a proposal; an exchange of drafts or abstracts will determine whether the sponsor understands the idea; a meeting with other workers in the field will reveal whether your company has the necessary expertise, background, or facilities. In competitive bidding, of course, there will always be good proposals turned down in favor of better proposals, but you can increase your success rate by paying careful attention to the early stages of the proposal process—in short by communicating well. The burden of proposal writing should be placed on preliminary discussion, not on the final document. As one expert consultant wrote concerning one of the proposals reproduced in this chapter, "This proposal is rough because I wrote it in one night. But that didn't matter since (as you know) the whole thing had already been settled through these last three months of conversations."

The Odds

It should be clear by this point that proposal writing can be a time-consuming and costly process. Large engineering firms can spend over $100,000 on the development of a major proposal. As a result, bidders choose projects carefully and try to avoid major commitments of time and money until they have some indication that a proposal will at least receive careful evaluation. For this reason, too, many agencies issue Requests for Qualifications (RFQ) in order to choose two or three finalists. The finalists can then incur the cost of proposal development knowing that their chances of success are 33 to 50 percent. Even with this winnowing process, a success rate of 20 to 30 percent is common, and a rate of 50 percent should be considered excellent.

SUMMARY

1. A proposal is a sales document that answers the questions: What is the service (or product) and why is it needed? How and when will the work be done? How much will it cost? How will the results be judged?

2. A proposal abstract is used to develop a project inside your company; a proposal prospectus can be used to solicit comment from your client.

3. Proposal writing and sales activities within a company can be organized according to one of three patterns: individual entrepreneurship, group work, or central planning. You should know which organizational model your employer uses.

4. The chief proposal types are the multipart, the narrative, and the letter proposal.

5. Multipart proposals are appropriate for technical projects; narrative proposals are useful for projects with less-concrete objectives.

6. A multipart proposal can have up to four major subsections: the technical, management, cost, and evaluation proposals.

7. Proposal writing requires extensive communication, both inside your company and with targeted clients. Proposals fail because of poor communication.

Proposal Case 1

THE AMERICAN INSTITUTE OF SWIMMING

After completing your B.S. in mechanical engineering, you have gone into business with four friends, forming Solar Associates, which specializes in designing and building solar systems for domestic and industrial use. Your chief business so far has been, surprisingly enough, solar-heated swimming pools. You have successfully installed solar heaters on seven residential pools, ranging from 450 square feet (15 × 30) to 800 square feet (20 × 40). But your last job was the biggest yet—the pool at the Thunderbird Country Club in Tempe, Arizona, which measured 1200 square feet (50 × 70).

This job has given you your best lead. The manager at Thunderbird tells you that the American Institute of Swimming in Corpus Christi, Texas, is considering a new competitive facility and suggests you check it out. You write a letter of inquiry and receive the following reply.

Ms. ________
Solar Associates
225 Pinon Drive
Phoenix, Arizona

Dear Ms. ________

In answer to your recent inquiry we are indeed planning a new facility that will require heating for part of the year. We would be interested in hearing from your company.

As you may know, we are a private foundation sponsoring training camps and conference competitions for high school and college swimmers. We also sponsor an active Masters Swimming Program. Our new facility will be a standard competition size (50 meters × 20 meters).

Our principal concerns regarding pool heaters fall into three categories: efficiency, price, and aesthetics. We are interested, most of all, in a dependable heater that will maintain a constant temperature during the colder months. But we are also concerned with costs, both for installation and maintenance, and with preserving as much of our natural setting as possible. We trust that your design will consider all of these factors. Since we are currently finalizing our architectural plans, decisions about pool heating must be made by September 20.

Sincerely,

Harmon Fulbright
Managing Director

ASSIGNMENT 1

Write a short prospectus to be sent to Mr. Fulbright. Include a statement of the needs of the American Institute of Swimming, a list of your objectives and procedures, a description of your personnel and facilities at Solar Associates, and an estimate of costs. Remember, the prospectus is not a full-fledged proposal; you're trying to get some feedback from Mr. Fulbright about his reaction to your plans. Here are some specifications for your solar heater.

Components

Pump
Solar collectors (2)
Differential thermostat
Pool filter
Auxiliary heater

Procedure

Collectors are plastic coils embedded in dark concrete deck surrounding pool.

Water circulating in coil absorbs heat from deck's surface and reduces surface temperature to comfortable level.

Pump circulates pool water through coils and back to pool again.

Differential thermostat turns pump on when coils are hot enough to benefit pool.

Filter keeps debris out of coils.

Auxiliary heater operates if solar heat insufficient to raise pool water to desired temperature.

Notes

Coils are concealed in concrete of deck: no eyesores.

Coils are plastic polymers that are stabilized against effects of heat and ultraviolet rays.

Plastic is inert; cannot be corroded by pool chemicals or eroded by fast moving water and does not allow scale to build up in water passages.

Black color of coils is permanent; does not have to be repainted, like metal collectors.

Deck will be dark colored.

Pool is in full sunlight; thus system is workable.

ASSIGNMENT 2

Mr. Fulbright reacts positively to your prospectus; you're on target with your design and your cost estimates. Now you must write the proposal to build the solar heater for the American Institute of Swimming. Since you're a small company, you decide to use a simplified format with the following sections: Introduction, Statement of Need, Goals and Objectives of the Project, Project Description, Task Statement and Schedule, Budget, and Staff and Facilities.

Proposal Case 2

The Departmental Computer

In the year and a half since you joined the Walden College chemistry faculty you have become increasingly frustrated by the department's lack of modern equipment. Your chair, Dr.

Martha Callaghan, shares your concern, but thus far the money for new equipment simply has not been available. Now Dr. Callaghan has a new idea. A private foundation, the Lutcher Group, has announced the availability of funds for instrument grants to small colleges. She suggests you write a proposal for a small computer, something the department has needed for some time.

Your letter of inquiry brings the following response from the Lutcher Group:

Dear Dr. ________

In response to your letter of 12 July, we are indeed interested in funding equipment proposals. Please note our closing date of 1 December. We look forward to your proposal.

Sincerely,

Robert Welch
Executive Secretary

Your first step, you realize, must be to assess the computer needs of the other members of your department: Dr. Eric Gustafson, Dr. Joan Langtry, Dr. Ed Pringle, and Dr. Larry Tzu. Following your interviews with these people you assemble these notes.

Undergraduate: Need computer that can be used in physical and analytical laboratories (Tzu); also one that can be used for lecture-demonstration; and labs in General Chemistry (Pringle). Must have software for teaching activities in chemistry.
Graduate: Computer must be capable of interfacing with various lab instruments (Tzu). Need dedicated instrument programs, at least for gas and liquid chromatography (Langtry).
Professional: Biggest use will be number-crunching calculations (Gustafson, Langtry, myself). Should also be useful for collecting and analyzing lab data (Tzu). Software: molecular orbital calculations, model building, analysis of spectra would all be useful (Callaghan). System should have large number of scientific programs available.
Technical: Should have at least 128K memory with capability of expansion to 512K. Should handle both Fortran and either BASIC or Pascal. Should have graphics printer. Needs laboratory interface system. Should provide training program for faculty users, plus service warranty.

After studying several available computers, you discover one—the Comstat company's IBIX—which will meet most of these requirements. It will also run the necessary software. The

IBIX costs $2000; an expansion card to increase its memory to 512K would be another $1800. A Univax dot matrix printer would cost $2300.

ASSIGNMENT

Go through these notes and develop criteria for the Walden departmental computer. Write the proposal to the Lutcher group to get funding for the IBIX plus peripherals.

CHAPTER • 12

Instructions and Manuals

INTRODUCTION

Manual-writing skills are in demand; look in the classified advertisements of any large newspaper. After all there's no point in designing or manufacturing a product if the buyer can't assemble, operate, or maintain it. Many technical writers spend most of their time writing manuals, instructions, descriptions, and classifications. Even if you don't plan to work steadily as a manual writer, you should understand the steps required for writing descriptions and instructions because engineers, executives, and writers must work together to produce effective product documentation

This chapter presents guidelines for defining mechanisms and processes, for classifying parts and steps, and for writing formal technical descriptions. Defining, classifying, describing, and directing are the skills required for writing instructional manuals. The chapter closes with rules for writing instructional manuals; we do not explain how to write manuals that are not instructional, such as codebooks or manuals of regulations. While you won't often be asked to write a formal definition of a product or even to develop a systematic classification of the parts of a mechanism or process, defining and classifying are necessary first steps.

DEFINING

Before you describe a product, you must be able to define what it is. Definitions set limits; they explain what is included in your understanding of a process as well as what you are excluding from analysis or discussion. The introduction of this chapter, for example, defines what we mean by manuals and also states what types of manuals we don't intend to discuss.

Definitions of mechanisms and processes commonly take a three-part form, which can be written as a simple equation:

species = genus + differentia

For example, a hammer (species) is a tool (genus) used for striking and pounding (differentia). More specifically a framing hammer (species) is a hammer (genus) with a 16- to 20-ounce head, a smooth or cross-hatched striking surface, and a claw for drawing nails (differentia). A definition can be simple or highly technical, depending on its purpose. For some definitions, you might need several paragraphs of specifications and background information.

The species = genus + differentia formula is a formal definition, useful for concepts that are either very complex or completely unfamiliar to your audience. In situations with simpler concepts or concepts that are similar to ones already familiar to your audience, you might prefer to use a simpler, more informal type of definition. It might be possible, for example, to provide a common synonym for an unfamiliar term or use a description or an example. In Figure 12-1, the definition uses an example (a sample of a simple program) rather than a formal definition.

Before you describe any object or process, you should normally provide a definition that points out what makes it distinct from others of its type. In Figure 12-1 the "programming" process is defined in the sense in which it is relevant to a specific hand-held calculator. Notice how the writers provide a definition that is tailored to a particular nonexpert audience. The company's concern was to define "programming" in a way that is clear; nonthreatening to someone who is just beginning to work with calculators; and not misleading to experienced computer users, who need to understand that this particular calculator is not "programmable" in the same way that more advanced calculators and computers are.

The last important point to remember about defining is that a definition should never be circular or tautological: that is, a definition should never attempt to define an object merely by

FIGURE 12-1
Definition using an example. Courtesy of Texas Instruments Incorporated.

A *program* for your calculator, then, is just a list of the series of keystrokes in the order needed to perform a particular calculation. Once you know these keystrokes you can program your calculator to remember them. To do this you simply press the [2nd] [Lrn] (Learn) key sequence. When you do this, you "turn on" a special memory in your machine that remembers the keystrokes that follow. You're telling the calculator "please remember the keystroke instructions I enter next".

repeating the name of the object, as in "A framing hammer is a hammer used for framing."

Once you have defined a mechanism or process, you are ready to divide it into parts and to classify those parts in a way that will help you write a clear and useful description. In practice you may sometimes want to start with a classification and use that mental exercise to define what is distinct about an object or process.

CLASSIFYING

Classification is a type of ordering process. When you classify a group of items, you identify their common qualities. Thus if you had a box of screws, you could classify them by length, thread size, diameter of the screw head, or even whether they had slotted or Phillips heads. By using any of these classifications, you could divide a large group of miscellaneous screws into several small groups of clearly defined screws. Notice that as you classify, you are also developing a preliminary description. Both classifying and describing are naming processes.

Before you can describe any part of a mechanism or process, you must carefully classify the parts in a way that includes every part and clarifies the relationships among the parts. As you will see in a later example, it is disastrously easy to jump in and start describing without laying down a foundation of definitions and classifications. Sometimes, even a simple mechanism can be hard to describe because the parts can be classified in several ways, and you may need to use two systems of classification simultaneously. Two models of a common framing hammer will demonstrate this point.

Hammer 1 (Figure 12-2) has three physically distinct parts: a wooden handle, a forged head, and a steel wedge that holds the head onto the handle by expanding the end of the handle.

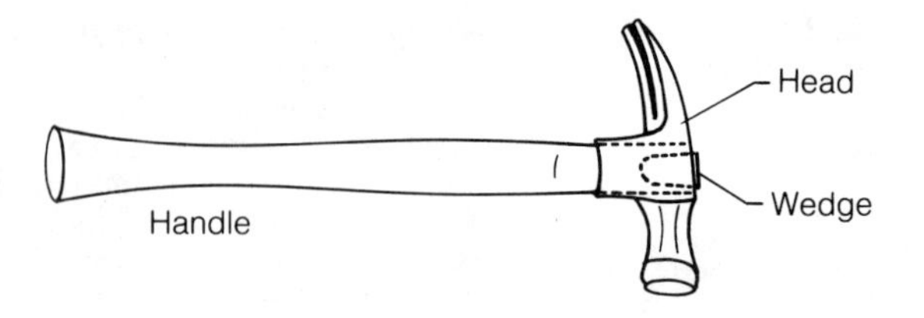

FIGURE 12-2
Hammer 1

You probably own a hammer like this. Hammer 2 (Figure 12-3) is built according to another common design. The head and shaft are forged from one piece of steel; a rubber grip fits on the end of the shaft; and a simple maple wedge is glued into the head to help dampen vibrations.

One way to describe these hammers is to consider each *physically separate* part in turn. Hammer 1 has a handle, wedge, and head; Hammer 2 has a grip, a wedge, and a head and shaft unit formed as one piece. Another, more abstract way to describe the two hammers is to classify *functions* of the parts. In the case of these two hammers, you might think of five general functions served by the parts: striking, drawing, fastening, leverage, and vibration dampening. In Hammer 1, the head performs the striking and drawing functions, the handle provides leverage and dampening, and the wedge fastens the other parts together. In Hammer 2, the grip and wedge dampen, the shaft provides leverage, and the head strikes and draws nails. Each hammer head is one piece, but in functional terms, each head has three subsections: a striking area, a drawing tool, and a hole for a wedge. This simple example shows how important it is to classify systematically and to be aware of the perspectives from which you classify parts. Consider how confusing it would be if you classified Hammer 1 by physical part and Hammer 2 by function, or if you described the handle of Hammer 1 in terms of function and the head in terms of composition. In the next section, you will learn six ground rules for classifying.

Characteristics of Classification

According to Edmond Murphy (*Logic of Medicine* (Baltimore, Md.: Johns Hopkins University Press, 1978), pp. 100-117), there are six basic characteristics of a good classification. A classification should be natural, exhaustive, disjointed, from the same perspective, useful, and simple. We'll consider each characteristic in order.

Natural

A classification should be based on natural characteristics of the object or process being described. In the case of the hammers, both classifications are natural. An unnatural classification of

FIGURE 12-3
Hammer 2

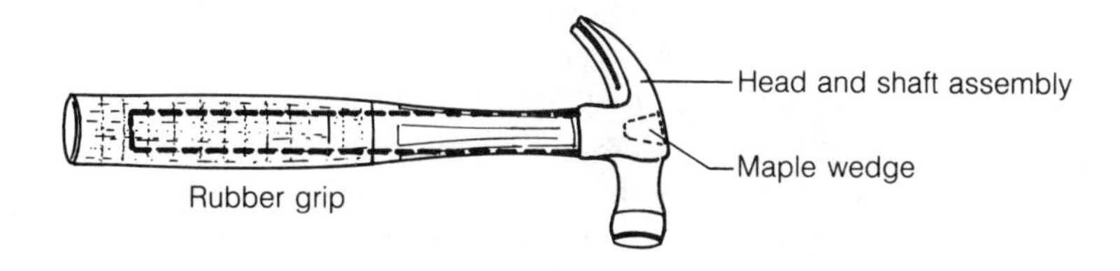

hammer parts might be front and back or right side and left side or a system that required discrete parts to be arbitrarily divided into sections, such as the top half and the bottom half of the handle.

Exhaustive

Every part or item should fall into one of the classes or groups you use. An ancient encyclopedia classified animals in the following far-from-exhaustive way: "(a) those that belong to the Emperor, (b) embalmed ones, (c) those that are trained, (d) suckling pigs, (e) mermaids, (f) fabulous ones, (g) stray dogs, (h) those that are included in this classification, (i) those that tremble as if they were mad, (j) innumerable ones, (k) those that have just broken a flower vase, (l) those that resemble flies from a distance" (J. L. Borges, *Other Inquisitions* (New York: Washington Square Press, 1966)). Most animals seem to fall into an "others" category of this system. Classifications often fail to be exhaustive because they leave out "insignificant" parts. Some people might consider the wedge, for example, an insignificant part of a hammer, but try using a hammer without one.

Disjointed

No part of a classification should fall into more than one class. The functional approach to classifying hammer parts runs into trouble here, because a wooden hammer handle is both a lever and a dampener; other parts would also need to be discussed under several functional classifications. In the example immediately above, mermaids fall into at least three classes: mermaids, fabulous animals, and animals in this classification. They might also resemble flies from a distance.

From the Same Perspective

A good classification should also be based on one viewpoint or on a clearly organized set of perspectives. In describing the hammers, we found that the major parts could be classified in terms of their being physically separate parts. We also found that a functional perspective would be useful for describing the subparts of parts that are one piece, such as the head and the handle. Just as the head could be divided into striking, pulling, and wedge zones, the wooden handle could be divided into the grip, the shaft, and the head.

Some classifications have no clear principle of classification, although they appear to have one. Murphy cites this example: alcoholics, problem drinkers, and social drinkers. Intuitively this classification of drinkers seems to be useful, but in practice the categories overlap so extensively that the classification is useless. Alcoholics are problem drinkers, and social drinkers are often alcoholics. The terms themselves are vague.

Useful

A classification must also be useful. One researcher classified a rare anatomical malformation—*talipes equinovarus,* or club foot—by enumerating all the possible combinations of its features. The resulting classification included over 200,000 groups, hardly a useful classification for rapid clinical evaluation and diagnosis. Most surgeons who work with that abnormality find it sufficient to rely on four basic categories that have been used since ancient times: varus, valgus, equinus, and adductus, which define the four principle ways that the human foot can deviate from a normal shape. Utility is closely tied to purpose. If you were ordering parts for a new mechanism, it might be useful to classify all the screws and bolts in one group. An assembler or installer, however, would probably prefer to have the bolts and other fasteners classified with the parts they hold together.

Simple

A good classification is memorable and easy to use. One of our students classified the parts of an ordinary latch key by observing that a key has two functional parts: a part that transmits information (the ridges) and a part that transmits force (the tab and shaft). That's simple, elegant, memorable.

Naming Parts

Classifications are not difficult to develop if you give them some thought. Definitions and classifications go awry, though, when they are developed too haphazardly or too quickly, or when parts are assigned arbitrary names. Asked to describe a common office stapler, one of our more stubborn students decided that simple names and a simple classification of parts were best, so he classified the parts of the stapler as the top, the bottom, and the hinge. Sticking doggedly to his chosen terminology, he soon found himself with something like Figure 12-4.

Describing the operation of the stapler, then, he explained how the top of the top pressed the bottom of the top toward the top of the bottom, where a staple was pressed out of the inside of the bottom of the top. . . . We draw the curtain on this scene.

FIGURE 12-4
The wrong way to name parts

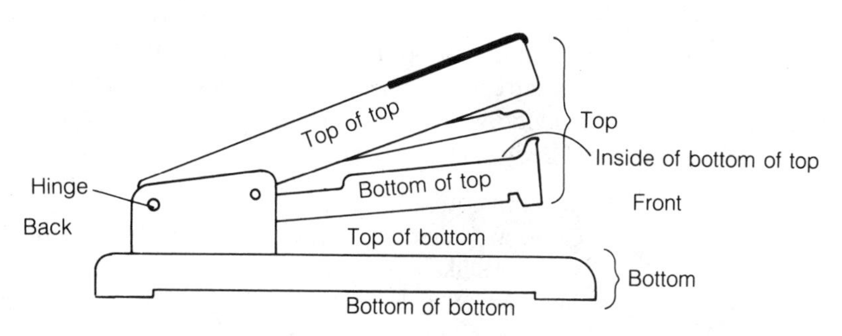

When you have defined a mechanism or process and have named and classified its parts, you are ready to write a description of a mechanism.

Describing Mechanisms and Processes

The first step in describing a mechanism or process is to determine your audience. How much does the audience need to know? What is their level of expertise in this field? Next, you should consider the purpose of the description. Descriptions of mechanisms are commonly used to direct the reader to initiate some kind of action. But what action? Do you need to reproduce all the technical specifications and tolerances for manufacturing? Or are you writing a less-detailed description to be used for assembly, operation, maintenance, or repair? Most descriptions are fairly short, because the detailed specifications are not relevant to the purpose. We will discuss mechanisms first, then processes.

Describing a Mechanism

A description of a mechanism has three main sections: the overview, the part-by-part description, and the description of one cycle of operation.

The Overview

The overview introduces the mechanism, defines it, and explains what it is used for. In this introduction, you should also name the main parts and orient the reader to the object with the help of an illustration. You should explain how you are going to use such terms as *top, bottom, left, right,* and other directions.

The Part-by-Part Description

The part-by-part description is the body of your presentation. Taking each part in turn, you describe its features and then divide the part into its subparts. Keep in mind that whenever you start describing an object, you should systematically work in one direction from that point, explaining parts as you go. Work from the left side of the illustration to the right; that order is most familiar. Should it become necessary to use a different order, be sure to explain the shift in perspective to your readers. Clearly identify the sections of your description by using subtitles or headings. Here is a general outline to follow:

1. Name of major part.
   a. General shape. An analogy is often helpful here (i.e., "Looks like a half moon").
   b. Material. What is it made of? Consider your audience's knowledge of materials.
   c. Dimensions. Either approximate or exact, depending on your purpose.
   d. Color or finish.
   e. Relationship to connecting parts or subparts.
   f. Main subparts.
2. Subpart One
   Repeat a to e, add a paragraph explaining how subpart one is connected to subpart two, and proceed to subpart two.
3. Continue this process until all parts and their subparts are described.

For example, after giving an overview of the parts of a framing hammer, you might begin with the head, describe it in general terms, and then give more detailed descriptions of the claw, the striking surface, and the hole, always being careful to orient the reader to the location of each subpart in relation to the others. When you have described either a major part or the complete mechanism, you can cap your analysis with a description of one cycle of operation.

One Cycle of Operation

At the end of the part-by-part description, you should describe one full cycle of operation so that the audience can understand how all the parts of the mechanism work together to perform its functions. In the case of a simple mechanism like a hammer, you could explain the operation at the end of the entire description. In the case of more complex mechanisms, you might find it more effective to stop at the end of each subpart to explain a cycle of operation. Keep in mind how your description will direct the reader's actions.

In practice manuals do not always provide a full-scale description of parts, because drawings show spatial relationships more effectively than words. Exhibit 12-1 is typical in this respect. Read this example carefully and note how and where the authors have included all the sections of a description of a mechanism. The names of the sections are different from those we have used—the cycle of operation is called "operating instructions," for instance—but all the content is there. We will return to this example when we discuss process descriptions and manual writing later in the chapter.

Describing a Process

Process descriptions are similar to descriptions of mechanisms. First you provide an overview of the process, and then you supply an inventory of the tools and materials that are needed. A step-by-step description follows, and the process description closes with a presentation of one full cycle of operation. In addition to these basic parts, a description of a process can include performance specifications, troubleshooting instructions, and warnings against dangers or shortcuts that will spoil the result. These topics are discussed under Manual Writing.

The Overview

The overview defines the process and explains its purpose and use. You might also want to include a brief discussion of the basic theory and principles of the process. For example, if you were explaining basic welding procedures, you might decide that a beginning audience needs to be reminded of the important differences between brazing and welding. The overview closes with a brief presentation of the principal materials and chief steps of the process.

EXHIBIT 12-1

Draft pages 12–15. Keuffel & Esser Portable Instrument Stand Manual N. 71-5030. (Courtesy of Keuffel & Esser Company, Parsippany, New Jersey)

KEUFFEL & ESSER COMPANY

Portable Instrument Stand
Catalog No. 71-5030

GENERAL DESCRIPTION: The K&E Portable Instrument Stand is a stable instrument support that has the advantage of portability. The stand provides a firm, rigid support for optical alignment, surveying, and other optical instruments. Stability is excellent despite the compactness and light weight of the stand.

HEAD ASSEMBLY: The column swivel head can be rotated 360 degrees in any position. It has a mounting ring at the top with U.S. standard 3½ × 8″ threads for attaching brackets or instruments. The column and swivel head are hollow to provide a 7/8" diameter clear aperture for downward sighting.

The tripointed hand wheel at the head assembly permits fine adjustment of the column height within a 3" range. A positive hand clamp prevents accidental height changes.

BASE ASSEMBLY: Triangular bracing of the legs assures stability of the base assembly. The legs are made of aluminum tubing; the lower legs slide in nylon tubing to prevent fretting and wear. Assembly of the base is facilitated by wing nuts and wing screws and requires no tools.

PHYSICAL CHARACTERISTICS: The net weight of the assembled stand is approximately 30 pounds. The height is variable from a minimum of 30⅝" to a maximum of 51½".

ASSEMBLY INSTRUCTIONS: The K&E Portable Instrument Stand is shipped in a single carton containing the following components.

- Head Assembly
- Leg Assembly (3 Each)
- Leg Brace (3 Each)
- Stainless Steel Point (3 Each)

All assembly hardware is installed in its proper location. Proceed as follows to assemble the stand. (NOTE: Numbers appearing in parentheses refer to parts identified in Figure 1 and the parts list.)

1. Loosen the wing nuts (8) on the upper ends of the legs (13) and mount the legs to the head assembly (7) in the slots provided. Hold the legs (13) in place with a snug fit by slightly tightening the wing nuts (8). See Figure 1.
2. Install the three leg braces (19) and tighten the six wing nuts (21).
3. Tighten the wing nuts (8) securing the legs to the head assembly (7).
4. Assembly of the instrument stand is now complete.
5. The swivel pads (16), provided for setup on a hard surface, may be replaced with the stainless steel points (27).

OPERATING INSTRUCTIONS:

1. Remove the plastic tripod cap (1) and mount the instrument or adapter on the threaded adapter ring (2) on top of the column.
2. To raise or lower the tripod, loosen the three wing screws (15), extend or retract the lower legs (14), and relock the three wing screws (15).
3. To raise or lower the instrument, release the clamp screw handle (11), then rotate the tripointed hand wheel (5) to raise or lower the center column (22). When the desired height has been reached, lock the center column (22) by tightening the clamp screw handle (11).
4. To rotate the instrument, release the thumb screw (23) and rotate the instrument by hand. Retighten the thumb screw (23) to lock the position of the instrument.

FIGURE 1
Portable Instrument Stand

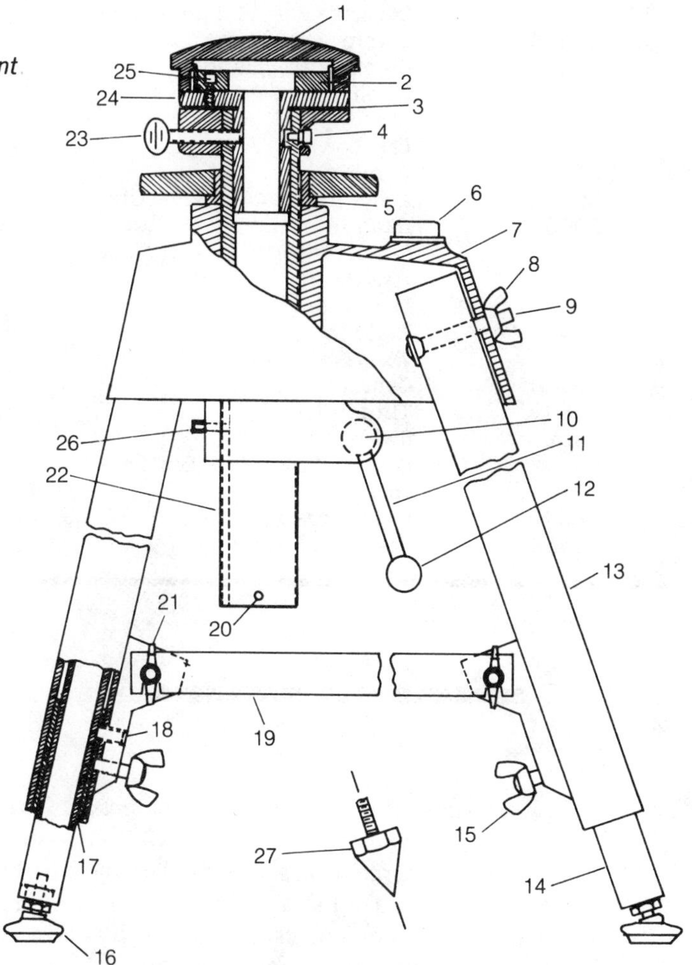

TABLE 1. Parts List—Portable Instrument Stand (71 5030)

Index* Number	Part Number	Description	Quantity Per Unit
1	A11697	Tripod Cap	1
2	A46635	Adapter Ring (3½ × 8 Thrd)	1
3	B3993	Washer	2
4	A46872	Guide Screw	1
5	C14401	Hand Wheel Assy.	1
6	941085	Bubble Level	1
7	D46584	Tripod Head	1
8		Wing Nut (3/8-16 N.C.)	3

9		Carriage Bolt (3/8-16 × 2-3/4″)	3
10	A46588	Clamp Screw	1
11	A46589	Clamp Screw Handle	1
12	939707	Knob	1
13	C3995	Leg, Upper	3
14	B3991	Leg, Lower	3
15	966083	Wing Screw	3
16	B3989	Assy., Adjustable Swivel Glide	3
17	B3990	Bushing	3
18		Setscrew (5/16-18 × 1/2″ Socket Head)	3
19	C3961	Leg Brace	3
20	952422	Roll Pin	1
21		Wing Nut (3/8-16 N.C.)	6
22	C46932	Center Column	1
23	B3959	Thumb Screw	1
24	B3992	Swivel Head	1
25		Screw (7-32 N.C. × 5/8″ Socket Head)	1
26	A46945	Guide Screw	3
27	B14415	Stainless Steel Point	3

*See Figure 1

Equipment and Materials

Next comes a complete list of equipment and materials. In the Keuffel and Esser manual (Exhibit 12-1), the assembly instructions list the three main subassemblies that must be attached to make the tripod. A full list of parts is not needed for assembly of the major components, so the parts list is placed on the last page of the manual. Generally, it is a good idea to group parts within a list according to a meaningful principle of classification. In the Keuffel and Esser manual, the numbering of parts is somewhat arbitrary; a clockwise pattern of numbering has been followed, with the result that the tripod head includes parts numbered 1 through 7 and 23 through 25. If the parts list had been broken into three subtitled sections and the numbering done top to bottom, the parts of the head would be assigned numbers 1 through 10 and so on, making it easier for readers to locate and order individual parts.

Step-by-Step Instructions

The step-by-step instructions are organized in the same way as a description of parts:

1. Introduce a main step.
2. Describe substeps in detail.

3. Summarize the expected result.
4. Lead into the next main step.

If an operator performs the process, describe what the operator does as well as what happens to the machinery or chemicals or whatever else is involved in the process. Computer manuals sometimes describe programming as if it happened automatically; a manual might say "Power on. Log in." rather than "Turn on the 'on' switch at the back of the CRT; log in by typing your user code." On the other hand, if the process is entirely automatic, don't hesitate to use the passive voice or to leave out any mention of an operator: "As soon as the main power is turned on, the main clock is activated and sends a signal that cues all internal programs. . . ." In every case, readers will appreciate information about what the process should produce at various stages if everything is going as it should. "After you log in, the computer should print out 'Ready.'" When you have described all the steps, you are ready to explain one full cycle of operation.

One Cycle of Operation/ Summary

The description closes with a review of one full cycle of operation. This is often done in the form of an example as in Exhibit 12-2.

Exhibit 12-2 is taken from the Texas Instruments T-I 55 manual. Notice that this manual assumes that you know what the "mean," "variance," and "standard deviation" are and what they are used for. (A separate text that accompanies this calculator does, in fact, explain the meaning of these statistical terms.) Consequently, the description of the calculating process skips the overview and starts with definitions of the tools, in this case the calculator keys used to perform the calculations.

The step-by-step description starts with clearing the calculator of any previous data. The data entry process is explained next, and the example shows one cycle of operation. Several short concluding paragraphs explain how additional data can be retrieved from the memory banks for use in other calculations.

Now that you have the guidelines for writing definitions, classifications, and descriptions, you are prepared to consider the more complex task of manual writing.

MANUAL WRITING

Manual writing draws on your ability to define, classify, illustrate, and describe, but the main focus of most manuals is *instruction.* Instruction manuals can be divided into several types:

EXHIBIT 12-2

Courtesy of Texas Instruments Incorporated.

MEAN, VARIANCE, STANDARD DEVIATION

[Σ+] **Sum Plus Key**—Enters data points, y_i, for calculation of mean, variance and standard deviation and for the linear regression routines.

[2nd] [Σ-] **Sum Minus Key**—Removes unwanted data entries for mean, variance, standard deviation and linear regression calculations.

[2nd] [Mean] **Mean Key**—Calculates the mean of the y array of data. Mean $= \bar{y} = \frac{\sum_{i=1}^{N} y_i}{N}$, i = 1, 2, 3...N

[2nd] [Var] **Variance Key**—Calculates the variance of the y array of data using N weighting.

$$\text{Variance} = \frac{\Sigma y_i^2}{N} - \frac{(\Sigma y_i)^2}{N^2}$$

[2nd] [S.Dev] **Standard Deviation Key**—Calculates the standard deviation of the y array of data using N – 1 weighting.

$$\text{Standard Deviation} = \sqrt{\text{Var} \times \frac{N}{N-1}}$$

All calculations here must begin and end by pressing [2nd] [CA] to totally clear the calculator. There are 4 pending operations available between entries and calculations of statistics, linear regression and trend line. However, arithmetic operations cannot be pending while actually entering data points. When doing trend-line problems, the implied x value must be reentered with the [x:y] key if arithmetic calculations are performed prior to entry of all data points. Statistical values are stored in memories 1 thru 7, so external values cannot be stored here without destroying the statistical data.

Data points are entered by pressing [Σ+] after each y_i entry and removed by pressing [2nd] [Σ-] after reentry of an incorrect point. The entry number N is displayed after each entry, N = 0, 1, 2...

Once entered, the data can be used to calculate the mean, variance and standard deviation by simply pressing the necessary keys.

Example: Analyze the following test scores: 96, 81, 87, 70, 93, 77

Enter	Press	Display	Comments
	[2nd] [CA]	0	Clear
96	[Σ+]	1.	1st Entry
81	[Σ+]	2.	2nd Entry

97	[Σ+]	3.	3rd Entry (incorrect)
97	[2nd] [Σ-]	2.	Remove 3rd Entry
87	[Σ+]	3.	Correct 3rd Entry
70	[Σ+]	4.	4th Entry
93	[Σ+]	5.	5th Entry
77	[Σ+]	6.	6th Entry
	[2nd] [S.Dev]	9.8792712	Standard Deviation
	[2nd] [Mean]	84.	Mean
	[2nd] [Var]	81.333333	Variance
	[RCL] **5**	504.	Total of Scores

Note that the standard deviation can be calculated first even though the mean is used to determine the standard deviation.

The data are accumulated in the memory registers with Σy_i in 5, Σy_i^2 in 6 and N in 7. The values stored in the memory registers can be recalled and used in other calculator operations.

For your convenience, the option has been provided to select N or N−1 weighting for standard deviation and variance calculations. N weighting results in a maximum likelihood estimator that is generally used to describe populations, while the N−1 is an unbiased estimator customarily used for sampled data.

Standard deviation and variance can be obtained with N or N−1 weighting. **The variance key uses N weighting and the standard deviation key uses N−1 weighting.** Variance is the square of the standard deviation. So, variance with N−1 weighting is obtained by pressing [2nd] [S.Dev] [x^2] and standard deviation with N weighting results from [2nd] [Var] [$\sqrt{x}$]

LINEAR REGRESSION

[x:y] **x Exchange y Key** – Enters the x values for linear regression calculations. Also used in conversions, roots and powers and certain arithmetic operations.

[Σ+] **Sum Plus Key** – Enters the y values for linear regression calculations.

[2nd] [Σ-] **Sum Minus Key** – Removes undesired data entries.

[2nd] [Slope] **Slope Key** – Calculates the slope of the calculated linear regression curve. If the line is vertical, the display will flash because the slope is infinite.

[2nd] [Intcp] **Intercept Key** – Calculates the y-intercept of the calculated linear regression curve. If the line is vertical, the display will flash because there is no y-intercept.

[2nd] [x'] **Compute x Key** – Calculates a linear estimate of x corresponding to a y entry from the keyboard.

installation, operation, assembly, maintenance, and repair manuals. Many manuals explain how to do several of these jobs. This section emphasizes instructional manuals. There are also several types of manuals we will not discuss, including reference manuals or codebooks that define performance levels and laws. The first steps in writing a technical manual are to assess the reader's level of knowledge and to determine the reader's needs. In other words, you must begin with audience analysis.

Analyzing a Manual's Audience

You learned general principles of audience analysis in Chapter 1; this section explains a few audience rules that are particularly important for writing instructional manuals.

The Reader's Knowledge

Manual writers often fall into the trap of writing for themselves, not for an audience that has less expertise and needs to learn and be taught. As a result, to cite one common example, there are computer manuals written in jargon and abbreviations that only another computer addict can understand. Your task as a teacher is to simplify, explain, and define terms.

The Reader's Need to Know

You also must ask what the reader needs to know. How much information does the audience need? What level of detail is necessary? How complex should your presentation be? Is it adequate, for example, to say that the mechanic should look for a part the size of a one-quart oil can, or is it necessary to give exact dimensions?

The Reader's Receptivity

Where and when a manual will be used are highly important considerations. Think about what you do when your car stalls. Do you open the hood immediately and start poking around in the carburetor and the wiring? Do you calmly consult a manual's troubleshooting guide? Or do you read a chapter on tuning an engine? Most people want to be productive immediately. As a manual writer, you should write and organize your instructions so that readers can quickly find the information they need and get right to work. One study showed that only 20 to 25 percent of readers completely read instructions before getting to work. Several important rules follow from these facts about your audience's limited receptivity.

- Place critical information and warnings in a prominent position and set them apart graphically so that they cannot be missed.
- Count on having no more than 10 to 20 minutes of the reader's time.

Writing Instructions

Having analyzed your audience, you can appreciate their need for clear instructions. Your audience should be able to get the information they need quickly and easily. There are four guidelines for you to follow in writing your instructions.

Use Simple Style

You should keep the sentences in instructions as short as possible, using the simplest vocabulary that will get your point across. As a rule include only one operation per instruction (i.e., "Turn the release lever," rather than "Turn the release lever; set the control switch to Pause.")

Use Imperative Mode

Instructions are written as commands in active voice (i.e., "Turn the release lever" rather than "The release lever should be turned").

Use List Format

Instructions are most useful when they are written as numbered lists. Never make your readers search through a paragraph looking for the next step.

Use the Rule of Seven

As we mentioned in Chapter 6, readers only have five to seven "spaces" for information in their short-term memories. This fact, often called the Rule of Seven, is crucial when you are writing instructions. A long, unbroken list of instructions can seem daunting to the reader. If you cannot limit the instructions to seven or fewer, subdivide the instructions into sets, using headings to indicate what each set covers (in Exhibit 12-1 notice that the instructions have been subdivided into assembly and operating instructions).

In Exhibit 12-3 from a manual for an Aerial Film Viewer, note the brevity, the simple style of the instructions, and the placement of warnings. Once you have considered your audience's knowledge, needs, and receptivity, you can organize an effective manual.

Organizing a Manual

Although the exact ordering of sections will depend on the audience and purpose, manuals typically include the following sections: overview, purpose, scope, principles of operation, tools and materials, descriptions of mechanisms and processes, warnings, performance criteria, troubleshooting guides, and illustrations.

Overview, Purpose, and Scope

Manuals should be organized to make the best use of the reader's time. Start with an overview, which can be either a page of introductory material and forecasting, a table of contents, or some other form of summary. Then, if it is appropriate, discuss the purpose and scope of your manual. John Muir's famous *How to Keep Your Volkswagen Alive and Well: A Manual for Compleat Idiots,* for example, explains its scope by defining four levels at which the manual can be used. The nonmechanical reader can use the book to evaluate used VW's on the market. More mechanically inclined readers can learn to do various repair jobs at three different levels, which Muir calls Phases I, II, and III repairs. Early in the book, Muir also describes the different sets of tools required for each of the three phases.

Principles

A discussion of principles comes next. In Muir's book, one chapter is devoted to explaining how an internal combustion engine works, so that novice readers will understand what they are dealing with. General purpose auto manuals often cover principles by means of a set of cutaway drawings of a working engine. Discuss only the principles your particular audience needs to understand: the hows and whys your audience does not already know.

Tools and Materials

If materials or tools are required, provide a well-organized list. Be sure to group or classify items so that materials and tools needed for one part of a process, or at one time, are listed together rather than at random locations in your list. List *all* the parts; don't assume that readers will understand what may be obvious to you: that you need welding rod to weld, for example, or glue to construct a model, or gasket sealant to attach new gaskets.

Descriptions of Mechanisms and Processes

The descriptions of mechanisms and processes are the heart of a manual. The critical factor for readability will be your ability to group steps or parts so that the reader can visualize a task as four or five major steps rather than as an endless number of disconnected minor steps and tasks. In the exhibits in this chapter, note how subtitles, graphics, and paragraphing are used to group tasks. The writers of Exhibit 12-1, for example, have separated assembly from operation, and the writers of Exhibit 12-2 have used separate paragraphs for principles of operation and data storage, data entry, calculations, and comments about mathematical details of the calculations.

Warnings

Warnings should be placed so that the reader notices them immediately. Exhibit 12-3 demonstrates one excellent way to make warnings and notices stand out. The Department of Defense defines three levels of warning:

EXHIBIT 12-3

Instructions. Courtesy of Keuffel & Esser Company, Parsippany, New Jersey.

INSPECTION AND ROUTINE CARE

The K&E Portable Aerial Film Viewer has been designed to assure years of dependable service when given routine care. Under normal working conditions the viewer requires little attention, however, certain commonsense precautions should be observed.

PRECAUTIONS

- When using the viewer, do not place any extra weight on the black aluminum rollers that are located on the top right hand and top left hand ends of the viewer.
- Take extra precaution to protect the Plexiglas easel from scratches.
- Keep liquids away from the viewer to prevent spillage into the lamp housing.

MECHANICAL CHECKS

NOTE:

Upon receipt of the Portable Aerial Film Viewer, a complete mechanical check should be made.

- Examine the easel to ensure it is undamaged.
- Operate all controls and check for smoothness of operation.

WARNING

THE VIEWER HAS TAPE HOLDING THE LAMPS IN PLACE DURING SHIPMENT. THIS TAPE MUST BE REMOVED PRIOR TO OPERATION.

- **Warnings** refer to dangers to personnel.
- **Cautions** refer to dangers to equipment.
- **Notes** refer to contingencies, including common problems with operation and maintenance.

Figure 12-5 makes highly effective use of graphics in a warning.

Performance Criteria

Place performance criteria at the end of each major step. Explain how the user can tell that the work is being done properly. At the end of Exhibit 12-3 users are instructed to test controls to feel if they run smoothly. If you were teaching a beginning camera class, you might explain that film is probably loaded

***Figure* 12-5**
Precautions. Courtesy of Panasonic Company, Division of Matsushita Electric Corporation of America.

PRECAUTIONS

Please read these precautions before you operate this unit.

AVOID SUDDEN CHANGES IN TEMPERATURE

If the unit is suddenly moved from a cold place to a warm place, moisture may form on the tape and inside the unit. In this case, the DEW Indicator will appear and the unit will not operate.

HUMIDITY AND DUST

Avoid places where there is high humidity or dust, which may cause damage to internal parts.

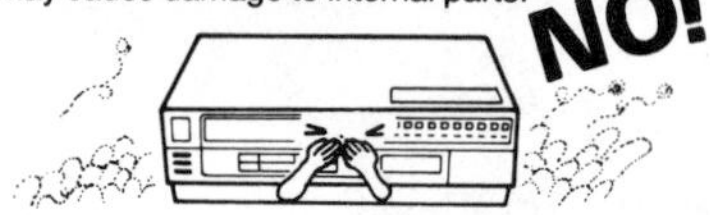

DON'T OBSTRUCT THE VENTILATION HOLES

The ventilation holes prevent overheating. Do not block or cover these holes. Especially avoid soft materials such as cloth or paper.

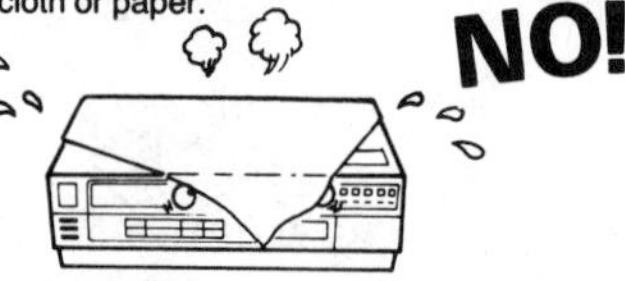

KEEP AWAY FROM HIGH TEMPERATURE

Keep the unit away from extreme heat such as direct sunlight, heat radiators, or closed vehicles.

KEEP MAGNETS AWAY

Never bring a magnet or magnetized object near the unit because it will adversely affect its performance.

NO FINGERS OR OTHER OBJECTS INSIDE

It is dangerous, and may cause damage, to touch the internal parts of this unit. Do not attempt to disassemble the unit. There are no user serviceable parts inside.

KEEP WATER AWAY

Keep the unit away from flower vases, tubs, sinks, etc.
CAUTION:
If any liquids should spill into the unit, serious damage could occur. In that case, please consult qualified service personnel.

CLEANING THE UNIT

Wipe the unit with a clean, dry cloth. Never use cleaning fluid, or similar chemicals. Do not spray any cleaner or wax directly on the unit or use forced air to remove dust.

STACKING

Use the unit in a horizontal position and do not place anything heavy on it.

REFER ANY NEEDED SERVICING TO QUALIFIED SERVICE PERSONNEL

correctly if the feed reel moves and if there is no ripping sound inside the camera when you cock the film advance lever. In other cases, you might refer to meter readings, the color or consistency of a product, or still other indications of proper or improper operation.

Troubleshooting

Your manual may also include a troubleshooting section. Often these are placed at the end of the manual or with the performance criteria at the end of each major section. Troubleshooting guides focus on troubles that arise after a task is performed. They explain probable causes for common problems. Troubleshooting information, however, is not a substitute for taking the time to teach correct procedures in the body of a manual.

Graphics and Other Aids

Graphics and other aids should be placed wherever words will not do the job. Graphics should always work with the text, not alone. They are particularly useful when you need to explain complex spatial and temporal relationships. When you need an illustration to explain such a relationship, though, take the time to design or find an illustration that does precisely what you want. Graphics are covered in detail in Chapter 8; it is worth repeating here that poorly conceived illustrations can be worse than useless because they confuse and frustrate the user. Remember some of the graphics that have frustrated you. One of our favorites was a photograph of the greasy underside of a car; everything in the photo was either black or dark grey, and the caption simply said, "The transmission drain plug is in the center of the illustration."

The manual excerpted in Exhibit 12-2 makes good use of layout to explain how to perform statistical calculations. Other ways to help the reader along include clear tables of contents, tables of illustrations, a glossary of technical terms, and appendices for additional information.

Teaching, Not Telling

Above all, your manual should teach; it should explain how to do a task, not just describe what should be done. A novice mechanic who is facing a frozen nut does not need to be told that the nut must come off; he or she wants to know how to deal with the situation. Your task as a manual writer is to know what emergencies are likely to arise and to be there with helpful suggestions. For example, Exhibit 12-1 tells the readers the steps for assembling the legs of the tripod. Telling is appropriate in this case because the parts are made to exacting specifications and assembly is simple. A contrasting example will show what we mean by teaching and anticipating problems. Weber Barbeques also stand on a tripod, which consists of three steel tubes that are pressed into cylindrical receptacles spot welded to the bottom of the barbeque's fire bowl. All parts are heavily coated with a rust-and-temperature-resistant paint, with the result that the parts fit together very snugly. In the instructions the Weber company cautions: "Make sure the leg is touching the bowl. If you have trouble, don't hammer, wiggle, or twist. Have a helper brace the bowl. Give a stronger push." (Weber Stephen Products, Publication 30411/1-83.) The company anticipated a common difficulty in assembling, and the manual teaches a solution.

A Sample Manual

Here is an example of a set of instructions using some of the sections described above. This part of a printer manual describes the installation and operation of one of the printer accessories: a forms insertion guide.

Section	***Manual Text***
Overview	*Introduction*
Definition	The Forms Insertion Guide, Model 8290, is one of the printer paper-handling options. It makes single-sheet paper feeding easier by automatically aligning each sheet of paper.
List of Materials	In addition to the forms insertion guide, you will need the clear plastic sound panel, which is placed across the platen to reduce noise.
Teaching	This sound panel should already be installed on the printer; if it is not, please refer to the instructions on page 23 of this manual.
Description	*Product Specifications* These are the dimensions of the forms insertion guide. Width—15 inches Height—9 inches Depth—2 inches Weight—2 pounds
Graphic	[Figure 2, Forms Insertion Guide Diagram]
Cautions	*Product Requirements* The forms insertion guide should be oper-

ated only at temperatures between 50° and 100°F (10° to 38° C) with a noncondensing humidity of 20 to 90 percent.

The forms insertion guide can be attached and detached without turning off the printer.

Warning

Paper containing staples or paper clips should never be fed into the printer.

Instructions

Installation Instructions

1. Open the sound panel (see Figure 1, Printer Diagram).
2. Open the paper bail (see Figure 1).
3. Hold the forms insertion guide so that its front, which shows the paper scale, faces the front of the printer. (See Figure 2, Forms Insertion Guide Diagram.)
4. Tilt the guide forward approximately 10° and position the attachment arms over both ends of the platen shaft.
5. Push the guide back and down until it rests firmly on the platen shaft, then slide the guide backward until it rests on the printer cover.
6. Move the left and right sides of the guide's paper chute to the desired position by loosening the two locking knobs. (See Figure 2, Forms Insertion Guide Diagram.)
7. Retighten the locking knobs when the unit is in position, allowing for a slight margin on either edge of the paper for smooth entry into the printer.
8. Close the sound panel.

Operating Instructions

1. Place a sheet of paper into the chute of the forms insertion guide.

Performance Criterion

2. Press the Form switch. The printer should then feed the sheet around the platen, stopping at the top-of-form position.
3. Enter the appropriate command at the workstation. Printing should begin.
4. Place each subsequent sheet into the chute when the current sheet is ejected by the printer.

Summary

1. Definitions commonly take a three-part form: species = genus + differentia. For example, a hammer is a tool that strikes.

2. A good classification of parts or steps should be natural, exhaustive, disjointed, based on one principle of division, useful, and simple.

3. Parts of mechanisms and steps in a process should be named carefully.

4. The description of a mechanism has three main parts: an overview, a part-by-part description, and a description of one full cycle of operation.

5. A process description should include an overview, a list of equipment and materials, step-by-step instructions, and a description of one cycle of operation. In addition, the description can include sections common in instructional manuals, such as performance criteria and warnings.

6. Manuals explain how to assemble, install, operate, maintain, or repair.

7. A manual typically includes an overview, a discussion of purpose and scope, a review of important principles of operation, lists of tools and materials, descriptions of mechanisms and processes, warnings of dangers to persons or equipment, performance criteria, troubleshooting information, and graphic support.

8. Manuals should teach rather than tell; they should explain how to do a task, not just state what should be done.

Exercises

1. Keeping audience and purpose in mind, define one of the following for an audience of college-educated magazine readers (your instructor may suggest a different audience or purpose).

   a. Brazing
   b. A dynamic load
   c. A catalytic reaction
   d. Vise grips
   e. Software
   f. Plate tectonics
   g. Gene splicing
   h. A quasar
   i. A polymer
   j. A quark

2. Write a description of a simple mechanism, including a definition and two illustrations. Choose one from the following.
   a. A stapler
   b. A retractable ballpoint pen
   c. A paper punch
   d. A manual pencil sharpener
   e. Another mechanism (chosen with the assistance and approval of your instructor)

3. Write a description of a process, including two illustrations. Choose from one of the following, or select a topic with the assistance and approval of your instructor. The process should have at least ten steps.
   a. Straightening a bicycle wheel
   b. Installing new engine points and timing an engine
   c. Preparing a difficult recipe, such as a soufflé
   d. Performing an editing operation in a word-processing program.
   e. Setting a videotape recorder to record a program with an automatic timer.

Instruction Case 1

The Production Design Division

The instructor in your mechanical engineering class has asked you to form groups of five to six students. He gives each group a bag of children's construction toys and the following assignment:

Gigantic Industries

TO: All members of the Production Design Division
FROM: Division Director
SUBJECT: New product models

As you know, there has been some concern of late over the lack of new product creation by this division. In an effort to stimulate productivity among the staff, I am asking you to do the following:

1. Use the enclosed materials to create a model of a product suitable to our expanded product line. Remember, with our recent acquisition of Megalith, Inc., we now have a virtually limitless field of production: recreational equipment, business machines, computer hardware and software, arms and materiel, scientific instruments, agricultural equipment, etc. Use your imagination.

2. Write a set of instructions for the assembly of your model so that my assistant, Mr. Haberman, can put it together at our sales meeting next month. Please remember that Mr. Haberman is far from the most coordinated member of my staff and that he has problems with polysyllabic words.
3. Present both model and instructions to me at the end of the class period.

ASSIGNMENT

Note: Your instructor will supply you with a bag of construction materials. Using the materials your instructor has provided, create a model and write the instructions for assembly. Be sure to consider graphics and effective format.

Instruction Case 2

THE PRINTING MANUAL

As part of your technical writing job at Digisystems Incorporated you must write operating manuals for all of the various instruments the company produces. Your usual procedure is to have the engineering department write a rough version of the operating instructions and specifications, which you then polish into an acceptable manual. However, the company's newest instrument—a small dot matrix printer—is going into rush production and the engineering department has no time to write anything.

Your solution is to corner one of the engineers (using appropriate force) and to ask him to explain how the printer is operated. This is what he tells you:

This thing is a great idea. It's designed for small offices—low speed, uses a grid and ribbon. But it shouldn't be used for high volume; just low volume.

What the operators really need to know about is loading the paper, because somebody always messes that up. You've got two kinds you can use, cut sheet and pin feed. Cut sheet? Well, it's sheets, you know, like regular paper. And pin feed is the stuff with the little pinholes on the sides. Listen, that stuff is a bear to load. You set the paper release lever to front position, yeah, o.k., open position, and you pull up the tear-off bar, then you set your forms thickness control lever. But the thing is they've got to line up the pin-feed holes in the paper with the pins on the platen in the printer, because they can't adjust it after the printing starts. So they feed in the paper from behind the platen until it won't go any further, and they have to line up the holes with the pins then. And then they turn the platen by hand until they've got the paper around the platen and under the tear-off bar. Then you close the tear-off bar again. Oh say, there's a cover, an access cover, over the platen and everything. They've got to take it off before they can get to the platen and then put it back on before they print.

The cut sheets are easier. You set your forms thickness control and pull up the tear-off bar. Yeah, o.k., you've got to take off the access cover. Then you just feed the paper around the platen like you would a typewriter, turn the platen by hand. If it's not straight? Well, there's this paper release lever up by the platen; you just move it to the left and it releases the paper and you can adjust it. Then you roll the paper over the platen with the platen knob to where you want the top to be. It's got to be above the tear-off bar, though. Of course you put the paper release lever back to the right before you position the paper; the platen won't turn unless you do. Then when the printer gets to the last four or so inches on the page, it stops and they have to press the continue switch to print any more lines. If they go to the last inch, the lines may slant, though, because the paper slips. Sure, yes, you have to put the access cover back on before you run the printer.

Forms thickness? Okay, you can only get to that by taking off the access cover. It's for setting the impact of the matrix. If you're printing a single sheet, you set it on 1, that's to the right. If you've got two sheets it can be right or left, 1 or 2. But if you've got three, it has to be set on 2, left.

That's something they should know about: all the controls. You've got your indicators, for one thing. They're right on the front of the printer, to the right. There's a power one that shows the machine is on, and a paper one that flashes when the paper supply gets low with the pin-feed. Then there's the attention one that tells you the pause switch has been hit. That lets them know if the pause is on accidentally, but it flashes too if they've hit the pause on purpose, of course. They're all red LEDs right above the operating switches.

The pause? Well, if they want to stop the printer for something, say they want to reset the length of the paper or something, they hit the pause switch, o.k, press the pause switch, and the printer will stop after it finishes the last command. The attention light flashes to let them know the pause is on. Then they hit, press, the continue switch to get the printer going again. You've got a bunch of other switches, too. That continue switch, for example—it not only gets the thing going after a pause, but you can press it to get the printer to go on printing one line at a time at the bottom of the page, after you've gone beyond four inches at the bottom.

The there's the page length switch. It's a dial on the front. You turn the arrow on the dial to the right number for the page length you want, that is, the amount, the length, of the copy you want on the page. I've got a table that gives the settings; I'll send you a copy.

What? Of course, there's a power switch. How else would the thing run? Why would you have to tell them to turn it on? Any idiot knows that. Okay, okay. It's on the back, lower left, next to the power cord. It says on and off—up for on, down for off.

So, how does it work? It's simple. You load your paper, get your length and your thickness, and let 'er rip. Then if you want to run the copy down to the bottom of the page, you hit the continue switch, and if you want to change anything, you hit the pause.

Dangers? Oh gee, nothing really. Well, I suppose you could tell them not to touch the print head when it's hot. And they should make sure the power's off when they hook up the interface cable with the computer. And it should be plugged into AC, of course. Oh yeah, also, the printer has to be turned on before the computer gives a print command or they may lose their data. You know what else? Sometimes people put stuff, coffee cups and stuff, on a printer—crazy. They shouldn't put anything on top of the printer. They shouldn't even lean on it. And if they drop something into it, they should turn the thing off before they try to get it out. That's a great way to lose a finger. And if they spill coffee or something, liquid, you know, they have to call a service rep.

Simple, right?

Two days later the engineering department sends you the following material:

Specifications for the 231 Matrix Printer

160 characters per second, 80-column format
Bidirectional printing
Nine-wire grid against inked ribbon
Dot matrix

Page Length Table

Dial Number (Page Length Switch)	Copy Length (In inches)	Lines at 6 lines/inch	Lines at 8 lines/inch
0	3	18	24
1	3.5	21	28
2	4.0	24	32
3	5.5	33	44
4	6.0	36	48

Dial Number (Page Length Switch)	Copy Length (In inches)	Lines at 6 lines/inch	Lines at 8 lines/inch
5	7.0	42	56
6	8.0	48	64
7	11.0	66	88
8	12.0	72	96
9	14.0	84	112

They also send you the illustrations below.

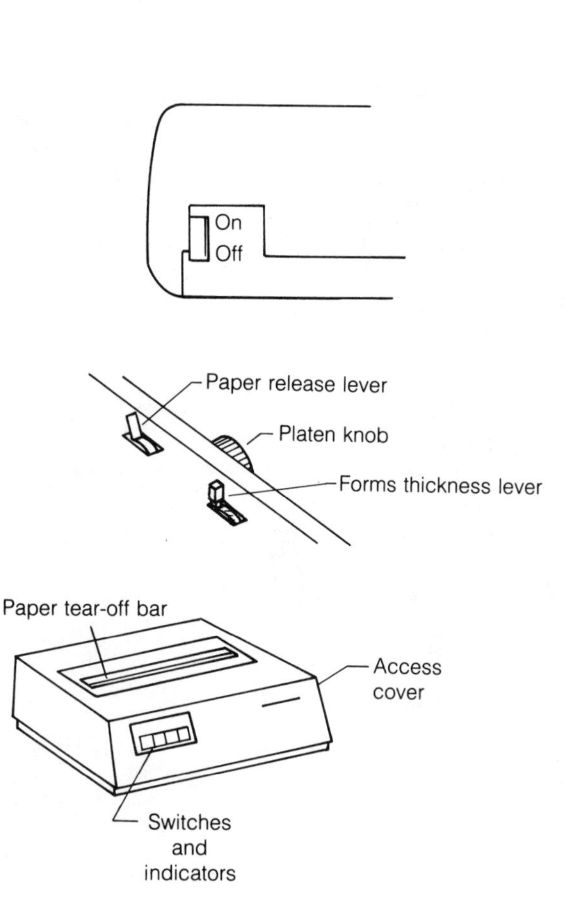

ASSIGNMENT

Write a set of operating instructions for the Digisystems 231 Matrix Printer. Be sure to include a description and classification of parts and all appropriate warnings and cautions. You should include the drawings the engineering department sent you.

P · A · R · T

Communicating at Work

CHAPTER • 13

Business Writing

Introduction

In your career, you will often write letters on your own behalf and for your employers. This chapter presents the basic principles and forms of business correspondence, with an emphasis on the forms you are most likely to use. This chapter covers the basic conventions of business correspondence; the principles of business correspondence; formulas for writing memos, request, and bad news letters; and correspondence forms used in the job search.

The Conventions of Business Correspondence

You want your correspondence to look professional. This section presents the two most common letter formats and examines the conventions of correspondence, including topics, such as how to address a reader whose name and gender are unknown to you, that often puzzle experienced writers. We'll begin with the two most common layouts.

Full Block Format

The full block letter format has become popular since the introduction of the electric typewriter. In full block, every line is justified on the left margin; you do not indent the date, the signature, or the heading. Exhibit 13-1 is an example of the full block format.

EXHIBIT 13-1
Full Block Format

COMPANY LETTERHEAD ON TOP, WITH ADDRESS AND PHONE

Date
Addressee's Name
Street Address (Notice the "open punctuation:" no commas after the
City, State ZIP code name or the street address.)

Salutation:

Body of the letter starts here. As you will see later, you can also insert a subject line before the salutation.

Notice that you double-space between paragraphs, but do not indent the first line of each paragraph.

Complimentary close,

Signature
Typed Signature

Additional lines can come after the signature for initials, enclosures, references, or information about the disposition of copies.

Semiblock Format

In semiblock format, the date, complimentary close, and signature are indented. Exhibit 13-2 is an example of semiblock format. Note that both formats include the same parts. These parts are discussed in the next section.

Parts of the Letter

The following sections explain the use of each of the parts of the formal business letter, from heading to additional pages.

Heading

Your company letterhead provides the heading. When you type your own letters on blank paper, type your address two inches from the top of the page and omit your name, which will appear in the signature line at the bottom of the letter. In full block

EXHIBIT 13-2
Semiblock Format

COMPANY LETTERHEAD: NAME, ADDRESS, PHONE

Date (indented)

Addressee's Name
Street Address
City, State ZIP code

Salutation:

In semiblock the paragraphs start on the left margin; do not indent the first line of each paragraph.

In semiblock, as in full block, double-space between paragraphs.

Complimentary close, (indented)

Signature (indented)

format, type your address as a block on the left margin, above the addressee's name and address. In semiblock, indent your own address.

Date

Type the date two lines below the letterhead. All of these distances are suggested, not cast in concrete. You want the final product to look good on the page. If your letter is short and you are using 8½-by-11-inch paper, you may need to start the date farther down the page, use a few more spaces between the inside address and the salutation, and stretch other distances.

Inside Address

The inside address includes the addressee's title (Ms., Mr., Mrs., Miss, Dr., Professor, etc.), job title, and full address. If you are writing to a department, use the name of the department in place of a personal name.

Subject Line

The subject line is optional. If you wish to attract attention to your letter or clearly indicate the subject of the letter, you can include a subtitle in the form, "Subject: Program Development." This line should appear before the salutation.

Salutation

The salutation (greeting) consists of the conventional "Dear" and an appropriate name or title. Unless you are a close friend of the addressee, do not use a first name. If you do not know the sex of the addressee, use the job title in place of the name: for example, "Dear Director." Do *not* write "Dear Sir or Madam." If you are writing to a department, you may replace the salutation with an attention line, "Attention: Credit Department," or you can write, "Dear Credit Department."

Body

The body of your letter should always be single-spaced, with double-spacing between paragraphs. Organization patterns are discussed later in this chapter.

Complimentary Close

As a rule, if you are writing to a superior, you should use a formal close. "Sincerely" and Sincerely yours" are always appropriate. "Respectfully" should be reserved for formal occasions. "Cordially" and "Warmly" are more intimate.

Signature

The signature section of the business letter includes both your typed name and your handwritten signature. You can also include your title and the name of the company (or the name of your division if the letter is directed to another division within the company).

References

The reference line indicates who dictated and who typed the letter. Typing forms vary, but the writer's initials always come first, followed by those of the typist, for example, "HWB/tr." When the signer did not write or dictate the letter, but merely approved it, the formula may (but need not) include the author's initials between those of the signer and the typist, for example, "MWB/KH/tr."

Enclosures

When you include other documents with your letters, you should note the enclosures on a separate line beneath the reference line. If you do not mention the enclosure in the body of the letter, you may follow "Enclosure" with a brief description of what you are enclosing. If there is more than one enclosure, note the number, for example, "Enclosures (2)."

Copies

If copies of the letter are to be sent to other persons, their names should follow the notation "cc:" below the enclosure line. The notation stands for "courtesy copies." There are also formulas for instructing the typist to send copies without letting the chief recipient know this. If you need to know these protocols, consult a full-scale letter-writing text.

Additional Pages

When your letter extends onto additional pages, those pages should carry the following information at the top of each page: full name of recipient, page number, and date.

Exhibits 13-1 and 13-2 provide you with an outline of these parts and the basic layout. Now that you have these graphic conventions in mind, you are ready to move on to four important communication principles that you should remember each time you draft correspondence.

Four Basic Principles of Business Communication

Four basic principles apply to all of your correspondence, no matter what your topic or what kind of letter you are writing.

1. Be responsible for every detail of any letter you sign.
2. Address the reader with the words *you* and *your* whenever you can.
3. Keep your language as simple and conversational as you can.
4. Use positive words, even when you are sending a negative message.

Be Responsible

Every document you sign is your responsibility. As far as the recipient is concerned, you *are* the company. Write your letters clearly and proofread them carefully.

Address the Reader Personally

Whenever you can, use the words *you* and *your*. Your audience will read more rapidly, stay interested, and be more sympathetic to your message. Some people dislike letters that use their personal names repeatedly, especially since word processor "mail merge" programs can automatically insert a name throughout a letter.

Use Simple, Conversational Language

Readers appreciate direct communication. You should avoid overly formal and outmoded phrasing, such as "regarding your letter of the nineteenth," or "it has come to my attention that. . . ."

Use Positive Language

Show trust in your reader's fairness and good judgment by expressing all of your messages—even bad news—in positive language. By the same token, avoid using words such as *unfortunately* or phrases such as "we regret to inform you," or "we are sorry to deny." While writers who have a talent for honest expression can make these phrases seem sincere and caring, this kind of negative language often makes readers bristle with irritation, with the result that they stop paying attention to your explanations.

The letter formats presented in the next sections are elaborations on these four principles.

MEMORANDA

Memoranda means "reminders." The best use of memos is to remind your coworkers of some fact or occurrence: a decision reached at a meeting or the date a new policy or procedure goes into effect. A well-designed memorandum can also be used for *minor* final reports. The most common abuse of the memorandum is to use the informality of the format as an excuse for writing a long, disorganized report on a significant subject. Memos should always be short; don't let yours sprawl over several pages. The memo is an *in-house* form. Memos can be written according to the various persuasive strategies discussed in the business letter sections that follow.

Memo Organization

Exhibit 13-3 shows a well-designed memo form. Notice the basic parts labeled TO, FROM, DATE, and SUBJECT. A memo is typically signed or initialled after the name typed on the FROM line. Also note the warning printed at the bottom of the reporting space: The company will proceed on the basis of this information unless corrections and additions are made immediately.

Memo Guidelines

Memos are best written immediately after a meeting or when you first recognize the need to inform coworkers of a new fact or policy. The virtue of the memo is its impromptu nature. Memos are ad hoc reports. The memo format should not be used to write reports longer than one or two pages.

DIRECT REQUESTS

Use the direct request format when your request is not likely to be questioned or challenged: in other words, when you do not need to persuade your reader (as in a sales letter) or break your news gently (as in most bad news letters). After we present the basic parts of the direct request and give you an example, we will offer specific tips on how you should state your requests.

Direct Request Organization

Direct requests have three parts: (1) the main request, (2) the explanation, and (3) the recapitulation and close. You will notice that most business letters follow this standard three-part format.

EXHIBIT 13-3
Memo-report form

COMMERCIAL
DESIGN
ASSOCIATES

4230 198TH STREET SW
LYNNWOOD, WASHINGTON 98036

CONFERENCE REPORT
TO:
FROM:
DATE:
PROJECT:
SUBJECT:
IN ATTENDANCE AT MEETING:

The following statements summarize our (telephone conversation / meeting) on the above date.

CDA is proceeding with the project based on the above statements being correct. Please notify CDA immediately if there are any corrections or additions to this report.

MAIL TO: ______________________ CC: ______________________

______________________ ______________________

______________________ ______________________

______________________ ______________________

Your Main Request Explain who you are and then state your request directly, being certain to specify everything you want. After you have stated the request, expand on it in your second paragraph.

The Explanation As you explain your request, stress the benefits the reader will obtain by granting your request, such as business good will, advertising, potential sales, public recognition, or simply the feeling of self-worth that comes of doing a favor. (Benjamin Franklin once observed that if you want someone to feel good about you, ask for a favor.)

Exhibit 13-4

Direct Request Letter

SIERRA INSTITUTE
P. O. BOX 1105
CLAREMONT, CALIFORNIA 91711
(714) 896-5673

January 31, 1986

C. W. Edwards, President
Public Design Forum
P.O. Box 18295
Dallas, Texas 78214

Dear Mr. Edwards:

Main request (implied)

The Sierra Institute, a nonprofit corporation that specializes in promoting information exchanges in environmental fields, is preparing a book that lists public space design agencies throughout the United States and Canada. We would like to include every public and private organization, agency, or action committee that is concerned with issues of historic preservation, open-space zoning, park design, and related issues.

Direct request

Would you please provide us with the following information so that the entry for your organization can be as complete and up-to-date as possible?

Numbered questions with implied benefits

1. Full name and date of incorporation of your group
2. Correct address, if address above is incorrect
3. Names of chief executives
4. Membership in regional or national consortia
5. Program emphases
6. Publications
7. Copies of your publications

Directly stated benefit

The Sierra Institute has published over 70 titles of this kind. Our reference books are found in most large libraries and in many government and private corporation reference collections.

Recap and reinforcement

Thank you for your cooperation. In order to meet our publication deadline, please reply by March 14, 1986.

Sincerely,

Oakmont Larch
Editor in Chief
SIERRA INSTITUTE

The Close

In a final paragraph, repeat your main points and thank your reader. Exhibit 13-4 is a short request letter.

Direct Request Guidelines

The letter in Exhibit 13-4 demonstrates several features of a well-designed direct request. Keep these points in mind as you write.

Specify All Details

Specify exactly what you want and, when appropriate, what you are prepared to offer. If you are asking someone to speak at a meeting, tell when and where the meeting will take place; whether an honorarium or consulting fee will be paid; whether meals, lodging, or travel costs will be reimbursed; and whether you intend the speaker to stay to answer questions, write an evaluation of the meeting, or participate in some other follow-up activity. If you ask for information, as in the letter above, explain what information you want and how you will make use of it.

Limit the Number of Requests or Questions

Limit the number of questions you ask, and be certain that each question is separate and distinct. Do not ask more than ten questions.

Number Questions or Requests

Number each question or request, and if possible devote a separate line or paragraph to each request. Your reader will answer questions that stand out on the page. If you run your questions together in a single paragraph, the reader will probably overlook some of them.

Ask Neutral Questions

Ask your questions in a neutral manner, without indicating your own preferences or opinions. In Exhibit 13-4 it would have been inappropriate for the author to state political opinions about conflicts between urban development corporations and park advocates.

Show Faith in Your Reader's Interest and Fairness

Believe in your reader's basic honesty and sense of fair play. You don't need to persuade readers to answer simple questions or provide lengthy rationalizations.

PERSUASIVE REQUESTS

The persuasive request format is used whenever you need to sell an idea to your reader. Most of the advertising you receive follows this format.

Persuasive Request Organization

Full-scale business-writing texts can introduce you to several popular variations of the persuasive request format. This section presents a simple four-part formula: (1) attract attention, (2) develop your reader's interest, (3) prove your point, and (4) specify an action to be taken.

Attract Attention

You can attract attention by a clever opening sentence; with promises, predictions, or offers; or by effective use of graphics and layout. Keep in mind that there is a distinction between attracting attention and demonstrating bad taste. For example, crass appeals to greed or gluttony—"Imagine a mountain of chocolate!"—are in bad taste.

Develop Interest

Next, develop your reader's interest by describing positive features of your proposal. Think of four or five points of interest that will sell your project, product, or idea. As you will see later in this chapter, employment application letters also follow this advice. The middle paragraphs of a good application letter always present your best credentials.

Prove Your Point

In a third section, offer evidence of the truth of one or more of your claims. In Exhibit 13-4 the Sierra Institute proved that it was worthy of receiving information by mentioning the number of their previous successful publications and places where they were distributed.

Specify Action

Finally tell the reader what to do. Advertisements specify action by closing with coupons, subscription forms, or cards to mark and send back. In an application letter, you specify action by asking for an interview or stating when you plan to call. Exhibit 13-5 exemplifies these features.

Persuasive Request Guidelines

When you first begin to use the four-point persuasive format, it might seem mechanical. You'll soon find that the method is highly effective and natural, and you will probably discover many reasons why it works so well. To learn the method quickly, analyze the advertising you get in the mail or see in magazines or on television. Notice how often persuasive material seeks

Exhibit 13-5

A Persuasive Request

Interpart Incorporated
P. O. Box 45
Bellevue, Washington 98456
(206) 368-3639
CABLE: INTERCORP

April 30, 1985

Ms. Roberta Williams
Director of Purchasing
Expert Engineering
Worthington Parkway 78
Industry, California 91040

Dear Ms. Williams:

Attract attention

The Interpart computerized on-line ordering system can save your purchasing department up to 20 percent of its operating costs and cut a week off the average time it takes to fill a parts order.

Describe and develop

Interpart is a new on-line computer network that links together 160 major industrial suppliers of equipment and materials. By using the Interpart system, you can compare suppliers' prices and be instantly aware of special purchase opportunities. You only pay for the system time you actually use for business transactions, and you can list your own products and surplus materials for a nominal fee.

Prove

Over fifty major industrial and engineering firms have become Interpart subscribers in the last six months. One user writes: "Interpart has enabled us to expedite rush orders and to take on contracts we would have had to pass by."

Specify action

If you would like more information about the services and materials available through Interpart, as well as a full list of subscribers, please fill out the attached card or call (206) 368-3639. We will respond immediately.

Sincerely,

Ellis Sinkiewicz
Director of Marketing

Enclosure: Interpart Action Card

to attract your attention, develop your interest, prove a point by testimonials or statistics, and then specify an action you should take.

BAD NEWS LETTERS

Bad news letters come in two types: the direct and the indirect. Use a direct bad news letter when the problem is minor, as with a slight delay in an unimportant shipment, or when the message is urgent, as with past due accounts. Use the indirect format when you want to break bad news gently and explain it fully.

Direct Bad News Letters

The direct bad news letter has three short sections: (1) an immediate statement of the bad news, (2) a brief explanation, and (3) a cautious close that includes some "resale" material designed to encourage return business. Remember one basic rule of business writing: Stay positive and avoid negative or discouraging language. Exhibit 13-6 exemplifies the direct bad news letter.

Indirect Bad News Letters

The indirect bad news letter, which is used when you want to buffer the shock, has four parts: (1) the buffer, (2) the explanation, (3) the implied decision, and (4) the positive close.

The Buffer

The buffer is a positive and encouraging statement that places your bad news in context. Start by thanking the reader for something, agreeing with the reader about some other matter, complimenting your customer for a past service, or offering some other good news. If you must complain about a late delivery, for example, compliment your supplier on a good delivery record.

The Explanation

When you explain bad news, do not blame or build webs of argument. State the relevant policies and facts clearly, and let the reader come to the logical conclusion.

The Implied Decision

In some cases, you might wish to state your decision directly, particularly if it includes many parts or significant details. If possible though simply imply your decision: Let the statement of facts and policies make your case for you. This is usually possible if the decision is a simple yes or no. Implied decisions are less confrontational and consequently less likely to give offense.

EXHIBIT 13-6
The Direct Bad News Letter

(Heading and Inside Address Deleted)

Dear Mr. Weatherby:

Your revised marketing software will be delivered on January 15, rather than January 13, 1986.

Our supplier was unable to deliver an adequate supply of the disks you prefer in time.

Thank you for the opportunity to work with you on this project. Please call if you have any questions about the delivery.

Sincerely,

S. Frind/Customer Relations

The Positive Close

Close your letter by affirming your belief in the reader's fairness, intelligence, or understanding. You can also suggest ways in which you might do business together at a later date. In simpler terms, close with a compliment and "resale" material, such as a brochure about a new product. Exhibit 13-7 provides an example of these techniques in action.

Bad News Letter Guidelines

Here are three points to remember about the bad news letter. Always show confidence in the reader's fairness, avoid blaming individuals, and use positive language.

Show Confidence in the Reader's Fairness

Fifteen years ago one of the authors of this text received a letter that challenged his truthfulness in making a complaint about a frozen dinner. He still remembers the firm's bad will. If you disagree with someone, don't challenge his or her fairness or intelligence. Instead, show that you trust the reader's judgment and then state the facts and criteria as you see them.

Praise Individuals, Blame Groups

If you wish to praise a person or a decision, use the person's name. Praise the individual. On the other hand, if you want to blame a person, blame a group instead. In other words, don't write: "Mr. Smith, you should have understood that you can't take a portable radio into the shower with you." Instead, write, "Portable, battery-operated radios are electrical appliances, just as are radios that are plugged into wall sockets. Neither 110AC or battery appliances should be immersed in water."

Avoid Negative Language

Avoid bluntness in your correspondence. Imply negative decisions rather than stating them. And don't use words like *careless, blame,* or *unfortunately.* These basic rules should make it easier to write effective business correspondence.

THE JOB SEARCH

The correspondence you write to contact employers, respond to offers, and change positions will be among your most important compositions. Letters of application, résumés, and the other important forms of employment correspondence can be used most effectively when their design and use are understood in the full context of the American job market. In this section, you will

- Learn key facts about the employment market
- Find out how to identify all of your marketable skills
- Practice how to write several types of résumés for different purposes
- Master the standard letter of application
- Learn strategies for writing effective after-the-interview letters

We'll start with a review of some key facts about the job market in America.

Facts About the Job Market

According to one recent study, about a third of the higher-level jobs in America are tailor-made for the skills of a particular applicant. The same study showed that when a job is built for *you,* you can expect to receive a substantially higher salary than when you fill an opening that was originally filled by someone else. The best way to look for work is to start by assessing your skills, your interests, and what you know about where you want to live, the kinds of people you like to work with, and other information you can use to decide where you would be most satisfied to work. Once you have compiled a profile of an ideal employer, you can begin to look for that employer in exactly

EXHIBIT 13-7
Indirect Bad News Letter

(Heading and Inside Address Deleted)

Dear Mr. Bridges:

Buffer	Thank you for applying for the position of Engineering Intern. We are highly impressed with your training and experience.
Explanation of criteria	Several hundred student engineers applied for the two advertised openings, many of whom shared your fine record. As we screened applicants, we considered the specific requirements of our software development program and narrowed the field to a few students who had taken advanced work in PASCAL and machine programming, and who had previous experience with manual writing.
Implied rejection	We will keep your application on file in case we have an opening that more nearly fits your qualifications and experience.
Compliment, trust in reader	Thank you again for applying. It was a pleasure to read such an excellent résumé.

Sincerely,

Mary Baker
Internship Coordinator

the same way you would look for an agency to fund a proposal. In the job search, you are actually looking for someone who will fund your life in return for allowing you to do the work you love and do well.

If you ask any group of people how to find work, you usually get four answers: the newspaper, personnel offices, private and government employment agencies, and contacts. However, contacts actually account for about 75 percent of all jobs. Newspapers, personnel offices, and employment agencies account for the remainder. The reasons for this are complex. Employers do not want to bear the costs of wide advertising, and they do not want to be besieged by hundreds of applicants when a network of personal contacts can find a suitable applicant more quickly and less expensively.

You can draw several conclusions from these facts:

1. You should develop your own network of contacts who know about your skills and will refer you to others for interviews.
2. You should focus your efforts rather than send out hundreds of résumés, devoting as much effort to researching companies as to looking for lists of openings.
3. Unless you are graduating in a field where there is enormous demand for your skills, you should start with some market research—a review of your own skills. Even if you are one of the lucky graduates who can count on several offers, you should take time to review your skills so that you can select the offer that will make the best use of your talents.

Skills Assessment: Gathering Data for a Résumé

Before you begin to write your résumé, you should put together a collection of information about yourself. This process will also help you decide exactly what you want to do. Most of us have preferences of which we are not fully aware and significant job skills that we have not used for some time, skills we enjoyed using in the past that we would love to use again if we had the chance. Gathering and organizing personal information is also an excellent way to prepare for interviews, particularly for those tough odd questions some employers love to pop, such as, What are your second and third choices for an occupation? There are four basic ways to gather information about yourself:

1. Biographical exercises
2. Skills identification exercises
3. Skills priority questions
4. Interviews with employers in which you discuss possible careers and possible uses for your skills

Biographical Exercises

You can begin to collect information for your résumé by writing out complete lists of (1) your educational experiences, (2) your work history, and (3) your participation in organizations and activities. Be sure to include learning you have done on your own as well as learning for school credit. And include volunteer work with your paid jobs. Once you have written all this information down, select the most important or relevant information for your current job search by doing skills identification exercises.

Skills Identification Exercises

As you scan your lists of work and other experiences, you might notice that you have important skills that have lain fallow for some time. To identify the skills you enjoy using, try this skills identification exercise. On a fresh sheet of paper, write down the date for every other year of your life back to the age of ten, earlier if possible. Then, for each line, try to write down two things you did that you were proud of having done. This exercise can help you to pick out your long-term strengths and might help you choose between careers or job offers. As you come closer to writing a résumé and drafting letters of application, you will also profit by doing the ideal job exercise, which asks you to imagine your ideal employer and working conditions. Employers often ask precisely these questions in interviews.

Ideal Job Exercises

To imagine your ideal job, ask yourself the following questions and any similar questions that occur to you.

- Do I want to work for a large or small company?
- Do I want supervision, or would I prefer a job in which I have considerable freedom to set my own schedule and work load?
- How important is job status to me?
- How much money do I need or want?
- What combination of my job skills would I most prefer to use?

Each of these questions can be broken into many more specific questions. When you are finished pondering your answers, try to answer this key question: If I had to trade off among job freedom, money, status, and the chance to use the skills I most enjoy using, what compromises would I be willing to make? You might decide that you would be willing to sacrifice a high salary for a chance to use certain skills, or you might be willing to accept less freedom in your job responsibilities in return for a higher salary. If you can decide what your priorities are, you will interview effectively.

Interviewing to Learn More about Yourself and Employers

If you decide to learn more about the marketplace and your salability in it by interviewing people in the field, follow these guidelines. Start with a friend, teacher, or former employer and get the names of a few people to talk with. Call to make an appointment well in advance. Be absolutely honest about the fact that you are looking for work, but be equally frank that you do not expect your contacts to have positions open. Take no more than twenty minutes of anyone's time. At the end of each interview, ask for the names of two more people you can call, using the name of the person you have been interviewing. Always write a thank-you note.

Identifying Your Skills

Now that you have collected some information about yourself, you can put it to good use in several ways. First, you can identify all your skills, including skills you have that you have not fully used for some time. Second, you can consider how to present all of your skills. Employers want to know about far more than your specific job skills, such as programming, circuit analysis, or knowledge of statics. They also want to know about your ability to direct and motivate yourself. Being on time is a skill; organizing your own workday is a skill. And employers want to know about your communications and personal skills. Your résumé and interview responses should show evidence of all of these types of skills. Once you have the data in hand, you are ready to write a résumé and practice for interviews.

Résumés: The Basic Parts

A résumé has three basic sections:

1. Information about your work experience
2. Information about your training
3. Information about how you can be reached

In addition, some résumés include references, personal data, and information about your hobbies and special interests.

A résumé is an attempt to present yourself on a single sheet of paper (some run longer, but as a rule a graduate should aim for one page). No matter how well you write your résumé, it will be a mediocre substitute for your personal presence. A résumé is *not* a security check: Its purpose is to sell you, not to present every fact about your existence. Use that page to highlight aspects of your experience that demonstrate your ability to fill a *particular* position.

Because the best résumés are tailored to specific openings, it follows that you should prepare a fresh résumé for every job you really want. People who send out hundreds of "general purpose" résumés get extremely poor results, unless they have rare skills. As a job seeker, you should identify a few places where you would like to work and develop those leads with care. One way to write fresh résumés quickly is to keep either a card file or a computer file of résumé sections that can be rearranged as you need them for specific applications.

One final point is highly important. You want your résumé to be memorable. People remember stories and numerical data, so mention incidents or figures whenever you can. If you increased production on an assembly line by 15 percent, list the figures. If you handled a cash flow of $2000, mention that. If you were commended for your suggestions, make a note of that fact when you describe that job. To write a good résumé, in other words, you need to consider how your audience is likely to read résumés, which is the topic of the next section.

How Résumés Are Read: Organizing an Effective Résumé

The top half of your résumé is the "hot zone." Most readers will read that first half page. After that, they may only skim subtitles and underlined information. For this reason, after your name, address, and phone number, the facts about yourself most relevant to the position should fill this space. In some cases, you will want to put your education first; at other times, you might want to list first your work experience, military service, or other facts that demonstrate how well you fit the opening. In the next sections, you will learn what to say—and what not to say—about your experience, education, and personal life. Remember to limit your résumés to one page if possible. Additional pages are justifiable as you gain some experience, but you should never fall into the error of believing that résumés are weighed and studied. The audience will only sample what you have to say. (The one exception to this is the after-employment résumé, which some government employers use to determine your rank on a salary schedule.) Now for the parts of the résumé.

Objective

Many standard résumé guides suggest that you open your résumé with a job objective. You can write an objective if you have a specific rather than a general goal. Avoid objectives that suggest that you merely want to learn more and then move on to a better firm. A statement that you wish to use your engineering or accounting or managerial skills is so broad as to say nothing. If you do not have a highly focused objective at this time, leave the section out. You need the space for more important information.

Education

As a college graduate or a student nearing graduation, your college degree and college or university name are important. High school training can be left out, unless there is a good reason to mention it. Good reasons to include high school might include evidence of teamwork, leadership, outstanding academic or community achievements, or a geographical location close to the employer. For example, if you attended college in Georgia and high school in California, a California-based employer will be pleased to note that you have lived in California and probably have roots and friends there. If an employer mentions specific job skills in the advertisement, you can list course work in those areas, particularly if your training is more extensive than usual. In addition to these basics, you can include your grade point average (GPA), information about leadership activities and honors, and information about how you paid for college. Your GPA can be presented in several ways, as long as you clearly label what you present. If your first two years were years of exploration and growing—and lousy grades—you can present your GPA for your last two years. Or you can present an overall figure and a figure for courses in your major. Employers are interested in well-balanced people. If your GPA is low, be sure to include information about memberships, sports activities, and the percentage of college expenses that you have earned. That information will help to explain the GPA and advertise your personal skills. By the same token, if you have a 4.0 GPA, present yourself as a well-balanced person by listing some activities. If you have trained yourself in an important skill or learned skills on the job, list these under education, too. Finally, if you list honor societies, explain what they are. A string of Greek letters means little to most readers. These principles are demonstrated in the exhibits later in the chapter.

Work Experience

Your work experience includes both voluntary and paid work. You don't need to indicate which is which. After all, as a college student, your voluntary work as a lab aid might say more about your skills and training than your summer job as a surveyor's assistant or hash slinger. If possible, you want to show that you have been given *responsibility* and have been *promoted* or have progressed from lower- to higher-level jobs. Don't be concerned if some of your work is not directly relevant to the job you want. If you were a shift boss at a fast-food restaurant, for example, you have demonstrated the ability to handle large amounts of money, cope with pressure, supervise people of diverse backgrounds, solve problems quickly, and maintain quality in a product. Any employer would appreciate these skills: your task is to point them out when you list the job. List your job title first if you can show an increase in responsibility or an impressive title. List the employer's name first if the firm is well known. Be sure to use one of these two formats consistently, though.

Activities, Personal Data, and References

Activities, personal data, and references are less-important sections of your résumé; you might want to leave them out altogether. Unless you know that an employer knows your references personally, you should simply refer the reader to your university placement service. A list of unknown persons does little to advance your case. References are often embarrassed because they get calls about people they do not remember well. Whenever you list a name or ask for a recommendation, secure the reference's permission and provide him or her with a copy of your résumé and a letter about your current career plans.

Personal data, such as your marital status, birth date, weight, height, and so on are not relevant to the employer's decision-making process unless the job is one where there is a legally *bona fide* reason for needing the information. Air Force pilots, for example, must be within a specific range of heights to fit into and operate standard aircraft. At the worst, this information will lead to inadvertent illegal interview questions. The same remark applies to your activities. If a hobby is directly related to the job, as hiking is related to forestry work or ham radio expertise to electrical engineering, you can mention it. Similarly, if a hobby reinforces an impression you want to create—that you are a good team player, a leader, or a sociable person—go ahead and mention the activity. When there is no good reason to mention activities, don't. This section is typically placed at the end of a résumé sheet, where a nervous interviewer's eyes tend to rest. As a result, you can find yourself being asked endless questions about running marathons or chess when you need to be talking about job skills. Never list six or seven hobbies; you'll give the impression that you are only interested in play. And if your hobbies include skydiving, spelunking, deep-sea diving, and ski racing, the employer may begin to wonder how long you are going to live and how often you will be on crutches.

All of the résumé sections can be combined in several formats: the chronological, the functional, the combined, and the creative. In the next section, you will learn about these formats and study examples of three of them.

Résumé Formats

The three most common résumé formats are the chronological, the functional, and the combined. There is also a creative format, which is appropriately used in art and design fields. This section will cover only the first three formats. Before you begin to write your own résumé, find out what is popular and expected in your field. The chronological and combined formats are always safe; the functional might be judged strange in some fields.

The Chronological Format

The chronological format organizes each of the two main sections—education and work experience—in reverse chronological order. For example, your work section would list your most recent job first, then the job before that, and so on. In the most traditional chronological format, the education section precedes the experience section. Exhibit 13-8 is a typical chronological résumé.

Notice that Bob Sproul has provided a short description of what he did on each job using strong verbs to introduce each descriptive phrase. The progress from engineering intern to assistant engineer and the fact that he was rehired by Ohio Computer suggest that he is a good worker. His student and restaurant jobs also suggest that Bob is highly responsible. Notice that the dates of employment are buried in the job description. Many résumé guides place the dates along the left margin, but we believe that this merely attracts attention to any holes in your career. If you list jobs in the summers of 1984 and 1986, but not 1985, some interviewers will immediately ask something negative, such as, "Why weren't you working in the summer of '85?"

The Functional Format

The functional format focuses on skills and is written in prose paragraphs rather than in short entries. Functional résumés often leave out dates, employers' names, and even job titles. You choose what you want to present. This résumé format is appropriate when you have had a variety of jobs, none of which was particularly significant or impressive in itself. For example, all of your work might consist of short jobs or voluntary work, and yet through those jobs you may have learned new skills and demonstrated the ability to manage responsibilities. Exhibit 13-9 is a functional résumé written using the same personal data presented in Exhibit 13-8. Notice how this format allows the student to focus attention on his achievements.

The Combined Format

A combined format joins the virtues of the chronological and the functional formats. You can use functional headings, downplay dates, and list jobs out of chronological order to group them more effectively. The purpose of the combined résumé is to help the reader pick out significant patterns of achievement.

EXHIBIT 13-8

A Chronological Résumé

Résumé of Robert Sproul (centered title optional)

(centered, or right, or left) Robert Sproul (name either here or in title)
1618 Britten Pl.
Columbus, OH 43210
(614) 571-9364

EDUCATION

Bachelor of Science in Electrical Engineering (expected June 1986)
The Ohio State University, Columbus, Ohio
Research assistant to Professor William Conrad, Circuit Design Project
Advanced courses in computer circuit theory
Related course work: nonlinear analysis, technical communication
Languages: Pascal, Fortran, APL, BASIC
GPA 3.86 in major; 3.2 overall (4 years)
75 percent of college expenses earned
Vice President, Engineering Honorary

EMPLOYMENT

Assistant Engineer. Circuit Manufacture Division, Ohio Computer Corp., London, Ohio. June-September 1985. Redesigned timing circuits and reviewed quality control procedures.

Peer Advisor. Lincoln Dormitory Complex, The Ohio State University, September-May 1984, 1985, 1986. Counseled students, enforced dorm policies, checked students in and out at the beginning and end of each academic year.

Engineering Intern. Software Division, Ohio Computer Corp., London, Ohio. June-September 1984. Designed an interface program for three Ohio Computer software packages.

Shift Supervisor. Speedy Burger, Worthington, Ohio. June-September 1983. Supervised a staff of three cooks and five servers during the dinner shift. Responsible for a cash flow of $2000.

REFERENCES

References available on request from The Ohio State University Placement Office, 32 Dormer Hall, Columbus, Ohio 43210.

EXHIBIT 13-9
A Functional Résumé

Robert Sproul
1618 Britten Place
Columbus, Ohio

(614) 571-9364

EDUCATION

Bachelor of Science in Electrical Engineering, The Ohio State University, Columbus, Ohio. Advanced course work in computer circuit design, research assistant to Professor William Conrad in computer circuit design project. Related course work in nonlinear analysis, technical communication. Languages: Pascal, Fortran, APL, BASIC. GPA, 3.86 in major, 3.2 overall. I earned 75 percent of my college expenses.

ENGINEERING EXPERIENCE

For two summers I have worked for a small computer corporation, where I was promoted from engineering intern to a full-time assistant engineer in the circuit manufacturing division. I have designed timing circuitry, maintained a quality control program, and written computer interface software for three packages.

MANAGERIAL EXPERIENCE

As a computer engineer, I improved record-keeping procedures for a quality control program that tested 1000 units per week. As a paid student peer advisor, I managed a fifty-person wing of a dormitory, enforcing social and property policies, arranging tutoring, and advising students with problems. As a dinner shift supervisor at a popular fast-food restaurant, I directed the work of eight employees, substituted at all positions, called in alternates, hired new shift personnel, and was directly responsible for the evening accounts and cash inventory.

REFERENCES

Available on request from the Ohio State University Placement Office, 32 Dormer Hall, Columbus, Ohio 43210

In Exhibit 13-10, we will use the data on Robert Sproul. For the sake of variety, we will place Mr. Sproul's experience first this time.

Notice that this résumé format forces the reader to notice the continuity of Bob's engineering experience and allows the reader to see the two supervisory jobs side by side.

Summary: Evaluating Your Résumé

Remember that no résumé is perfect. Every résumé is an attempt to put a human being on a sheet of paper, and that can't be done. You will have done a good job of writing and organizing if you manage to convey the main facts about yourself in the top half of the first page and in the subtitles and underlined information. If you apply this test to Robert Sproul's last résumé (Exhibit 13-10), you'll see what we mean. The top half presents Bob's two engineering jobs and shows that he was promoted and given more responsibility. The remaining subtitles say that Bob has management experience as a peer advisor and shift supervisor and that his engineering education will culminate in a B.S.E.E. That much information will provide an interviewer with material for plenty of questions that Bob will enjoy answering.

Keep in mind how and when résumés are used: once in an initial screening and probably again during one or more interviews. Every underlined phrase might trigger a question. If you underline the word *hobbies*, then you may spend thirty important minutes discussing skeet shooting or running shoes. Finally, your résumé should be neat and written specifically for each job. All-purpose résumés are great for all-purpose jobs, but we've yet to see an advertisement for a jack-of-all-trades. With your résumé in hand, you are ready to write a letter of application.

Letters of Application

A letter of application is a guided tour but not a full repetition of your résumé. Your task in the letter is to point out a minimum of three ways in which your credentials match the employer's needs. Some employers, in fact, have their staffs scan letters of application to see whether they mention precisely the qualifications listed in the advertisement for the job. (If the ad says "software engineers," don't call yourself a "programmer/engineer"; use the language of the advertisement.) The letter of application has three parts: introduction, body, and close. Each part has a specific function. After we discuss these parts, we'll provide an example and close with some tips on writing variants on the standard format.

EXHIBIT 13-10
A Combined Résumé

Résumé of Robert Sproul
1618 Britten Place
Columbus, Ohio 43210
(614) 571-9364

ENGINEERING EXPERIENCE

Assistant Engineer. Circuit Manufacture Division, Ohio Computer Corp., London, Ohio, June-September, 1985. Redesigned timing circuits and reviewed quality control procedures.

Engineering Intern. Software Division, Ohio Computer Corp., London, Ohio. June-September 1984. Designed an interface program for three Ohio Computer software packages.

MANAGEMENT EXPERIENCE

Peer Advisor. Lincoln Dormitory Complex, The Ohio State University, September-May 1984, 1985, 1986. Counseled students, enforced dorm policies, checked students in and out at the beginning and end of each academic year.

Shift Supervisor. Speedy Burger, Worthington, Ohio. June-September 1983. Supervised a staff of three cooks and five servers during the dinner shift. Responsible for cash flow of $2000.

ENGINEERING EDUCATION

Bachelor of Science in Electrical Engineering (expected June 1986)
The Ohio State University, Columbus, Ohio
Research assistant to Professor William Conrad, Circuit Design Project
Advanced courses in computer circuit theory
Related course work: nonlinear analysis, technical communication
Languages: APL, Fortran, Pascal, BASIC
GPA 3.86 in major; 3.2 overall (4 years)
75 percent of college expenses earned
Vice President of Engineering Honorary

REFERENCES

References available on request from the Ohio State University Placement Office, 32 Dormer Hall, Columbus, Ohio 43210.

The Opening Paragraph

Your opening paragraph must do two jobs: (1) name the job you are applying for, and (2) mention where you heard about the job. Companies often have several positions open; announcing your objective immediately allows your letter to be routed to the correct office. Reporting where you heard about the job serves two purposes. Companies spend large amounts of money advertising openings; they appreciate information that will help them decide which means of advertising are most cost effective. You do the company a favor by naming the source. Second, you can drop a name if you have been referred by someone known to the company's management. Do not drop names if you have no reason to believe that they are known at the company. ("George Smith, the receptionist at my college placement office, suggested that I apply to General Engineering," won't cut any ice.) You may also state when you can start work or your date of graduation.

The Body: A Review of Relevant Credentials

In the main paragraph of your letter of application, present at least three credentials that precisely match the job specifications. As far as possible, use the language of the job description. Writing this paragraph requires that you carefully select the facts about yourself that best support the case that you are qualified for the job. Do not ramble, list qualifications that are not relevant to the position, or write more than one long paragraph unless your qualifications are extraordinary. You want the whole letter to be no more than one single-spaced page. Be selective.

The Close

Close your letter by asking for an interview or stating your availability and your interest in discussing the position and your qualifications further. Be sure to provide your telephone number and say when you are available at that number. If you have a separate message phone, give that number too. Do not refer the reader to your résumé for your phone number. Why make the personnel officer do extra work? Besides, résumés and letters often become separated. Avoid awkward thank-yous, such as "thanking you in advance for looking at my credentials." If you feel comfortable with being assertive, you can indicate that you know about the company, are strongly interested, and will call in a week or two to arrange an interview. If you are planning to be in the employer's vicinity in the near future, you can say that you will call ahead to arrange for a visit to the factory or offices. The sample letter in Exhibit 13-11 has three closes, one basic and two more assertive.

EXHIBIT 13-11
Robert Sproul's Letter of Application

1618 Britten Place
Columbus, Ohio 43210
January 31, 1986

George Williams
Director of Personnel
Micro-Comp, Inc.
P.O. Box 1109
San Carlos, California 94070

Dear Mr. Williams:

I am applying for the position of Assistant Software Engineer that was advertised in The New York Times, January 28, 1986. I will be available to start work after June 10, 1986, when I graduate with a B.S.E.E. from the Ohio State University.

As an engineering intern with Ohio Computer, I was given full responsibility for designing three software interface programs. While I am fluent in the use of several basic computer languages (APL, BASIC, Pascal, and Fortran), I have a special expertise in machine programming and have assisted in the development of new generations of circuitry and language development. At Ohio State, I assisted Professor William Conrad in his circuit modeling research and completed several advanced courses in circuit design and alternatives for problem solving. In my second summer at Ohio Computer, I was promoted to assistant engineer and was given responsibility for designing a new timing circuit and solving the associated software problems. Additional information about my work experience is included in the attached resume.

CLOSE 1 (Basic). I would like to learn more about the position; I am available for campus interviews. My telephone number is (614) 571-9364 (after 4 pm); messages can be left with the Ohio State Engineering Placement Office, (614) 584-3000.

CLOSE 2 (Assertive). I have read about Micro-Comp's product advances and am very interested in learning more about the position. I will call on February 10th to arrange an interview with your college representative. Should you wish to contact me for further information or references, my telephone number is (614) 571-9364 (after 4 pm EST); messages can be left with the Ohio State University Engineering Placement Office (614) 584-3000.

CLOSE 3 (Assertive/Visit). I am very interested in Micro-Comp's software line. In late February I will be visiting the San Francisco area. When I arrive, I will call to arrange a visit to your offices so that I can learn more about career opportunities with Micro-Comp. Should you wish to contact me for further information or references, I can be reached at (614) 571-9364 (after 4 pm, EST) or at my message number (614) 584-3000 (Ohio State University Placement Office).

Sincerely,

Robert Sproul

Enclosure: Résumé

Variations on the Standard Letter of Application

You can write a longer letter than the one in Exhibit 13-11 if you have good reason to do so. We'll mention three common strategies for longer application letters. If you have a strong desire to work for a specific company, you can state what you know about the firm, where you obtained your information, and how you think your training and experience will fit the firm's needs. This information must be conveyed with tact and precision, however. It won't do simply to say "I've heard that General Software is a wonderful company, and I think that someone with my computer training could fit in well." Let the company decide whether your qualifications are appropriate.

A more dangerous strategy is to offer to help solve a problem you know the company faces. Senior executives sometimes use this approach in changing jobs; it requires good research skills and strong persuasive abilities. This strategy can easily backfire because you are challenging the company's judgment. This method also works best if you plan a long campaign of letters, lunches, and interviews.

A third strategy is to pull out all the stops, use a clever opening sentence in your letter, present arguments, or detail your credentials. This kind of "guerilla warfare" letter should be a last resort to be used only when other methods have failed or when the competition is so great that routine applications will probably fail to make an impression. Your best strategy always is to apply for jobs for which you are fully qualified and demonstrate how well you match the job specifications. If your

EXHIBIT 13-12
An Assertive Interview Thank-you Note

Dear Mr. Dickerson:

Thank you for taking the time to speak with me about the computer programming position at Amcom Engineering. I was particularly pleased about your plans to expand your computer graphics department, because this is an area in which I have good training and a strong interest.

If you need any further information about my training or experience, please call me at 689-7779. I will call your office in two weeks to ask how the selection process is going. You are at the top of my list.

Sincerely,

Diane Montgomery

letter and résumé are successful, you will be invited to an interview. After that interview, you have at least one more writing task to perform.

Correspondence after an Interview

When you finish a job interview, your communication work has only begun. You should always write a thank-you note after an interview; later you will need to write either a letter of acceptance, a letter of refusal, a letter accepting a rejection, or a letter asking for more time to decide.

Thank-you Letters

As soon as your interview is over, sit down and draft a short thank-you note. This letter has three purposes:

1. To thank the interviewer for his or her time
2. To remind the interviewer of one of your skills or to remind him or her of the interview by mentioning something you learned about the company
3. To express continued interest in the position and leave an opening for further communication.

EXHIBIT 13-13
Letter of Acceptance

Dear Mr. Dickerson:

I am pleased to accept your offer of the position of assistant computer engineer at Amcom Engineering. In this letter I wish to confirm several details of the hiring conference I had with Mr. Byers of the Personnel Management Office on June 3. As I understand the offer, my initial salary will be $25,965 with full medical and dental benefits for myself and my family. I understand that I will have three weeks of vacation after my first year of service, personal life insurance, the Aetna annuity program, and, in view of my credits toward a Master's degree, a salary review at three months rather than six months.

I am eager to begin work on Amcom's new graphics programs and will report for company orientation on Monday, June 11. Thank you for your courtesy to me during the interview process.

Sincerely,

Diane Montgomery

If you are assertive, you can mention your phone number and even indicate that you will call or write in a specified period of time to learn how the selection process is going. You can also use this letter to add information to your file, but it is best to keep it short to avoid an impression of special pleading. Your task is to reinforce the good impression you've already made. Exhibit 13-12 is a brief but assertive thank-you note.

Letters of Acceptance

When you are offered a position, it is a good idea to send a letter specifying your understanding of the terms under which you have been hired. This letter is not always necessary; often it is sufficient to return a signed copy of the letter, offer, or contract sent by the company. If you have negotiated salary benefits, review schedule, or working conditions, however, it is in your best interest to write a straightforward, neutral letter stating the content of your understanding. Exhibit 13-13 provides an example.

The Letter of Refusal

A refusal letter is a bad news letter. When you refuse a company's offer, you want to maintain good will and leave the door open. Do not compare the company to others that were in the competition, and avoid words like *unfortunately.* Nothing hurts more or shuts off communication with more finality than a sentence like "Unfortunately, there were many companies that made far better offers than yours." Use a buffer and extend good will. Exhibit 13-14 is a sample of how you should turn down a job offer.

Notice that Ms. Montgomery does not need to say where she decided to work or what criteria affected her decision.

The Letter Accepting a Refusal

A letter accepting a turndown is not necessary, but if you wish to keep a channel of communication open, you can write a brief note accepting the company's decision. Mention something you liked about the company, your relevant skills, the interview, and if you wish your continuing interest. It is also acceptable to ask what winning qualifications the successful applicant had. Most people are not assertive enough to make this request comfortably, yet an honest answer to the question can help you in later applications and interviews. If for some reason that company needs to reopen the search or hire additional personnel, a tactful acceptance letter in your file can reinforce the good impression you have already made.

The Letter Asking for More Time

On some occasions you might need to ask for more time to make a decision. This typically occurs when you are waiting to compare another offer. It is best to handle this transaction by telephone; a letter will then confirm what was decided during your conversation. Be sure to emphasize your interest in the company and set a definite limit on the extension you are requesting. A week's grace is common: longer than a month is more than most employers are willing to wait.

A Last Word

As you change roles from applicant to employer or interviewer, you will need to write letters from the other side of this process. When that time comes, the same principles apply: Use buffers, tact, and good will. A full-scale business-writing text will include examples.

Tips for Survival

Looking for the right job is hard work. You should schedule your time and spend 20 to 40 hours each week interviewing, writing letters, and developing contacts. Studies show that many job seekers spend as little as five hours a week looking. No wonder it takes them so long to find work! Job searching is an

EXHIBIT 13-14
Letter Refusing a Job Offer

Dear Mr. Dickerson:

Thank you for offering me the position of assistant computer engineer. I was strongly impressed by Amcom's plans for an expanded computer graphics division. I have chosen, however, to accept another offer, which I hope will offer me as many opportunities to use my skills in computer graphics programming.

Thank you again for your consideration during the interviews.

Sincerely,

Diane Montgomery

ego-threatening process; every day you are being judged, classified, and sometimes rejected. Build up a network of friends and fellow job seekers who can talk with you about the occasional bad interview and boost you when you need a good word.

SUMMARY

1. Use conventional letter formats; readers are persuaded by familiar usages.
2. Address your readers directly.
3. Proofread all your work carefully; you are responsible, and in the readers' eyes, you are the company.
4. Use simple, conversational language; avoid stale formulas.
5. Use positive words; avoid words that will anger your reader, such as *blame* and *unfortunately*.
6. Use memos for short notes; a memo is not a report format.
7. Use the three-part format for all correspondence: introduction, explanation, and positive close.
8. Tailor your résumé for each important application.

9. Show how well you match the job description in your letter of application.
10. Write appropriate thank-you letters, acceptances, and refusal letters after your interview.

EXERCISES

1. Put the following dictated letter into correct format; use either full block or semiblock format.

   Mr. Coleman Blaylock, Merrieman Enterprizes, 241 Booker Street, Columbus, Maryland 18414. June 18, 1986. Dear Mr. Blaylock: We have processed your order of May 23 and are sending the 3-inch and 8-inch struts immediately. The 5-inch struts are temporarily out of stock, but a new shipment is expected on June 20. We will fill your order as soon as this shipment arrives. Thank you for your patience. Sincerely, Lorna N. Gridder, Production Department.

2. Consider a product or service related to your term paper project. Write a direct request letter asking for information.
3. Bring to class a recent advertising letter you have received—for book or record clubs, political organizations, or magazines, for example. Analyze the structure. Do the writers follow the basic persuasive request format? Can you see ways the letter could be improved?
4. Consider bad news letters you have received in the past. Which ones were effective? Which ones were not? Try rewriting one that struck you as ineffective, using the bad news format on page 313.
5. Write a résumé and letter of application for your ideal job using your current qualifications. Try going through the skills and needs identification exercises on page 317. Now locate a real job opening, either through newspaper want ads, your placement office, your professional association, or other contacts. Write a résumé and letter of application for this job.

Business Writing Case 1

THE HOME CARE PROJECT

For the past two years you've been involved in the Home Care Project at your local volunteer center. The idea has been to provide hot meals, errand service, and companionship for twenty-five to thirty elderly shut-ins in your area. Using a donated van, you and other volunteers travel to the homes of your elderly clients three times a week, cook a meal, and do anything that needs doing around the house. The program has been very successful, but it depends on outside funding to provide food, gasoline, and maintenance for the van. That's the source of your problem.

After three years your funding source, the Corbett Foundation has cut back your grant to two-thirds the former amount. Their reason was that increased demands on their resources required across-the-board reductions. The project steering committee has decided that, rather than dropping any of your clients, you'll cut back your visits to twice a week. That way no one will be completely cut off, although everyone will have to suffer reductions.

As chairman of the project you've been selected to inform everyone of the bad news.

ASSIGNMENTS

1. Write a direct bad news letter to the volunteers who have worked on the project. Remember, the project is not being canceled, it is being reduced in size.
2. Write an indirect bad news letter to one of your clients, Mrs. Hannah Wocjeck. You need to tell her that you'll be visiting her two times a week rather than three.

Business Writing Case 2

THE ULTRASONIC TEST

As a mechanical engineer your job with Amax Components has involved many products, but your main interest has always been with the steel pipe Amax produces for the construction industry. Amax has been in the business for over twenty years and deserves its good reputation. Still, you've been concerned with Amax's quality control procedures. Up to now the idea has been to select pipes randomly from each production lot and test them under stress conditions. The tests reduce the pipes to scrap, which is a waste of material, but, more importantly, they can be done only on selected samples. There is always a possibility that a defective pipe may slip through undetected.

Your solution has been to propose the use of ultrasonic testing equipment. This equipment would be installed directly into the production line. Each steel pipe would be examined by a search unit transmitting an ultrasonic beam. If the beam is intercepted by a defect in the elastic continuum of the material, it will be deflected back to the search unit and will show up on a CRT display. No pipes would be destroyed in the process and, since each pipe would be inspected, any defective ones would be detected before leaving the factory.

Your proposal was submitted to the vice president for production six weeks ago and you've just gotten the news: It's been accepted. The process will be phased in within the next six months. You have permission to investigate ultrasonic testing units and select the one you think is most effective.

ASSIGNMENTS

1. Write a memo to the support staff of the pipe production unit. Explain the vice president's decision and what will be done.
2. Write a direct request letter to LaForrest Electronics of LaForrest, Illinois. You are interested in their Bar/Tube test station. You need to know whether it is adaptable to 6-inch pipe and whether it can be installed on a conveyer system. You are also interested in cost, speed, and ease of operation.
3. Write a persuasive letter to Court Manning of Manning Construction Company, Lavesque, Louisiana, one of your biggest customers. You want to inform him of your new process and convince him that the quality of your product will be even better.

Business Writing Case 3

THE WRITING JOB

As a senior English major at Corona State University, you've long been aware of the employment problems you face. You don't want to teach, but you also don't want to switch majors. You like writing more than any other subject. You've decided your solution is to become a technical writer. You know the jobs exist in your area and you think you'd enjoy the work.

One day the director of the placement office sends you the following notice:

> *Position Available*
>
> Technical writer. To write, edit, and assist with composition of manuals and flysheets for electronic instruments. Writing and editing experience a must. Some computer background desirable. Send résumé and letter of application to Ms. Joanne Lorenzo, Communications Group, Digicom Corporation, 19152 Sierra Drive, City.

You know Digicom; they're a local company in the process of expanding. This looks like a good opportunity for you.

Your school has no technical writing major, but your advisor has suggested some helpful courses over the past three years. You've taken both basic and advanced technical writing, business communications, and advanced composition. You've also taken some journalism courses: layout and design, newswriting, and graphics. Your technical course work is a little more spotty. You've had fifteen hours of computer science, including programming in BASIC. You've also had a year of chemistry and a semester of algebra. Last year you took basic statistics, but you didn't do very well: You got a C+. It's really not your field.

You do have some good work experience. You've worked on the school newspaper for three years; this year you're assistant editor. You've had a lot of experience there with pasteup and typography since you were responsible for getting the final version of the paper in shape for the printer and seeing it through production. Last summer you had an internship at your local shopper's newspaper. You wrote copy for advertisements and edited the material turned in by your stringers. The editor felt you did very well; he gave you credit for increasing want ad revenue through a new system of telephone ad placement that you designed. You also worked for a year when you were a sophomore as a shift manager at your local Bonus Burger. Your shift was six o'clock to midnight and you and eight other employees handled around 300 customers a night. Your boss there wanted you to stay on, but you finally quit because the work interfered with your studies.

You have an overall GPA of 3.64 (drat that statistics course), with 3.75 in your major. You'll be graduating in June.

ASSIGNMENTS

1. Write a résumé for the Digicom job. Choose one of the three standard formats—chronological, functional, or com-

bined—that you think will show off your qualifications to the best advantage.

2. Write the letter of application to accompany your résumé for the Digicom job. Remember, Digicom is located in your city, so you can come in for an interview easily.
3. Congratulations, you've been offered the job. Write either a letter of acceptance, a letter of refusal, or a request for more time.

CHAPTER • 14

Oral Presentations

Introduction

Oral presentations are a major part of every job. They range from informal one-on-one explanations to a manager to formal research and marketing presentations before large groups. This chapter will take you step by step through the preparation and delivery of a standard presentation. We will discuss analyzing your audience, defining your purpose, organizing your materials, preparing visuals, and delivering your presentation. We will also follow the preparation of a typical presentation to demonstrate how this procedure applies to an actual situation. You will probably notice that the steps in preparing an oral presentation are similar to those in preparing a written presentation; these similarities are not accidental. Once you become accustomed to the standard procedures in organizing and presenting technical material, you will find that they are applicable to most communication tasks.

Why Presentations?

Harold P. Erickson has estimated that professionals spend up to 30 percent of their time on oral presentations (in *The Teaching of Technical Writing,* Cunningham and Estrin, eds., p. 156). These involve not only formal presentations at conferences and meetings but also presentations within organizations, such as those a manager gives to his staff, presentations to other companies, such as those given to potential customers, and informal presentations to anyone who wants more information about a project or product.

Presentations have many advantages over written communications. You can get immediate audience reaction and make sure that the audience understands and accepts your ideas. You can also discover problems in the audience's response and develop solutions. Presentations offer you more control over your audience than written reports; you decide how to order your material, unlike a report where readers may begin anywhere they like. You also determine pace and emphasis. At the same time presentations are often easier for the audience to understand. Oral communications are frequently simpler and more direct than written ones. Oral presentations are also more personal. Finally, it is simply easier to listen than to read.

There are disadvantages of course. Oral presentations are unsatisfactory when the information will be needed for reference. Even handouts accompanying a presentation will not have the reference value of a written report. Moreover, there is a definite limit on audience size for an effective presentation: The bigger the audience, the less likely it is to follow the speaker. Finally, one of the advantages for you is a disadvantage for the audience: They are "captives." The audience has no control over the presentation, which may severely affect their comprehension.

On balance, however, the advantages of presentations outweigh their disadvantages. Since you will be preparing and delivering presentations in almost any job, we will follow a typical presentation step by step to see how it is put together. First of all, you will analyze your audience.

ANALYZING YOUR AUDIENCE

Your first step in preparing your presentation will be to consider the people to whom you will be speaking. As with written reports, oral presentations must be audience centered. You must know as much as possible about your listeners in order to satisfy their needs. Consider four things about your audience: their knowledge, their interests, their attitudes, and their ability to act.

Knowledge

First you must consider how much your audience already knows about your subject and how much they need to know. Their knowledge will affect the amount of detail you present.

Interest

Your next step is to decide what aspect of your subject will interest your audience the most. This might be something involved with their day-to-day concerns or something reflecting a special situation, but it will be the primary interest of your audience. For example, management personnel might be chiefly interested in cost, while research personnel might be more concerned with theory. This aspect of your subject may not, of course, be the one that you yourself find most interesting, but it is the aspect you will want to emphasize since you want the audience to be interested in your presentation.

Attitudes

You must also consider the attitudes your audience might have about your subject, particularly with persuasive presentations. Consider group attitudes (does this organization, as a whole, have any reason to react strongly to your project?), as well as individual backgrounds. Also consider possible reactions toward you personally. Remember that in oral presentations, you become part of the material. In general, use the same care in assessing your audience's preconceptions and biases as you would in a written report.

Ability

Finally, you must consider your audience's ability to act. If you are suggesting a program or an idea, are these people the ones who can implement it? If you are giving information, can these people use it? If not, consider whether this audience is the right one for you to address.

Let's look at how one persuasive presentation was put together. Our speaker is a biomedical engineer at a large hospital. Each month his technicians made up to twenty-five current leakage checks on over 1500 pieces of equipment to make sure that there were no discharges ("leaks") of electric current. Even a slight discharge of current could result in serious injury or even death for some patients. The technicians used equipment "rosters," computerized forms that were printed by an outside firm under contract to the hospital. The rosters included the equipment type, serial number, and location. The results from these checks were handwritten into four different files. Three technicians worked together: one to run the equipment checks and the other two to deal with the paperwork.

This system was inaccurate and time-consuming; some equipment was missed each month, and some records were illegible. The engineer had researched a system that would use a computer both to make and to record these tests. It required a microcomputer—the Signus III—an analog to digital and digital to analog interface card, and a special meter for making the equipment checks. It also required a special program, which could be written by the hospital's programming division. Considering the potential saving in time and money the new system would represent, the engineer thought it was worth the investment.

The engineer decided to make a presentation to the hospital equipment board to try to convince them to implement this system. This was a five-member committee made up of the chief biomedical engineer, the chief of staff, the chief of surgery, the assistant financial director, and the assistant personnel director. The engineer's first step was to analyze this group.

The board's knowledge was mixed. Concerning the present situation, the doctors and the chief engineer all knew the need for leakage checks, but the administrators probably did not. The chief engineer was probably the only one familiar with the present procedures. Concerning the proposed procedures, the chief engineer and the administrators had some knowledge of the basics of computer systems, but the doctors might not. The chief engineer was probably the only one who might be familiar with the specific capabilities of the proposed system. Probably no one was aware of the specific costs involved.

As for interest, the chief engineer was concerned with a more efficient use of his staff, since the present situation required an excessive amount of time spent on paperwork. He was also interested in a more accurate system to avoid mistakes. The doctors were also interested in a more accurate system in order to reduce the possibility of equipment failures that could endanger patients. The administrators were primarily concerned with the costs of the new system compared with the old one, and the costs of implementation.

Attitudes did not seem to be a special problem. The engineer saw no reason for group hostility. The greatest change was for the chief engineer, but he was already dissatisfied with the present system. Moreover, since he had nothing to do with the design of the present system, he had no reason to be loyal to it. The administrators might be prejudiced against any new expense so the engineer would need to justify costs clearly. The doctors appeared to have no special prejudices to worry about. The only attitude toward the engineer personally was a possible suspicion of inexperience since he would be the youngest person in the room. The engineer therefore decided that he should appear confident yet respectful.

The easiest part of the engineer's analysis was the ability to act: The board made recommendations about equipment purchases to the hospital's board of directors. Any new procedure regarding equipment maintenance needed their approval, so they definitely had the ability to implement the engineer's proposal. The board of directors would, of course, make the final decision, but the chances were that on a relatively small project, such as this one, the board of directors would accept the equipment board's recommendation.

Having analyzed his audience, the engineer could use the information he had deduced to help him choose relevant material, organize his presentation, and decide on the best approach to the audience. His next step would be to define the purpose of his presentation.

DEFINING YOUR PURPOSE

After analyzing your audience, the next step in both written and oral communication is to consider exactly what you want that communication to achieve. What do you want the audience to do after they hear your presentation? The presentation itself is not your purpose; the presentation is always a *means* to some end. Your real purpose is always involved with the audience reaction you hope to gain.

Knowing your purpose can help you in several ways. It can help you decide how formal you want to be: As a general rule, the more important the presentation, the higher the degree of formality. Knowing your purpose can also help you decide on your scope—how much information you will cover—and your theme—the ideas you want to stress.

There are two basic purposes for oral presentations: to inform and to persuade. Informative presentations include professional conference papers, descriptions of new products and procedures, even classroom lectures. Yet even in informative presentations, the speaker needs to persuade the audience of the worth of the information. Persuasive presentations are those seeking approval or support for new ideas, projects, or programs. These include marketing presentations to potential customers. Yet the persuasive presentation usually also includes information. The line between these two types is not clear-cut.

In our hospital illustration, the engineer's purpose was to persuade. Although he wanted to inform the board of problems with the present system and the advantages of his system, he ultimately wanted to persuade them to implement the system he proposed. Thus when he began to organize his material, he emphasized the facts that were most likely to persuade the board to accept his idea.

Preliminary Considerations

Before you organize your material further, you should consider some outside factors that might influence your presentation. Timing is critical; if you have any control over the date of your presentation, give it serious consideration. Give yourself enough time to prepare adequately, and make sure that you avoid conflicts with other projects. If your audience has just funded a major project, now may not be the best time to ask them to fund something else. Consider whether your presentation would work best as one session or several. If you have a large or extremely mixed audience (some with a great deal of background and some with very little, for example), you might consider a number of presentations to different groups.

You should also do some networking before making your presentation. If you can, discuss your ideas with others informally; try to discover possible support and possible problem areas before you proceed. Pay particular attention to opposition; it is much easier to work with individuals than with groups. Perhaps you can defuse some objections in informal discussions or include answers to them in your presentation.

You should always include negative material, just as you would in a written report. This shows that you have considered possible objections, but you can then demonstrate why they do not change your conclusions. You can also be ready to deal with these objections in question-and-answer sessions, which we will discuss later in this chapter.

In the case of the hospital proposal, the engineer decided to discuss his ideas with his boss, the chief biomedical engineer, whom the engineer knew fairly well. He was thoroughly familiar with the system in use, and the engineer could determine during the discussions whether his chief had any objections to the proposed changes. If the chief engineer supported the new system, the engineer could feel more confident about his chances of convincing others.

Organization

Having analyzed your audience and decided on your purpose, your next step will be to organize your material. Keep in mind your audience's needs and your purpose, what you want the audience to do. You should also be aware of any time limits. Most presentations are between twenty and sixty minutes long; in some cases, such as presentations at professional conferences, time limits must be strictly followed. Make sure you time yourself during your rehearsals so that you will know approximately how much material you have.

You can divide your material into things the audience must know, things that are useful for them to know, and things they might like to know. Omit anything that does not fit into these categories. You can begin with the "must knows" and move through the "usefuls" to the "likes."

The Beginning

The audience must be aware of several things at the beginning of any presentation:

- Your name and title, assuming that they do not already know
- Your subject and your purpose, why they should listen to you
- Your scope, what you will and will not cover
- Your criteria, what you will base your judgment on
- Your goal, what you want them to do at the end of your presentation

It is also a good idea to get your audience's attention at the beginning. You can give them a surprising fact or troubling statistic, an example, even a joke if it is appropriate and relevant. And you should provide any background material necessary to orient your audience about your subject. You should know how much information is necessary after your audience analysis.

The Middle

Your statement of subject and purpose in the beginning should lead logically to your discussion. Basically, you present your major points along with the necessary subpoints and supporting data: You begin with the "what" and move to the "why."

Like everything else the order of your supporting points is determined by the needs of your audience and your purpose. The audience's most important need usually goes first. If you are dealing with controversial material, you should begin with your least controversial point and move on to the more controversial ones after you have established your credibility. If your subject is complex, begin with the basics and move to the more complicated material after you are sure your audience understands.

You can use any of the standard orders for information and persuasion. For an informative report you might use chronology, spatial order (e.g., left to right, east to west, etc.), causal analysis, or classification. For persuasive reports you could use any of the standard orders of reasoning: induction, deduction, or elimination (presenting all the possible alternatives and disposing of all but the one you favor).

The End

The end of your presentation is your last chance to impress your ideas on your audience. A summary of your major points is usually a good beginning. With an informative presentation you can simply restate your main ideas. With a persuasive presentation, you can summarize your argument. If you are making a recommendation, this too will be included in your conclusion.

Your conclusion should also have a "hook," a final wrap-up point that will get your idea across. This could be the next step you would like to see taken to achieve your purpose. It could be the consequences you foresee if action is *not* taken. Or it could simply be a final comment that wraps up your presentation and says "That's it; now you can applaud," something like "The next move is up to us: Let's get going."

Transitions

You should be careful to tie your presentation together with transitional material as you go. Indicate when you have finished one topic and are moving to the next (e.g., "Now we've seen the design of the new frimbus, let's consider its advantages over the old whatzit."). With a long or complex presentation, you can use what Thomas Leach (in *How to Prepare, Stage, and Deliver Winning Presentations*) calls "minisummaries" along the way to restate your key points before you introduce more material (e.g., "Now we've seen the design of the new fribus. It contains three dinguses, an advanced fribble, and an interconnecting thrip. Now let's look at the advantages of this design over the old whatzit.").

A Sample Organization

Now we can consider the organization of the hospital proposal. The engineer decided to begin with a couple of surprising statistics. In the present system over 50 percent of the technicians' time was spent on paperwork; moreover, each month between five and ten pieces of equipment "disappeared" and were not checked because of faulty rosters. The first fact was aimed at the chief engineer and the administrators since it represented an inefficient use of personnel; the second fact was aimed at everyone since it implied the possibility of faulty equipment.

The engineer decided to follow these facts with a clear statement of his purpose: to propose a system that would be more efficient, more productive, and, ultimately, more economical. Then he planned to explain his criteria in a forecasting statement that would cite his major points. Finally, he would express his goal, his hope that the board would recommend implementing his system.

For the middle section of his presentation the engineer had two areas to cover: the need for the system (which meant describing the old system and its problems) and the practicality of the system. He considered the primary interests of his audience: efficiency (the chief engineer), cost (the administrators), and equipment maintenance (the doctors). He decided to begin with the present system, demonstrating its shortcomings in these areas.

The engineer wanted to begin with equipment maintenance because, although it was a primary concern for the doctors, it was at least a secondary concern for everyone. He started by pointing out that the current equipment rosters giving equipment descriptions and locations were supplied by the outside firm each month. These rosters were frequently late; moreover, because changes were very difficult to make (they involved contacting the company and adding information to a computer program), the rosters were often inaccurate. They did not always list new equipment and they frequently included equipment no longer in use. Also, they frequently listed inaccurate locations for equipment since equipment was missed every month no matter how conscientious the technicians were (a point the engineer wanted to stress so that the chief engineer would not feel his staff was being attacked).

Efficiency was a primary concern for the chief engineer, and a secondary concern for the administrators since inefficient use of personnel meant increased costs. The engineer decided to begin by pointing out the number of forms the technicians had to fill out currently and the amount of time it took. He also stressed that three technicians had to work together to perform the tests effectively.

Cost would be the engineer's most speculative but most dramatic point, so he decided to save it for last. He could begin by reiterating that the excess labor required by the current system was costly and then mention the technicians' hourly wage. Then he would point out that the equipment maintained by the system was worth over $500,000 and would be even more costly to replace. Finally, he would point out that many lawsuits involved charges of faulty equipment. Currently the inaccurate rosters and sometimes illegible written records could be a problem if the hospital were trying to prove its competence.

Having outlined the problems, the engineer proceeded to his solution: the microcomputer system. First he would outline its components: the computer, the leakage meter, the analog to digital and digital to analog interface card (to connect the computer and the meter), and the computer program. Then he would return to each of the audience concerns in order.

He would next explain that rather than relying on the rosters, his system would store all data on floppy disks. The data could then be continually updated by the technicians themselves: new equipment added, outdated equipment deleted, and locations changed. When a technician wanted to check the equipment in a particular section, he could ask for a display of that section's equipment on the computer screen. At the end of each month, the computer would print a list of any equipment not checked and the technicians could locate it. The result would be fewer instruments missed and a constantly updated record of equipment maintenance.

The engineer would point out that with the new system the technician could hook up the equipment to be tested to the leakage meter; the computer would make the test and print the results on its screen. The technician could type in the identification number of the equipment and write the results once, on the equipment tag. The computer would print all the other entries. Not only could the computer save the technician paperwork, it would require only one technician for operation rather than three.

Cost would be the trickiest fact to deal with because the new system would cost more initially (a consideration for the administrators). The engineer needed to give the costs of each part of his equipment (computer—$2190; A-D/D-A card—$269; leakage meter—$1015; program—$1500), but he could point out that there would be a saving in the increased efficiency of the technicians. He could also explain that the program could be done economically by the hospital programmers, and that this would enable the engineers and technicians themselves to have more direct input into the program's design.

For his conclusion the engineer would begin by briefly summarizing his points: The present system was inaccurate, inefficient, and potentially costly; the new system would improve accuracy and efficiency and prevent expensive errors. Then he would suggest that the board recommend his system to the board of directors. Finally, he would give his hook: $4974 might seem to be a large investment, but if the new system saved one piece of equipment or prevented one lawsuit, it would have paid for itself many times over. (To see how this presentation actually works out, consult the model script in Exhibit 14-1 at the end of this chapter.)

FIGURE 14-1
A sample script

AUDIO	VISUAL
1. Purpose of this presentation: • Introduce new way to cut milching time in half. • The new frimbus.	1. Introducing The New Frimbus!
2. We'll cover • Design of frimbus • Advantages over whatzit • How it fits into production plans	2. Agenda • Design • Advantages • Your Plans
3. Frimbus features . . .	3. Diagram Frimbus Assembly

VISUALS

Having organized your material into a rough outline, your next step is to turn that outline into an effective presentation. There is a basic difference between a presentation and a speech. A speech appeals only to the sense of hearing; it depends on the dynamism of the speaker to make an impression on the audience. A presentation uses both sight and sound; it holds the audience's attention by the use of visuals.

Air Force studies have shown that a combination of words and visuals increases listener retention six times more than words alone and three times more than pictures alone (in M. B. Broner, "Stand Up and Be Heard," *A Guide for Better Technical Presentations,* R. M. Woefle, ed., p. 36). But visuals must be well designed, and they must complement the spoken portion of the presentation. Many professionals suggest composing a "script" or "storyboards," laying out the words and visuals side by side. With a **script** each page is divided lengthwise into two columns, one side for the words and the other for the accompanying pictures. Each column is then divided into squares or "frames," one for each visual and its accompanying audio (see Figure 14-1). Think of each frame as a paragraph, a complete unit. It is easiest to work with short audio sections, designing visuals to accompany each one. **Storyboards** use a similar format, but in this case you use notecards rather than a complete page. Each card has one visual and key words to suggest the accompanying section of the audio (see Figure 14-2).

TABLE 14-1 Some Visual Advantages and Disadvantages

Medium	*Cost*	*Advance preparation*	*Formality*	*Audience size*	*Audience interaction**	*Comments*
Chalkboard	0	None	Informal	Under 35	Poor	Requires preparation during talk. Time-consuming and little audience contact.
Flip charts and posters	Moderate to high	Requires time	Informal to semiformal	Under 20	Good	Artwork sometimes expensive. Not highly portable. Small audiences.
Models	High	Requires much time	Formal	Under 20	Good	Very expensive. Effective with small, influential audience.
Transparencies	Low to moderate	Can be done quickly	Semiformal	Under 50	Good	Inexpensive; fast to prepare. Does not require darkened room.
Slides	Moderate to high	Requires time	Formal	Over 100	Fair to good (depending on darkened room)	Can use color. May require darkened room.
Videotapes	Moderate to high	Requires time	Semiformal	Under 20	Fair	Less expensive than film for small groups. Allows more audience interaction than film.
Films	High	Requires much time	Formal	Over 100	Poor	High interest, but expensive and inflexible.

*Indicates the amount of contact the visual provides between speaker and audience (e.g., does the visual allow the speaker to face the audience, must the room be darkened?).

Figure 14-2
Sample storyboards

TITLE Introducing the New Frimbus	AGENDA • Design • Advantages • Your plans	(Design Explanation) Frimbus Assembly

Types of Visuals

Your visuals can involve anything from words to tables, graphs, and diagrams. There are two main categories of visuals, however: nonprojected and projected. Nonprojected visuals include chalkboards, flip charts, models, demonstrations, and handouts. Projected visuals include overhead transparencies, slides, videotapes, and films. Nonprojected visuals are suitable only for small groups and less formal situations; projected visuals are best for larger groups and formal presentations, but generally are more expensive and time-consuming to produce. Some advantages and disadvantages of each type are shown in Table 14-1.

The same general rules apply to all visuals:

- Keep visuals simple; include only one point per visual. Use no more than five words in a line and no more than eight lines in a visual.
- Avoid unnecessary numbers or words; you can provide details in your explanations.
- Make your visuals readable: use the largest type with plenty of white space.
- Do not simply take the graphics from your written report; simplify them so that they can be easily read and understood.
- Put your information at the top of the visual and orient the material horizontally rather than vertically so that the people in the back of your audience can see clearly.
- Use a consistent type style, art style, and scale for all your visuals.
- Use only high-quality visuals; the care you take in their design will convey its own message.

Most companies either have art departments or hire freelance artists to help produce visuals for written and oral presentations; you can use their advice for design. Remember: the audience should get the point of your visual within fifteen to twenty seconds; the more complex the visual, the more your audience will be distracted from you. (For more information about visual design, consult Chapter 8.)

To begin preparing the visuals for the hospital presentation, the engineer broke his outline down into segments he wanted to illustrate. He tried to limit himself to material that could be covered in two- to three-minute intervals. And he tried to get a variety of visuals.

For the beginning of his presentation, the engineer decided to use two visuals consisting of words only: one summarizing his surprising statistics and another giving the main points he intended to cover. (This is known as an "agenda slide.")

For the middle of his presentation the engineer decided to try a mixture of words and examples. Pictures included examples of current rosters to show the number of missing instruments and the illegibility of the notations, one flow chart to show the current procedure and another to show the new system, an example of the new computer listing of the equipment to show legibility, and a table of costs for the new system. Word visuals included a list of current potential costs (e.g., the technicians' hourly wage, equipment replacement, and lawsuits), the components of the new system, and a summary slide for each subpoint.

The engineer's ending used largely word visuals. He decided to have two summary slides: one giving the old system's drawbacks and another giving the new system's advantages. He also decided to design a final slide to dramatize the comparison of the cost of the new system and the high potential cost of the old one.

The engineer considered the potential range of visuals. He rejected nonprojected visuals as too casual for his purposes. Of the projected visuals, videotapes and films were clearly inappropriate; slides would work, but the artwork involved would be expensive and time-consuming. For these reasons he decided to use overhead transparencies that he could design and produce himself.

DELIVERY

Some people feel that they must become newscasters or political speakers when delivering a presentation. Actually your object should be to come across as the best possible version of yourself. Your language should be simple but appropriate, like the language you would use in conversation with the people to whom you are speaking. You should consider your audience's needs; avoid jargon, define terms they might find difficult, get straight to your point, and provide enough detail to make your ideas clear. Remember also to moderate your pace: fast enough to keep your audience's attention, but not so fast that they have trouble absorbing your information.

Voice and Appearance

Try to speak loudly and clearly enough that the audience can hear you in all parts of the room. Also try to be animated when you speak; a monotone will only put your audience to sleep. Avoid verbal distractions like repeated words or phrases (*ok, you know, uh, I mean,* etc.) or physical distractions like rattling keys or playing with eyeglasses.

Make eye contact with your audience throughout your presentation. Avoid staring at your visuals or your notes; try to look at several members of the audience in different parts of the room.

Try to stand reasonably straight (not "at attention," however) with your shoulders back and your head up; a good stance will help you project your voice. Hands seem to be a problem for some speakers; you can simply hold them at your sides, but don't be afraid to use your hands to emphasize your points.

Nervousness

Every speaker is occasionally nervous, even those who have delivered many presentations. Nervousness is a particular problem for beginning speakers, however. First of all, remember that generally the audience wants you to succeed; what seems like a major mistake to you will probably seem minor to them. Moreover, they have come to hear you; they know that you know more about the subject than they do.

The peak of nervousness usually occurs three to five minutes into the presentation; thereafter you should start to feel better. If you can, direct your nervous energy into the presentation itself: raise your voice, make emphatic gestures. If worse comes to worst and you either draw a blank or encounter other difficulties (missing part of your notes, for example), just try to keep going. Chances are the audience will not notice the problem.

Remember that the better prepared you are, the easier the presentation will be. Go over your presentation several times; deliver it to friends or colleagues and get their reactions. Time your delivery and make sure it is within your time limit. The more familiar you are with your material, the easier it will be to deliver. Finally, if you can, try to look pleased to be there; your enthusiasm will win over your audience.

Notes vs. Manuscript vs. Visuals

The ideal delivery situation is to use your visuals as the basis of your presentation. Let the visual provide you with cues; get the point across and use transitions to get from frame to frame. If you feel uncomfortable with this approach, however, it is best to work from notes or from an outline, anything that will keep you in contact with your audience.

Very few situations justify reading from a manuscript. Occasionally papers are read at professional meetings, but it is far more common to deliver material extempore, without text. Reading is by far the least interesting delivery; it easily becomes a monotone and provides little eye contact. Try to deliver your presentation without reading.

Using Visuals

The major point about using your visuals is to make sure that you are not blocking the view of someone in the room. Do not stand in front of the screen, chart, or board. You can usually stand beside an overhead projector and get your cues from the transparency itself, without watching the screen. With slides, you can stand next to the screen with the slide control in one hand and a pointer in the other. Remember always to face your audience; do not deliver your presentation looking at the screen and showing the audience your back.

Feedback

The best speakers are continually aware of the audience. While they are speaking they are also listening and watching. Adjust your pace to people's reactions to you: If they seem restless you can speed up; if perplexed, slow down. You should also vary your pitch and volume occasionally, to increase interest. Most of all, be aware of how your audience is behaving and be ready to react to it. Skip a section if people are yawning; add more details if they are fascinated. Do not be afraid to ask questions of your audience before giving information (e.g., "How many of you are familiar with how a frimbus works?"). If you stay in contact with your audience, you have a better chance of reaching and impressing them.

Final Considerations

Always try to visit the room where you will be speaking before the time of your presentation. Check the physical layout. Is the temperature comfortable? Will everyone be able to see your visuals? Are there distractions (traffic noises, for example)? Make sure that your equipment is there and functioning and that your visuals are in order and right side up. It is even a good idea to have a spare bulb on hand for the projector or to know where to find one in a hurry.

The Question-and-Answer Session

Questions provide the best direct contact between presenter and audience. They can arise during or after a presentation. Questions during a presentation can take care of problems immediately, but they can also break up a presentation and make it hard to maintain continuity. One solution is to give the audience paper and pencils and ask them to keep track of questions for the session after your presentation. (Always announce that you will have a question period.)

As you prepare your presentation, try to analyze it from the audience's point of view. Where might people ask for more information? What problems or objections are likely to arise? Try to be ready with answers for the most obvious questions.

Ask for questions. If none arise immediately, try asking one yourself (e.g., "One question that is frequently asked about this is . . ."). Keep the following guidelines in mind:

- Take your time. Give people time to recall their questions. Do not cut questioners off; let them finish the question even if you already know what it will be. If you think others in the room might not have heard the question, or if you want to make sure that you understand it, try repeating it before you answer.
- Be friendly. Treat all questions as legitimate, even if you suspect that some are not. Do not make a joke about a question (at the questioner's expense) or try to duck it by saying it was answered in your presentation—obviously it was not for this questioner.
- Be frank. If you do not understand a question, ask for an explanation. If you do not know the answer, say so, but offer to find out.
- Try to involve everyone. Take questions from all segments of the audience, not just the important people. Try to keep one person from monopolizing the session. (Suggest that you could give him or her more information after the session is over.)

With the hospital presentation, the engineer began by trying to pinpoint problem areas so that he could take care of them within the presentation. He needed to define *leakage*, for example. Then he considered areas that might inspire questions. The administrators might want to ask about possible alternatives, for example, such as finding another supplier for the rosters. There might also be questions about the particular pieces of equipment like the Signus III. Finally, there might be questions about the seriousness of the problem, whether equipment had, in fact, failed. The board might also like to know what other hospitals did about this problem. The engineer needed to research all of these questions, although the material might not go into the report itself. Knowing the answers helped him feel more confident about the presentation as a whole. Exhibit 14-1 is the final script for the engineer's presentation.

EXHIBIT 14-1

Sample Script for the Hospital Presentation

AUDIO	**VISUALS**
Subject: Leakage Check System Present System —50% of technicians' time on paperwork. —Time needed for equipment servicing —5 to 7 pieces of equipment missed monthly because of faulty rosters. Leakage checks —Needed to ensure that no equipment discharges current. Current discharges could lead to serious injury for some patients. Faulty rosters = risk of faulty equipment	**Slide 1** Did you know? ▪ 50% on paperwork ▪ 5 to 7 instruments missing each month
New Leakage Check System —More efficient —More productive —More economical in long run Comparison between systems in terms of: —Equipment maintenance —Efficiency —Costs First: outline of old system and shortcomings.	**Slide 2** Present System vs. Proposed System ▪ Equipment Maintenance ▪ Efficiency ▪ Costs

Second: new system.
Goal: you'll agree that new
system should be implemented

Present System
—Monthly rosters
Give equipment descriptions
and locations
—Generated by Universal
Datasystems
Based on info. we supply
—Frequently late
5 to 7 days behind schedule
—Hard to change
Universal must add info. to
their program
—Often inaccurate about
equipment
Don't always list new
equipment
Sometimes list old
equipment
—Often inaccurate about
locations
—Techs try to be conscientious
Equipment missed each
month because location
listed is inaccurate

Slide 3
Rosters—Universal Datasystems

- 5 to 7 Days Late
- Inaccurate Equipment Descriptions
- Inaccurate Equipment Locations

System uses techs' time
inefficiently
—Results of checks handwritten
Four separate files
—Three techs work together on
checks
One checks equipment
Two do paperwork
—50% of techs' time spent
writing up test results

Slide 4
[Flow chart showing typical leakage check procedure]

Costs
—Techs' salary: $9/hr., plus
overtime
—Equipment
$500,000 to replace
—Legal
Records currently inaccurate
Records occasionally
illegible

Slide 5
Costs (Real and Potential)

- Technicians' Wage: $9.00/Hour
- Equipment: $500,000/Instrument
- Legal: ?

Possible problems if there's a lawsuit	**Slide 6** [Reproduction of typical file to show legibility problems]
New System Components —Signus III microcomputer —Leakage meter From Billings Corp. Attached by Digital to Analog/Analog to Digital interface card New System Procedure —Meter attached to each piece of equipment Computer makes check Computer records results	**Slide 7** [Diagram of new system]
Advantages: efficiency —Equipment rosters eliminated All info. stored on floppy disks —Techs could update info. Add and delete equipment Change locations —Techs could see display of equipment in each section on monitor Computer could print out list of all equipment not checked at end of month. Techs could locate and check it —Continually updated list of equipment Fewer instruments overlooked	**Slide 8** [Sample of computerized equipment file]
Advantages: Less paperwork —Tech writes results once on equipment tag Computer records results of test in all other files Tech only needs to type in i.d. number of equipment. —Can be done by one tech instead of three	**Slide 8 Remains**

Advantages: Cost
—Initial cost: $4974
    $2190 for computer
    $ 269 for interface card
    $1015 for leakage meter
    $1500 for program
—Increased efficiency of techs should compensate for cost
—Program can be written in-house
    More economical
    More input from techs and biomed staff

Slide 9

Costs:

Computer	$2190
Interface Card	269
Leakage Meter	1015
Program	1500
Total	$4974

Summary
—Present system
    Inaccurate
    Inefficient
    Potentially costly

Slide 10

Present System
- Inaccurate
- Inefficient
- Potentially costly

—Proposed system
    Improved accuracy and efficiency
    Prevents expensive errors
Proposal
—Recommend system to Board of Directors for purchase

Slide 11

Proposed System
- Accurate
- Efficient
- Cost Effective

Hook
—$4974 may seem large
    Compare with cost of one damaged piece of equipment or one lawsuit

Slide 12

$4974 vs. ? ? ? ? ?

SUMMARY

1. Oral presentations have several advantages over written reports in some situations: They are more immediate, permit more audience control, and are usually easier for the audience to absorb.
2. Your first step in preparation is to analyze your audience in terms of their knowledge, interest, attitudes, and ability to act.

3. Your next step in preparing a presentation is to define your purpose. Most presentations are either informative or persuasive.
4. Before making a presentation, you should discuss your ideas informally with potential audience members to gauge probable reactions.
5. Presentations are organized with beginning, middle, and concluding sections basing the order on audience needs.
6. All presentations contain a visual component; the audio and visual sections of the presentation can be composed together using either a script or storyboards.
7. Visuals can be either projected or nonprojected. All visuals should be simple, readable, and high quality.
8. Delivery should be as natural as possible: both audible and relaxed.
9. Delivery should be from visuals or from notes, not from a manuscript; you should maintain eye contact with your audience and try to adjust your presentation to their reactions.
10. Question-and-answer sessions provide an opportunity for speaker and audience to have direct contact; you should prepare for questions by analyzing your audience's probable reactions in advance.

EXERCISES

1. Consult the Carlyle Windsor House Case on page 113. Assume you were going to present your report orally to the foreman and painting crew in order to ensure that this problem doesn't occur again.

   a. What points would you emphasize to this audience?
   b. What visuals could you use?

   Now assume that you were giving this oral presentation to Bill Morrasco and the crew foremen. What points would you emphasize for them? What visuals would you use this time?

2. Write out the script or storyboards for an oral presentation of any of the following cases: the Sikes Chalk Trend (p. ______), Supermarket Scanners (p. ______), Access for All (p. ______). Use the audiences specified in the case descriptions.
3. Write a script—audio and visual—for an oral presentation based on a topic that interests you. This could be a term paper subject, some-

thing from your school or city that concerns you, or even a hobby or special interest. Specify the audience to whom the presentation would be directed and then tailor it to their needs and interests.

Oral Presentation Case 1

Day-Care Development

As an assistant city planner you've submitted many written reports, but you've managed to escape a full-scale oral presentation. Now you're facing your first and it's an important one. For the past two years you have watched the development of the Southside Rejuvenation Project. Spearheaded by the Monterrey Neighborhood Association, the project is working on restoring several historic buildings, constructing new housing, and attracting new business to the Monterrey District, only two miles from downtown.

So far, as a result of a lot of hard work and dedication, the association has been successful. The buildings have been redone, a new plaza has been constructed, six new stores have already opened and five more are scheduled to open during the next six months. But the greatest success has been in convincing Midsouth Enterprizes to open a new office building housing their corporate headquarters on the plaza. The people at Midsouth estimate approximately 1800 employees will be working out of their new headquarters; the Monterrey District will be booming.

Which brings you to the problem. Currently there are only two small day-care centers in the Monterrey District, both of them housed in church basements. These two centers can handle only thirty children each and both are already filled to capacity. After studying census records, the neighborhood association anticipates a need for at least 150 more day-care spaces by the time the new stores and Midsouth Enterprizes are both in place. They ask you for help.

After studying the problem, you think you see a solution. The city already owns a building in the Monterrey District which would be perfect for a day-care center: it's a former elementary school, currently closed because of a school district consolidation. The Monterrey District Association would be willing to rent part of the building as their headquarters, if the city would administer the day-care center as part of its municipal day-care program. Federal funds are also available for approximately 20 percent of the costs if welfare and low-income parents use the center. The rest of the costs could be made up by fees from parents. The building already meets the federal and state standards for day-care centers.

You must convince two groups of the feasibility of this program. First, you must convince the school board to allow its building to be used for this purpose. Although the building has been empty for almost a year, there has been a suggestion recently that it could be used as a warehouse for schoolbooks and supplies. There is some precedent for using school buildings for other purposes: The former Benjamin Harrison Elementary School is now a center for performing arts on the city's north side. The Monterrey Neighborhood Association would pay rent to the school board for the use of the building; the day-care center would be administered by the city's social services department and thus would not pay any rent.

The other group you must convince about this plan is the city council. They must approve the decision to open another day-care center to be administered by the department of social services. The city already administers three other day-care centers, handling between twenty and thirty children each. This would be a much bigger center, but there are no commercial day-care centers in the area to compete with the city's center. The city would have to invest around $1500 to get the building in shape as a day-care center, but you figure that between the rent from the neighborhood association, the federal grants, and the fees from parents the $1500 would be repaid within eight months of the opening of the center. The center would probably make only a negligible profit, figuring salaries and overhead, but it would not lose money, and it would provide a necessary service to the south side.

ASSIGNMENTS

1. Write the script, using both audio and visuals, for your presentation to the school board. You do not, as yet, have the approval of the city council, but you need an indication that the board will allow their school building to be used for this purpose. The school board is elected at large, but one of the members comes from the south side.
2. You have the approval of the school board; now write the script for your presentation to the city council. Include both

audio and visuals. Remember that these are elected officials from various districts around the city; only one council member is from the south side. However, they will be concerned with the good of the city as a whole.

Oral Presentation Case 2

THE EIGHTH GRADE DIET

As a student you've given several oral presentations in your classes, but those were to your fellow students. Now you're going to do something altogether different. Ever since high school you have been interested in food and diet, both personally and professionally. Now you have almost achieved your goal: a degree in nutrition. You're planning on working as a dietitian when you graduate.

In the middle of your studies, however, your mother presents you with a problem. She is an eighth grade science teacher and, thanks in part to your interest in diet, she has become increasingly concerned about the eating habits of her students. Now she has hit on what she considers to be a great plan: She wants you to come and talk to her classes about diet and junk food. She thinks that if they hear it from you, a young person, rather than her, an old fogy, they just might listen.

You're at a loss about what to say to a bunch of thirteen-year-olds about diet and health. You remember the speeches you heard yourself about food when you were younger: a mish-mash about the four basic food groups that didn't appeal to you even with your natural interest in the subject. While you mull the whole thing over, you come up with a few points you might cover:

Balanced diet means variety: protein, dairy foods, fresh fruit and vegetables, and whole grains. No food has everything.

Additives: counting corn syrup, dextrose, annual per capita consumption of sugar in U.S. over 100 pounds. Annual per capita consumption of salt fifteen pounds a year.

Salt: most eat 10 to 15 grams a day; we only need 3 grams (about ¼ teaspoon).

Snacking itself not bad; many small meals may be better than three large ones. But snacks should be good food, not just candy and chips.

Sugar not as effective in providing energy as protein; pieces of cheese provide more long-term energy than pieces of candy.

Chocolate made palatable by high sugar content; also 52 percent fat and high in caffeine (20 milligrams in 1-ounce bar, same as some cola drinks). Try carob: no caffeine, 2 percent fat, high in natural sugars, low in starch.

Fast-food restaurants have high-protein meals, but also high in calories from frying and high sugar content of some drinks. Typical fast-food meal provides 100 percent of minimum daily carbohydrate requirement, 50 percent of daily calorie requirement, but little fiber and vitamins. Salad bars help. Have milk instead of shakes—sometimes milkless!

Pizza highest in protein of fast food and lowest in fats, but little vitamins and may be high in salt.

Fried chicken higher in protein than hamburgers, but also higher in fat.

Fried fish lowest in calories but also in protein.

All fast-food restaurants salt food: up to 1½ teaspoons per serving.

Peanut butter supplies protein, carbohydrates, vegetable fat; very nutritious. Some peanut butters have sugar, hydrogenated oil (heavy oil, hard to digest, keeps peanut butter from separating). Check label; "natural peanut butter" worth the money. Make your own in blender (1 cup peanuts plus 1 tablespoon oil).

Cereals: some over 50 percent sugar. May be highly vitamin fortified, low in fiber, overprocessed. Granola types sweet, high in calories because of honey and oil, also high in fiber and vitamins/minerals.

Bread: Switch from white to whole grain for fiber. May be more filling. Look for bread made with milk for extra nutrition.

Milk is good source of protein, calcium, some vitamins. Skim milk is good way to cut down on fats if you can take the taste.

Yogurt is high in protein and calcium; it is milk cultured with live bacteria. May be good for digestion. Commercial yogurts may use thickeners like gelatin; avoid citric acid or sodium caseinate. Fruit syrups add sugar, thus calories. Look for "natural ingredients," high bacteria count (some companies pasteurize yogurt, kill bacteria).

Eggs highly nutritious, also high in cholesterol. Conflicting studies on cholesterol. Rule of thumb: no more than four eggs/week.

Cheese good source of calcium and protein. Hard, natural cheese best food value; processed cheese has water, other additives, less cheese for money.

Ingredients on food labels listed by amounts: If food has more sugar than anything else, sugar comes first on label.

Anything called "drink" or "food" on label—"fruit drink" or "cheese food"—has lots of additives, mainly water.

Reviewing your list you see you have a lot of information, but not much shape. Obviously, you're going to have to work on preparation.

ASSIGNMENT

Write the script, both audio and visual, for your presentation to the eighth grade. First analyze your audience. What are they likely to know about your subject? What is likely to interest them? Then consider your purpose—what exactly do you want to accomplish? Finally, study this list of information: what facts will help you accomplish your purpose and be of interest to an audience of thirteen-year-olds?

P · A · R · T

Advanced Strategies and Formats

CHAPTER • 15

Writing in a Group: Managing a Team of Writers

INTRODUCTION

You will probably do much of your writing as a member of a group. This does not mean that you will write every sentence with others' help, but it is likely that you will be asked to write parts of reports that will be completed and edited by other professionals. This chapter explains how you can organize an effective writing team by following a simple system for planning and monitoring the writing process. This group management approach specifies three main checkpoints: (1) an assignment conference, (2) a reorganization conference, and (3) a review and editing conference. Even if you are not a group manager this information will help you be a successful member of a writing group. In this chapter you will learn how to schedule a writing project. You will also learn how to delegate responsibility and authority in a writing group, and how to coordinate report writing with technical or engineering work that is proceeding at the same time. A final section explains how you can revise other writers' work.

AN OVERVIEW OF GROUP WRITING

The basic principle of group writing is that planning for the final written products should start as soon as the project begins, even before any technical work starts. This planning should include a careful division of responsibility, as well as attention to the problems of gathering, classifying, and storing all the data that will be needed when it comes time to write reports. A writing group manager should schedule regular meetings to ensure that all the necessary writing and information gathering is going ahead on schedule, just as any good manager regularly reviews the progress of design, construction, or sales. As a writer within the group, you should also be aware of the tasks the group must complete and the best methods for completing those tasks.

The following story illustrates the number of factors you must keep in mind when you manage the writing, reproduction, submission, and evaluation of a report. Regional Coal, a large western electric utility, was planning to build a new coal-powered generation plant in one of the western states. In order to get local government permission to proceed with plans for the plant, the utility was asked to produce a preliminary report—similar to an Environmental Impact Statement—discussing the economic and environmental consequences of building the plant, the projected need for electricity in the surrounding region, and many other issues raised by the proposal. For one year, engineers and economists from the utility performed a variety of studies and gathered data to be included in a year-end report to the regional government.

With the December 15 deadline only a month away, the managers of the power project decided it was time to plan the final report. The head of the project sent a memo to each of the engineers doing the technical and economic analyses, asking for a final report and explaining that these final reports would be edited into a comprehensive study of the generation plant proposal. The manager then asked his staff to hire an outside technical editor who would be responsible for compiling the final report, editing all the sections so that their prose was of uniform quality and, most important, writing the introduction and the conclusion, which no one else had been assigned to draft. The manager estimated that the report sections would arrive within two weeks, leaving the outside editor something less than two weeks to compile and finish a report that was estimated to be 400 pages long. Notice that the most important task—writing the conclusions on the basis of the information contained in the subsections—was left to an outsider who had done none of the research and who was probably unfamiliar with the policies and expectations of the agency, not to mention the attitudes and composition of the audience. This amazingly compressed schedule also left no time for such important tasks

Figure 15-1
Regional coal's project management plan

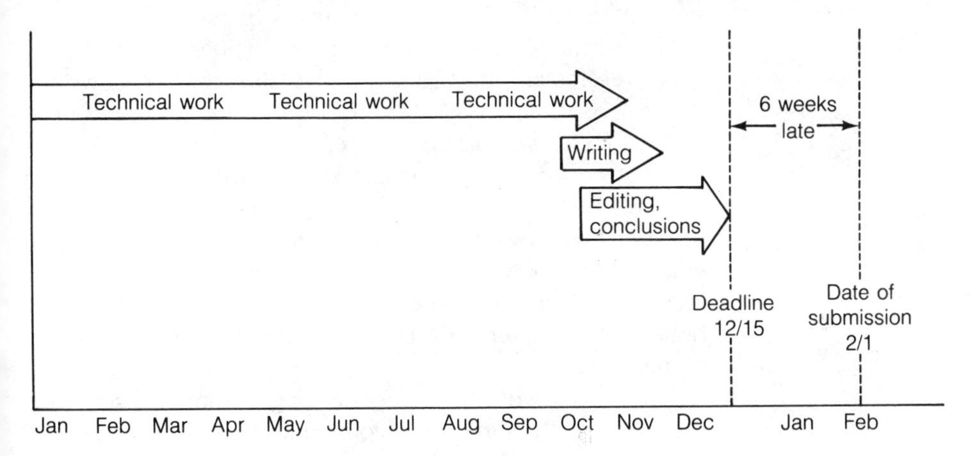

as a careful review, typing and reproduction, or graphic design. Nor were any plans made for a meeting among the various authors to establish guidelines for length, style, section numbering, use of subtitles, graphics, audience, common purpose, or persuasive strategy. As it turned out, the report was finally finished—over a month late and after many hastily arranged conferences, rewrites, and haphazard reviews.

Regional Coal's report writing "plan" is presented graphically in Figure 15-1. Time is plotted along the horizontal axis. Obviously this was a disastrous way to write a large report. The authors had no systematic way to meet and discuss common topics or to be certain that the conclusions of the various sections were in agreement. A much better approach to team writing is presented in the next section, which explains how you should view the relationship between report writing and the other tasks involved in a large project.

Effective Group Writing

In Figure 15-1, you saw how the staff of Regional Coal completed all of their technical work before they began to work on communicating their results to interested government agencies. A more effective way to manage the combined communication and technical task is to recognize that the communication job starts at the same time as the technical job, if not before. When a project team assembles, among the first questions to be answered are, How will this group establish procedures for communication and review? and Who will keep track of records and results so that they are available when we need them? Figure 15-2 presents an effective way to visualize the group writing process. As in Figure 15-1, the horizontal axis represents time; the major checkpoints are also represented. As you can see in the figure, the three main checkpoints are the organization meeting early in the project, a midcourse correction or regrouping meeting toward the end of the technical work, and an editing and review meeting shortly before the final draft report is submitted for evaluation and preparation for publication. The bulk of this chapter is devoted to a discussion of these checkpoints.

While the work team is organizing, communication tasks take most of the time. Later communication is still important but subordinate to the task of solving the technical problem. Toward the end of a project, communication tasks again become primary. Our main point in this chapter is that if you pay attention to the communication tasks throughout a project, you will never face the crises that plagued Regional Coal, and you will seldom find yourself falling seriously behind schedule or searching for records no one thought to keep.

Checkpoint 1: Organizing the Group and Assigning Tasks

When a project team assembles to divide the technical tasks, the communication work should also be organized and divided. At this checkpoint, you have three objectives: (1) selection of a management system, (2) assignment of general responsibilities, and (3) discussion of the documents that need to be written, including the final report and any progress report.

Selecting a Manager and Assigning Responsibilities

Projects normally have managers who assign the technical work, and often these managers also oversee the development of reports. Problems arise when the role of communication manager is not clearly assigned or defined. If someone other than the project manager is assigned the task of coordinating report writing, it is important to decide how records, interim reports, and notes will be kept and filed. If the record-keeping job is not assigned early in the project, the technical staff will keep some records, the communication staff will keep other records, and some important notes and documents may fall into a no-man's-land. One of the authors of this text worked on a project where it was important to keep a running list of editorial

Figure 15-2
Effective project management showing the three major checkpoints for writing management

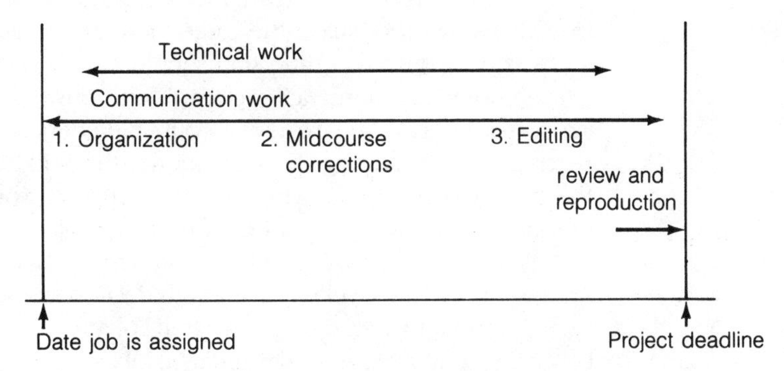

changes in a government document. Unfortunately, this task was not mentioned to the editor until the final draft of the new document had been accepted and was about to be placed on the agenda of the regional government for a public hearing. Although other members of the team had been aware of the record-keeping requirement, the task had not been discussed or assigned to any member of the group.

Once these basic management roles have been defined and assigned, the other team members can receive their assignments. If the members of the group have not worked together before, this is a good time for brainstorming and for open discussion of each person's skills and interests. Early in the project, when egos are not heavily involved, it is also critical to find out which members of the group have strong skills and motivation as writers and editors. As a writer this is your time to assess your own skills and interests. With what part of the project are you qualified to deal? What part would best reflect your interests? The manager should assign the final editing authority to a good writer at the beginning of the project to avoid bruised feelings at the end when the final report is edited for uniformity of content, style, and approach. At this meeting, the manager should also decide how decisions will be recorded and transmitted to members of the group. Will there be a system of memos? Will the managers issue periodic project reviews? Will each working group write a periodic progress report that will be edited into the final report?

Analyzing the Document

The third topic you should discuss at the first checkpoint is the document or documents that are to be written. The entire group should understand the audience or audiences. Will the final report be read by legislators? Investors? Executives? Other technical staff? These are the same decisions you make when you write reports on your own. In a group it is most important that all the writers share the same view of the audience. A tentative schedule and time line should also be worked out, leaving time for emergencies, clerical functions (such as word processor entry), and review. Group members should also reach an understanding about the money that will be available for graphics services, copying, and reproduction. Some corporations even issue style manuals for large projects, specifying details of usage, layout, section numbering, typefaces, and graphic style. This level of attention to detail becomes increasingly important when different sections of a report are written by groups that have little day-to-day contact. A major national study of nuclear waste was written over a period of years by three corporations: one on the East Coast, one on the West Coast, and one in the Midwest. Together, the project generated over a dozen enormous volumes of reports. In that case, a comprehensive manual of style was necessary; for a smaller project, it is useful to agree on a single published manual of style (e.g., *The Chicago Manual of Style* or *Words into Type*).

At the same time, the communication director should find out if the agency or corporation has uniform policies for report formats, binding, or graphics. In the authors' experience, graphics guidelines can surface just as a report is going to press, forcing last minute design changes that require additional time and money.

Scheduling

Scheduling deserves extra emphasis. In Figure 15-2 the deadline is at the far right, some distance from the point marked "final draft." Writers often plan to have the final draft ready *on* the deadline, forgetting that there are many other tasks to complete before the final draft is ready to be turned in for evaluation. It is a good practice to schedule a project by starting with the deadline and then subtracting all the time that will be needed for typing or word-processing, review by supervisors, typesetting, proofreading, preparation of final graphics, making changes, and printing. Throw in some time for unavoidable emergencies, and you can calculate when your final draft needs to be ready. For example, if you had a term paper due next Monday, you would be wise to schedule one day for typing, one day for review, and perhaps another day for foul-ups. Your deadline for the final draft, then, would be Thursday, not Sunday night.

Outlining

The first group meeting is also the time to draft a tentative outline for the progress reports and for the final report. If you are going to keep records for several months, or even years, a tentative outline will provide you with a set of categories for classifying and storing records. While the beginning of a project might seem an odd time to outline the final report, there are several reasons why it should be possible to do so. Often the members of a team will have completed similar projects before, or they will have access to reports written by other groups. At the very least, you are bound to have some valuable preconceptions about the direction the project will take, the questions you need to answer, the technical problems that must be solved, and the difficulties you will probably encounter. An outline can help you agree on the major points to be covered and any supporting points that may go along with them. Finally, writing a preliminary outline this early helps clarify confusion about organization and the division of responsibility. (Multiple outlining techniques, a related topic, are covered in Chapter 5.)

Communicating

The project team should agree on a method for communicating feedback and recommendations while the project is continuing. How will memos be distributed? What measures will be taken to assure that all concerned personnel receive each memo? Who will be responsible for this task? How will liaison be maintained between the technical and communication managers? After all of these issues have been discussed and settled, the communication manager can issue a memo that records all the decisions reached at the meeting, especially those concerning the division of work, the schedule and tentative outline of progress reports, and the final report. Figure 15-3 is an example of a brief project organization memo. A checklist of all Checkpoint 1 activities is printed at the end of this chapter.

Checkpoint 2: Preparing for the Final Report

The second team meeting focuses on midcourse corrections. You might want to have more than one of these meetings, depending on how well the project is going. If there is only one midproject checkpoint, it can be scheduled for about the time that the technical work is winding down, at the beginning of the writing, or even after the first draft has been completed. Two agreements need to be reached at this meeting, one about the

FIGURE 15-3
Memorandum of understanding and project outline for service manual review

Systems Software Manual Revision
Division Manager and Project Head: Margaret McKinley 528-9000
Administrative Assistant to Manager: Ellen Orford 528-9000
Engineering Liaison: Art Cranmer 528-9120
Consulting Editor: Jonathan Redding 232-1664 (2-5 pm only)
Technical Editorial Service
1949 East Lake, Seattle 98112

Audiences:	(1) Technicians, (2) consumers, (3) large buyers (in order of priority).
Purposes:	(1) Reference for technicians and buyers (to coordinate with information they already have) and (2) Step-by-step instruction for consumers (to direct action).
Graphics:	Limit changes in this area. Graphic work will be done by the internal arts department to standard specifications for drawings and blueprints.
Length Limit:	100 pages.
Index:	Editor's task.
Record Keeping:	Engineering liaison will maintain records, particularly all notes from the engineering division concerning technical explanations and the technical accuracy of suggested revisions.
Meetings:	Every two weeks. The division manager will be represented by her assistant who will have final authority over changes and revisions.
Outline:	Separate page (appended to original).
Schedule:	A separate time line is appended.

content and style of the document, and one about the scheduling and use of limited resources, such as clerical time and money for graphic artists and reproduction.

Reviewing and Editing

The first goal of the midproject meeting is to establish common guidelines for revision and editing. As a group member you should review your understanding of the audience and purpose of the report and then go over the preliminary outline to make any changes that are necessary. You should also reach agreement about the details of the physical format of the final report:

size, layout, binding, technical vocabulary, graphics, and so on. As the technical work has gone forward, the different work groups have probably begun to drift away from the guidelines established in the first meeting. One group may have found it necessary to make use of special techniques for illustrating research findings: Will this throw the report out of balance? Or as the work has developed, various members of the group may have redefined common technical terms in ways that are relevant only to one part of the project. In addition to these issues, the group should discuss the level of writing quality for which team members should strive. Some members of the group are bound to be nervous or simply blocked; open discussion of quality, audience, and purpose will help them to finish their work on time. (Levels of writing quality are discussed in Chapter 6.)

The second topic of this meeting is more mechanical, but equally necessary. As the project draws to a close, there will be a rush to use limited clerical and support services. The group should establish fair limits for the use of these services and, if possible, develop a schedule of staged or progressive deadlines for the report sections, so that support services are used evenly. Before the end of this meeting, the manager can remind the group that final editorial authority was delegated to one person at the first meeting.

Checkpoint 3: Reviewing the Document

The third checkpoint is placed far enough before the deadline to allow time for revision, reproduction, and distribution. In some cases, there will be a need for two review points, one before the final report is submitted for evaluation and a second after the report has been returned for additions and corrections. In either case, the meetings should be scheduled to leave enough time for necessary revisions.

Reviewing Audience, Purpose, and Organization

At this meeting, the main question is whether the report has achieved its purpose, and this question, of course, calls for review of the audience analysis and organization of the draft. If you originally planned to reach an audience of executives, is the language appropriate, or have you included too much technical discussion? If you planned to write a proposal, does the draft report make a persuasive case or is it merely informative?

Your second item of business at this meeting should be a review of the format. Are headings and graphics uniform in style and design? Do all the parts fit together into a whole? Have you included the necessary introductions, transitions, and conclusions?

At this time the designated editor may take over and complete the final draft. Alternatively, the group may prepare the final draft, leaving the editor to do the final text editing for style, grammar, and spelling. Too often, as in the case of Regional Coal, the group thinks that a technical editor called in at the last minute can make up for all the planning that should have been done months before and correct every flaw. In fact, if audience analysis and purpose were never settled, the resulting organization is likely to be so far off target that nothing short of a complete reorganization and rewriting will produce a successful document. On the other hand, if a group has successfully dealt with audience, purpose, and organization from the beginning, the editor will have a relatively simple and manageable task to perform.

Revising Other Writers' Work

If you have been assigned the task of editing either your colleagues' or your supervisor's work, follow four rules: (1) reach an agreement, (2) consider the level of writing needed, (3) focus on organization, and (4) be tactful, particularly with your boss.

Reach an Agreement

Some writers only want to hear praise; they have no interest at all in impartial editing. If you are asked to edit, explain that you will do your best to provide honest, constructive criticism and changes. Should the subject matter be outside your knowledge, explain that you can edit for style, logic, and organization, but not for the accuracy of the content.

Whatever limits you agree on, be certain that the author understands them clearly. You don't want to be blamed for failing to correct a fact or equation you don't understand, and the author should not submit a final draft believing you have polished every detail when you never intended to do so. If you are asked to edit a long report, find out whether you are to provide a one- or two-page list of suggested revisions, whether you are to make the revisions yourself, or whether time permits only a rapid scanning for major grammatical errors and misspellings. Above all, be sure that you understand the audience and purpose of the document.

Consider the Level of Writing

Disagreements over editing commonly develop when the editor does more than the writer wants. Some editors have an urge to see everything written in their own style and marked by their own idiosyncracies, but your job as an editor is far different from this. Your basic task is to bring the entire document up to the *author's* level of competence. (If the author wants to hire you to prepare an entirely new draft, OK, but that job goes well beyond the limits of basic editing.)

When you edit someone else's prose, you want the writing to be uniform in quality and representative of the author's best work. (We assume that you have been asked to edit because you are more competent than the author in some way: faster, a better speller, a clearer writer, etc.) In some cases, you might improve the writing to a quality beyond the reach of the author. You are not expected, however, to bring the report to the level of your personal best because that effort would probably require more time than you can afford, and you would more likely produce a final draft that would be too obviously your own work. Just how much you revise in any individual case will depend on the agreement you reach with the author.

Focus on Organization

By focusing on organization (rather than prose style) you will accomplish more and you will also avoid conflicts over stylistic questions. Start by summarizing the argument; as you read, jot down a rough outline and then check it for completeness and logic. This approach is easier than it sounds, and it yields excellent results. Read carefully, too, for subtitles, introductions, and transitional sentences. Many reports can be greatly improved by adding introductory sentences and subtitles that the author probably considered too obvious or too time-consuming to bother writing.

Be Tactful

Whenever you edit, be tactful and willing to compromise. Some colleagues will fight over every comma, while others will permit you to throw out whole pages without comment.

If you make extensive changes, one common approach is to present the author with a clean copy of the revised text, rather than with the heavily marked original. Many writers will accept the new draft as their own work without noticing how much you have changed. Your boss may prefer this approach to being faced with an extensively annotated copy of his or her work. Be extra careful not to change the author's meaning if you're going to use this approach.

When you correct grammar, be prepared to explain or justify your conclusions, especially to superiors. And when you edit for style, change only those sentences that absolutely must be corrected; you are not being asked to rewrite the report as if it were your own.

If you are asked to edit rather than advise, make all the corrections and revisions yourself; by all means do not add marginal notations such as "vague," "awk.," and "sentence structure needs work." On the other hand, if you have agreed to provide comments rather than editing and rewriting, add a cover letter that explains your recommendations, particularly any changes in organization that you advise. Remember to praise what you like, too. Constructive criticism should be written in the form of a bad news letter (see Chapter 13), with a buffer, without direct condemnation or negative language, and in a spirit of cooperation.

SUMMARY

1. A little time spent on communication early in the project can save enormous amounts of time, money, and grief when the deadline draws near.
2. Plan at least three checkpoints depending on the length and complexity of your project.
3. At the beginning of the project, meet to establish guidelines and divide responsibility.
4. Toward the middle of the project, meet to make corrections in the outline and to review your understanding of audience, purpose, and details.
5. Before final editing, review how well the document achieves its purpose and make necessary revisions.
6. If you are responsible for editing the final document, follow four rules: reach an agreement, consider the necessary writing level, focus on organization, and be tactful.

CHECKLIST FOR CHECKPOINT 1

1. Have you selected a communication manager for the project?
2. Have you investigated the skills and interests of the group members?
3. Has a final editor been selected?
4. Has the group reached consensus about the audience and purpose of the report?
5. Have you drafted a preliminary outline?

6. Has someone been assigned the task of keeping all records?
7. Has the group established a realistic schedule that allows time for production and review?
8. Have you established clear lines of communication and decision making? How will all group members be informed of decisions and problems?
9. Are all members of the group aware of company or project guidelines for graphics, format, style, and the use of support services?
10. Is everyone aware of the budget for this report?

EXERCISES

1. As a group project, meet to discuss one of the assignments listed below. Write a memo of understanding that covers all the issues that should be discussed at the first checkpoint. Assignments: Day-Care Development, page 361; The Swimming Pool Proposal, page 263.
2. If you have already completed a group project in this or another course, write a three-page analysis of your group's effectiveness. Did the whole project go smoothly? Did any misunderstandings arise? How could you have improved the organization of the project? What changes would you make?
3. Exchange final group reports with another group in your class. In your group, write a three-page analysis of the other group's project, focusing on ways that additional group communication could have improved the final project.
4. Reorganize and rewrite the group report in Exercise 5, below, or any other student group report that needs careful editing.
5. Read the following group proposal. In a two-page paper, discuss the proposal's strengths and weaknesses. How could additional meetings have improved this proposal? What issues did the group fail to discuss? What issues did they address successfully?

PROPOSAL TO PLANT SERVICES DIRECTOR
CENTRAL UNIVERSITY
<u>STATEMENT OF PROBLEM</u>

One of the most attractive features of Central University Campus is the parking loop at the campus entrance. Surrounded by the school's oldest and most attractive buildings, and shaded by oak trees over one hundred years old, the loop area is a favorite gathering place for students and alumni, and the focal point of many campus events throughout the school year. At night, however, the loop is a nightmare. Because the sidewalks are cracked and there is no lighting, except for the lights on the sides of nearby buildings, it is easy to trip and fall, especially on dark and rainy nights. Security is another issue. Many female students, including myself, are hesitant to walk across the loop at night. I think that by adding streetlights on poles at several strategic locations in the loop this problem could be solved and the loop could become a student gathering place in the evening.

SECTION 2: PROPOSED SOLUTION

A lighting analysis of the loop area was performed on May 13, a night when there was no moon and light cloud cover. These conditions approximate what this group believes to be average night lighting conditions in the loop area. Using a Weston Number 703 Foot-Candle meter, candlepower ratings were taken at ten points throughout the loop, particularly along the major walkways through the area. The readings indicate that lighting is far below the minimum suggested by national lighting standards for walkways in an area deemed to have potential security hazards.

The quantity of light could be increased to near or equal the national minimum standards for walkways by placing five equidistant street lights along the margin of the loop. It is estimated that this light placement would afford optimum lighting along the parking area and near optimum lighting in the center of the loop. The suggested placement of the streetlights is indicated in figure 1, attached.

SECTION 3: BUDGET

A candlepower analysis of the campus loop area indicated that the central loop area received only 50 percent of the minimum light recommended by national standard for walkways. A discussion with Jim Williams of the Plant Services Division revealed that last year there was a $4000 surplus in the plant services budget and that plans are now being made to spend these funds on additional flowering shrubs for the east campus area. We feel that it would be far wiser to spend this money on additional lighting for the loop than on more shrubs. Some of the additional costs—beyond $4000—might be recovered from reductions in insurance costs related to the reduction in hazards in the loop area because of the new lights.

We estimate that the cost of the new lights would be

5 lamps (see figure 1, below, for lamps)	$3400 or $3750
Installation at $200 per lamp	$1000
Trenching and trench wiring	$ 800
TOTAL	$5200 or $5550

These estimates are based on conversations with University Hardware and Supply Corporation. From the available types of lamps we selected two models that have the correct range of candle-power for this application.

Park Lamp Co. #1540	$750 per unit
Zone Lite AC2000	$680 per unit

Figure 1: The two lamps judged appropriate for this application

The Zone Lite AC2000 is a basic aluminum-cowled street lamp with a mercury vapor element. The more expensive Park Lamp #1540 is an imitation gas light with a mercury vapor lamp inside. The committee members believe that the Park Lamp would fit in better with the campus environment.

SUMMARY

If these lamps are installed for the minimal cost of $5550, the campus security division can probably count on a reduction in crime in the loop, where two robberies and one assault occurred last year during the regular school session.

Group-Writing Case 1

The Home Windmill

You and four friends have decided to go into business part time while you finish your college work. The four of you are interested in energy alternatives, but what you're really concerned with is wind power. It strikes you that windpower is an underrated home energy source. A windmill generator system would be less expensive than a solar system and would be the perfect alternative for people who want to augment the power system they already have. The five of you have decided to market a small windmill setup for the average home.

Your first customer comes more quickly than you expect: One of your professors hears about your project and expresses interest in having you install your windmill at his home. You explain that the windmill couldn't power his house completely, but that it could serve as a supplemental power source. The professor agrees. His home is a conventional three-bedroom ranch-style design on a half-acre lot. He wants to use the windmill system to power his yard lights and also the lights in a garage which has been converted into a work area for automotive and household repairs.

ASSIGNMENT

Working in a group of five, research windpower possibilities. Decide on the best specifications and design for a small home wind generator. Follow the guidelines for group work in this chapter:

1. Decide on a project manager.
2. Decide on a record keeper to keep track of notes and interim reports.
3. Decide which member of the team will accomplish each task.

When you have assembled enough information about your system, write a short proposal for your professor with the following format:

Introduction
Discussion (include system design, capabilities, and cost)
Budget
Summary

Again, follow the group-writing guidelines in the chapter:

1. Assign specific writing tasks based on the skills and interests of group members.
2. Select an editor.
3. Reach a consensus on the purpose and audience of the report.
4. Draft a preliminary outline so that each group member is clear about individual tasks.

Group-Writing Case 2

Access For All

As part of your course work in architecture, you have undertaken some research on access to buildings for the handicapped. The more you read about this problem, the more you begin to realize that your own campus presents a number of problems for handicapped students, particularly the wheelchair-bound students. Doors are hard to open, some buildings lack elevators or ramps, restrooms do not have proper toilet facilities, etc. Your professor realizes this, too; she divides your class into groups of five to study the problem.

Your assignment is to survey your campus to assess how well the buildings and physical layout meet the needs of the wheelchair-bound student. Are there barriers to access that need to be overcome? How could they be dealt with? What are the applicable local and federal regulations about handicapped access? What solutions to the access problems have been tried at other institutions?

ASSIGNMENT

In a group of five, write a recommendation report for the school administration regarding campus access for wheelchair-bound students. Remember the guidelines of group writing.

1. Decide on a project manager.
2. Decide on a record keeper to keep track of notes and interim reports.
3. Decide which member of the team will accomplish each task.
4. Assign specific writing tasks based on the skills and interests of the group members.
5. Select an editor.
6. Reach a consensus on the purpose and audience for the report.
7. Draft a preliminary outline so that each group member is clear on individual tasks.

CHAPTER • 16

Writing Environmental Impact Statements

INTRODUCTION

An environmental impact statement is a carefully documented analysis of what might happen to an environment if a specific project is allowed to proceed. The term *environment*, as used in these documents, includes not only the physical and biological environment, but also the social, human, and aesthetic environments. Environmental impact statements are required for many types of projects by an assortment of federal, state, and local laws and regulations. In every case, an EIS is meant to be used in deciding whether to issue permits to allow a proposed project to go forward.

Environmental impact statements are usually written by teams of specialists, often with the assistance of an editor and project manager. If your career plans include work with government agencies, planning or architectural firms, or consulting firms, you can expect to have at least some contact with these statements. As a format, the EIS can be classified as a special form of recommendation.

Because the details of environmental impact statements are so complex, and because federal, state, and local guidelines differ, this chapter is designed to familiarize you with (1) EIS vocabulary, (2) the basic stages of the process, and (3) the most common EIS format. This chapter also introduces the history of the EIS. For a full guide to the EIS process, you should consult the regulations and references listed at the end of the chapter and call the nearest office of the Environmental Protection Agency for information about recent updates of EIS policies and regulations.

History

Environmental impact statements were mandated for all "major" federal government projects by the National Environmental Policy Act (NEPA) of 1969 and by several additional laws concerning clean air and water. Under that legislation, an EIS is "to be included in every recommendation or report on proposals for legislation and other major federal actions significantly affecting the quality of the human environment" (*CEQ Regulations,* p. 10). The NEPA established the Council on Environmental Quality and gave the CEQ broad powers to assess the environmental consequences of federal government plans. The law defines the EIS as an "action-forcing device"; Federal officials are to use the EIS to plan future action and make decisions (*CEQ Regulations,* p. 9). Under the provisions of NEPA, federal agencies are not permitted to take any irreversible actions on a project until the impact process is complete. (The limited exceptions to this rule can be found in the *CEQ Regulations.*) Additional impact statements, such as the Federal Energy Regulatory Commission Exhibit E, are required by other statutes.

At the state level, environmental impact statements have been required by some jurisdictions since 1970. At a 1973 conference, the Council of State governments put forward a suggested State Environmental Policy Act (SEPA), and by 1974 fifteen states and Puerto Rico had enacted environmental policy acts. Since that initial burst of legislative activity, environmental impact legislation has been debated in many states, but few provisions have been enacted. The state laws extend the EIS requirement to state actions and, in many cases, to private projects for which federal, state, or local building permits or licenses are required. In some states that do not have environmental policy acts, the same function is performed by land use regulations that mandate documented environmental assessments. Since the mid-1970s, many jurisdictions have also accepted a "functional equivalency rule," whereby similar documents would be accepted in place of a formal EIS. Under a similar policy, projects requiring both federal and state impact statements can often submit one document for both review processes. In addition to the federal and state regulations, some regional and local governments (such as coastal commissions and county governments) have established impact statement requirements. At this time requirements vary widely from state to state; some areas still have none. The Council on Environmental Quality can provide information about current requirements and legislation in your area. More information can be obtained from the Association of Environmental Professionals (2321 P Street, Sacramento, California 95816), the principle organization of EIS specialists. We will limit our discussion to current federal guidelines.

THE EIS PROCESS

The environmental impact statement process includes five main stages:

1. An environmental assessment to determine whether a full impact statement is required
2. A draft EIS
3. A period of hearings, written evaluations, and comments on the draft statement
4. A final EIS that includes responses to the comments on the draft
5. A formal record of the decision that is reached, including discussion of any mitigating actions that will be implemented.

The Assessment Stage

If it is obvious that an EIS will be required, the initial assessment can be skipped. An EIS is required if the proposed action is of a type that normally requires an EIS. Section 1501.4 of the 1979 *CEQ Regulations* covers the full range of situations that require an EIS. An EIS is also required if the proposed action is without precedent. In practice, the rules leave a limited gray area of possible projects for which an EIS is not automatically required. In these cases, the initial assessment involves a conscientious but rapid survey of the possible environmental consequences of a proposed action. If this initial assessment reveals no environmental impacts, a finding of "no significant impact" is prepared. This statement must be made available for public review if the proposed action is similar to actions that normally require an EIS.

In some cases, a community needs assessment process precedes environmental assessment. Community needs assessment is a method for finding out what the members of the community perceive as their wants and needs. The process can involve public hearings, discussion with government officials and citizens groups, surveys, and many other information-gathering techniques. If a community does not want a project, or wants a different kind of project, everyone will benefit by finding this out before any public or private funds are spent on plans and environmental assessments. Some commentators have expressed the hope that the EIS legislation will gradually bring about more community needs assessments and at the same time discourage the proposal of projects that have been developed without public participation.

The Draft EIS

The draft EIS includes (in the recommended order)

- A cover sheet
- A table of contents
- A discussion of the purpose of and need for action
- A discussion of all the alternatives, including the proposed action
- A definition of the affected environment
- The environmental consequences
- A list of preparers
- A list of agencies, persons, and organizations who are to receive copies of the draft
- Supporting documents

Notice that the proposed action is *subordinated* by being placed within a discussion of alternatives. At the same time, the EIS format emphasizes the community need or problem that needs to be solved. For example, instead of proposing a new stadium in a specific location and then discussing the environmental consequences of such a project on the surrounding neighborhoods, the EIS format requires authors to discuss first the various dimensions of the community's need for an athletic facility. Only after the need has been defined and explained are alternative solutions to be presented and considered. Under the current rules, the alternative of *no action* must be thoroughly evaluated. All reasonable alternatives must also be covered, together with a discussion of actions that might mitigate the negative environmental consequences of each alternative.

The affected environment is defined through a process called "scoping," which requires public notice and consultation with other agencies that could assist in identifying "significant issues" (*CEQ Regulations,* p. 7).

The discussion of the environmental consequences is the most technical section of an EIS. This section can include direct and indirect effects; interagency conflicts over land use and preservation policies; energy use; natural and depletable resource issues; and recreational, historical, architectural, social, human, aesthetic, and spiritual values. Historically, the earliest EISs emphasized cost-benefit analyses. Over the years, biological, historical, architectural, social, human, and spiritual values have been added to the list of criteria. In the EIS for the proposed Copper Creek Dam in Washington State, for example, a section was devoted to the spiritual value of bald eagles to both Native American religion and the American national identity. Under current rules any cost-benefit analysis must be placed in the *appendix* so that cost findings are not mixed with the discussion of purely environmental or "value" issues.

Within these sections, other reports can be "incorporated by reference," a method of citation called "tiering." Tiering allows the agency preparing the EIS to focus on specific, significant issues without repeating discussions, supporting facts, and policy statements that are readily available elsewhere. The agency can simply refer the reader to the other reports that present this material.

The discussion of environmental effects must also consider methods for mitigating or reducing the adverse environmental impacts of the various alternatives. In the EIS excerpt in Exhibit 16-1, for example, the effects of dumping acid waste into the ocean are mitigated by a plan to limit the quantity of acid that can be dumped at one time.

A list of preparers and their qualifications follows the discussion of impacts. Early EISs placed less emphasis on the names and qualifications of the specialists who wrote the documents. The current regulations encourage authors to express their judgments fully and carefully; the regulations also discourage specialists from making judgments in areas outside their expertise.

Commenting

The agency preparing an EIS has an obligation to invite comments from other government agencies, concerned organizations, and the general public. Government agencies that have legal jurisdiction or special expertise concerning any environmental impact "have a duty to comment" (*CEQ Regulations,* p. 16). Comments can be presented at hearings or by submitting written responses and letters.

The Final EIS and Supplemental Statements

The final EIS is made up of the draft statement, all the comments received, and responses to the comments. In responding to the comments, according to *CEQ Regulations* (1979, Section 1503.4, p. 16) the preparers can:

***Exhibit* 16-1**

One Section of an EIS

ENVIRONMENTAL IMPACT STATEMENT (EIS)
FOR
NEW YORK BIGHT ACID WASTE
DISPOSAL SITE DESIGNATION

August 1980

Prepared Under Contract 68-01-4610
T.A. Wastler, Project Officer
for
U.S. ENVIRONMENTAL PROTECTION AGENCY
Oil and Special Materials Control Division
Marine Protection Branch
Washington, D.C. 20460

CHAPTER 4

ENVIRONMENTAL CONSEQUENCES

The release of acid waste at any of the alternative sites would produce similar environmental consequences. There will be minor, short-term, adverse effects on the plankton and minimal effects on bottom-living organisms. Effects on the benthos are most probable and would be easiest to demonstrate at the Southern or Northern Areas; effects (if present) at the Acid Site are obscured by the multiple contaminant sources, while no effects are expected at the 106-Mile Site, which is located in water depths of 2,000 m.

Adverse effects from acid waste disposal on the public health and water quality will be minimal except for a site in the Southern Area, where acid waste disposal might interfere with development of exploitable shellfish resources. Demonstrable, adverse effects on the ecosystem are most probable at a new site in the Northern or Southern Areas since wastes have never been released in these regions.

Most importantly, 32 years of study at the existing New York Bight Acid Waste Disposal Site have not demonstrated any adverse effects resulting from the disposal of these wastes.

This chapter details environmental effects of waste disposal at various alternative disposal sites outlined in Chapter 2. Included are unavoidable environmental consequences which will occur if the proposed action takes place. The effects discussed first are environmental changes which directly affect public health, specifically, commercial or recreational fisheries and navigational hazards. Secondly, the environmental consequences of acid waste disposal at each alternative site, are discussed, which cover effects of short dumping in nondesignated areas. Finally, the chapter concludes with descriptions of unavoidable adverse effects and mitigating measures, the relationships between short-term uses of the environment, maintenance and enhancement of long-term productivity, and any irreversible or irretrievable commitments of resources which will occur if the proposed action is implemented.

Much data were examined to evaluate potential effects of acid waste disposal at the sites. The principal data sources for each area are:

- Acid Site: NOAA-MESA studies of the entire Apex beginning in 1973. NMFS/SHL study from 1968 to 1972. Site-specific studies sponsored by NL Industries, Inc., beginning in 1948. Routine monitoring surveys sponsored by the permittees.
- 106-Mile Site: NOAA surveys, starting in 1974. Waste dispersion studies and monitoring of short-term disposal effects sponsored by the permittees. Public hearings concerning relocation of sewage sludge disposal sites and issuing of new permits.
- Southern Area: NOAA survey in 1975. Public hearings concerning the disposal of sewage sludge in the Bight.
- Northern Area: NOAA and Raytheon surveys in 1975. Hearings concerning the disposal of sewage sludge in the Bight.

Information from these and other sources was collected and compiled into an extensive data base entitled Oceanographic Data Environmental Evaluation Program (ODEEP; see Appendix D). The following discussion is based on an evaluation of the available data.

EFFECTS ON PUBLIC HEALTH AND SAFETY

A primary concern in ocean waste disposal is the possible direct or indirect link between contaminants in the waste and man. A direct link may affect man's health and safety. An indirect link may cause changes in the ecosystem which, although not apparently harmful to man, could lead to a decrease in the quality of the human environment.

COMMERCIAL AND RECREATIONAL FISH AND SHELLFISH

The most direct link between man and waste contaminants released into the marine environment is via consumption of contaminated seafood. Shellfishing, for example, is automatically prohibited by the FDA around sewage sludge disposal sites, or other areas where wastes are dumped, which may contain disease-producing (pathogenic) microorganisms. Thus, the possibility of consuming shellfish which may be contaminated by pathogens, is eliminated or minimized. Harmful effects caused by eating fish containing high levels of mercury, lead, or persistent organohalogen pesticides have been documented. Certain compounds (e.g., oil) have made fish flesh and shellfish unhealthy and unpalatable. Therefore, wastes containing heavy metals, organohalogens, oil, or pathogens must be carefully evaluated with respect to possible contamination of commercially or recreationally exploitable marine animals.

Foreign long-line fisheries exist on the Continental Slope, but U.S. fishing in the mid-Atlantic is mostly restricted to waters over the

Continental Shelf. Commercial fishing and sportfishing activities on the Shelf are widespread and diverse; finfish and shellfish (mollusks and crustaceans) are taken. The Bight is one of the most productive coastal areas in the North Atlantic, and the region may be capable of even greater production as new fisheries develop.

Important spawning grounds and nursery areas lie within the Bight, but critical assessments of the effects of man-induced contamination on fish and shellfish populations are lacking. Many factors complicate the collection and assessment of these data. For example, normal short-term and long-term population cycles are not well understood, catch data may not be adequate, and the complete life cycle and distribution of the stock may be unknown. Natural population fluctuations, overfishing, and unusual natural phenomena may have greater influences than man-induced contamination on the health and extent of the fisheries' resource. Therefore, assessment of the effects of ocean disposal includes uncertainties due to the weaknesses of existing fisheries information.

NEW YORK BIGHT ACID WASTE DISPOSAL SITE

There is an extremely low potential for endangering public health from continued acid waste disposal at this site. The site location was chosen 32 years ago because it was not a point of concentration for fish or fishing and because the sandy sediments of the site are seldom associated with productive fishing. Ironically, the site has become a sportfishing area because the discoloration of the water caused by acid-iron waste disposal apparently attracts bluefish, a prized sport fish, to the area (Westman, 1958). Fishermen in the New Jersey and Long Island areas claim that the discolored water hurts the fishing for other pelagic sport fish. In winter, a commercial whiting fishery exists near the Acid Site, and lobstering may occur northeast of the site.

Effects of acid waste disposal on these resources are practically nonexistent. No health problems associated with sport fish caught at the site have been reported. No tainting or harmful accumulations of waste components in the flesh of fish taken from the area have been reported. Longwell (1976) observed adverse effects on mackerel eggs taken near the site, but did not suggest that dumping caused these changes, which also appeared at stations halfway to the edge of the Continental Shelf. Long-term damage to the resources by the acid waste disposal have not been documented (EG&G, 1978b; ERCO, 1978a,b). The area nearer shore is closed to shellfishing because of the material released at the Sewage Sludge and Dredged Material Sites.

Acid waste contains only small amounts of tainting substances, e.g., oil and grease. Relative to total inputs of oil and grease to the Bight, acid waste is an insignificant contributor of these contaminants. Waste constituents may reach the bottom and be assimilated by organisms, but other sources of contamination are probably more significant. (The Sewage Sludge Site is only 2.8 nmi from the Acid Site.)

106-MILE SITE

Commercial or recreational fishing is infrequent at this site; consequently, acid waste disposal will not directly endanger human health. NOAA-NMFS resource assessment surveys do not extend beyond the Shelf, however, densities of fish eggs and larvae are low beyond the edge of the Shelf. Foreign fishermen are near the site in the later winter, but usually catch highly migratory fish. The probability of fish accumulating toxic levels of contaminants from the waste is extremely unlikely.

A small fishery for the deep sea red crab (Geryon quinquedens) exists near the Shelf-Slope break in the mid-Atlantic. Immediately north of the 106-Mile Site, crabs are found in moderate abundance (33 per half hour otter trawl), but the water depth is much shallower than at the site (311 to 732 m). At a station 70 nmi (130 km) northeast of the site, at a comparable depth, no crabs were taken (Wigley et al., 1975). The site is within the range of smaller crabs, yet none of commercial size were taken deeper than 914 m. As with finfish, the probability of liquid wastes affecting a benthic animal is extremely low. Therefore, disposal at this site does not directly endanger human health by contaminating edible organisms.

SOUTHERN AREA

Numerous surf clams, ocean quahogs, and scallops are found in the Southern Area. Most commercial shellfishing is presently to the west, near the New Jersey coast. However, declining harvests may cause the Southern Area to be exploited in the future (EPA, 1978a). Recreational fishing is unlikely at the site due to the distance offshore and the competition from more attractive sportfishing areas closer inshore. If the area were used as a disposal site for wastes, similar to those presently being released at the Acid Site, a real but low potential for an accumulation of waste constituents in the flesh of shellfish would exist.

NORTHERN AREA

Disposal of aqueous acid wastes in the area would probably not directly endanger public health. The site is not in a known commercially or recreationally important fishing or shellfishing area. Resource assessment surveys show that this area has a similar, or lower, density of fish eggs and larvae compared with other Shelf sites (NOAA, 1975). Shellfish are not abundant in the area. The area supports no commercially or recreationally exploitable finfish or shellfish, thus a health hazard from eating animals contaminated by waste materials is unlikely.

NAVIGATIONAL HAZARDS

Navigational hazards may be separated into two components: (1) hazards resulting from the movement of transport barges to and from a site, and (2) hazards resulting from barge maneuvers within the site.

If an accident occurred involving the release of wastes, the effects from the dumped waste would probably be equivalent to a short dump. The effects from the other ship would depend on the cargo, and could be severe if the barge collided with an oil or liquefied natural gas (LNG) tanker. There is the possibility of loss of life in any collision. Sites further offshore have a longer search and rescue response time than sites closer to shore.

With respect to all sites, barges must pass through the Precautionary Zone centered around Ambrose light (see Chapter 3, Figure 3-10), where traffic is densest and hazards are greatest. Once through the Precautionary Zone, potentials for problems increase with increasing distance from shore. Table 4-1 shows the distance and estimated transit time for the four alternate sites.

ACID SITE

The Acid Site is across the outbound section of the Hudson Canyon Traffic Lane from New York Harbor, but the barging operations are designed to minimize interference with traffic. In 32 years of use at the Acid Site, no collision between a barge discharging waste and a ship has ever occurred. In April 1976, a collision did occur near the Acid Site between a ship and a barge outbound for the 106-Mile Site. The permittees now using the site barge wastes about once a day. Any accident would be close to New Jersey or Long Island beaches.

106-MILE SITE

Barges in transit to the 106-Mile Site from New York Harbor use the Ambrose-Hudson Canyon Traffic Lane for most of the journey. There is a slightly greater possibility for problems during the round-trip transit to the 106-Mile Site than to a site closer inshore.

Hazards resulting from maneuvers within the site are negligible. The site is extremely large, and permittees are required to use different quadrants of the site if there are simultaneous disposal operations. The frequency of existing barging is only two or three times per week. Increased frequency of use would not significantly increase navigational difficulties.

SOUTHERN AREA

The Southern Area lies outside traffic lanes for New York Harbor, thus its use would cause few navigational hazards. Barges could use the Ambrose-Barnegat Traffic Lane for most of the trip. However, increased ship traffic resulting from offshore oil and gas resource development would slightly increase the hazards. The degree and extent of such hazards would depend on the speed and magnitude of oil and gas development in the area. Any accidents would likely occur in the heavily fished coastal waters off New Jersey.

NORTHERN AREA

The Northern Area lies outside traffic lanes for New York Harbor thus its use would cause few navigational hazards. Barges could use the Ambrose-Nantucket Traffic Lane for most of the trip. Mineral resources have not been found in the area, so there is no possibility of increased hazards from future resource development. Any accidents would be near coastal waters off Long Island.

TABLE 4-1 Distances and Transit Times (Round Trip) to Alternate Sites

Site	Distance*		Transit Time (hours)**	
	nmi	(km)	5 kn (9 km/hr)	7 kn (13 km/hr)
Acid Site	17	(31)	7	3
106-Mile Site	113	(209)	46	32
Southern Area	53	(98)	22	16
Northern Area	50	(93)	21	16

* Measured from the Rockaway-Sandy Hook Transect
**Does not include time in transit from the loading dock to the Rockaway-Sandy Hook Transect (New York Harbor), nor time spent at the sites.

1. Modify alternatives including the proposed action
2. Develop and evaluate alternatives not previously given serious consideration by the agency
3. Supplement, improve, or modify the analyses
4. Make factual corrections
5. Explain why the comments do not warrant further response

When the changes are substantial, the preparers may be required to rewrite the entire draft statement. Under other circumstances, supplemental draft and final statements can be appended to a final EIS. If a supplemental statement is written, it must pass through the draft and commenting stages. The final EIS is limited to either 150 or 300 pages, depending on the complexity of the issue. The excerpt in Exhibit 16-1 gives you an example of a final EIS.

The Decision and the Record of Decision

Once the lead agency (the agency chiefly concerned in the issue and the one that prepares the EIS) reaches a decision, it must file a public document stating what the decision is and how it was reached. The document must specify the alternative or alternatives that are considered preferable. An agency can discuss its preferences among alternatives based on factors such as economic and technical considerations. And the agency is required to identify and discuss all the factors, including any considerations of national policy, which the agency weighed in making its decision (*CEQ Regulations,* Section 1505.2, p. 19).

The decision record must also specify what mitigating measures were adopted and why others, if any, were rejected. Until a record of decision is issued, no action can be taken on the proposal that would have an adverse effect on the environment or limit the choice of reasonable alternatives (*CEQ,* Section 1506.1, p. 20). For example, an agency that planned to build a dam is not permitted to contract for several million yards of concrete and then claim that it must proceed in order to avoid a contract cancellation penalty. The *CEQ Regulations* explain special exceptions to this rule whereby special equipment can be ordered in advance of final project approval. After the decision record is filed—and often before—remaining disputes are settled in the courts.

Organizing the EIS

Organizing the comparison of alternatives is often a problem in writing an EIS. One famous EIS evaluated four alternative sites for dumping hazardous chemical wastes at sea. (Part of the fourth chapter of this EIS is reproduced in Exhibit 16-1.) The authors chose to compare the four sites by discussing each class of environmental impact (biota, water quality, etc.) in a single main section, with the four sites in subsections. For example, under the general heading of BIOTA, the report is organized like this:

- Plankton
  - Acid Site (present dumping ground)
  - 106-Mile Site
  - Southern and Northern Areas
- Nekton (fish and swimming mammals)
  - Acid Site
  - 106-Mile Site
  - Northern and Southern Areas
- Benthos (animals living in or on sediments)
  - Acid Site
  - 106-Mile Site
  - Southern Area
  - Northern Area

Notice that this method of organization places the emphasis on the biota, rather than the total comparison of sites. One alternative would have been to use the following method of organization:

BIOTA
- Acid Site
  - Plankton
  - Nekton
  - Benthos
  - Summary
- 106-Mile Site
  - Plankton
  - Nekton
  - Benthos
  - Summary
- Northern Area
  - Plankton
  - Nekton
  - Benthos
  - Summary
- Southern Area
  - Plankton
  - Nekton
  - Benthos
  - Summary
- Summary of Biota Evaluation

Yet another organizational scheme would place the emphasis on the sites themselves by using them as the main headings.

ACID SITE
- Biota
- Water Quality
- Sediment Quality, and so on

It is not easy to choose among these alternatives; the issues and values discussed and compared in an EIS are often so complex that they will not fit well into any organizational scheme. For these reasons, the authors must be particularly sensitive to differences in emphasis that can be created by organization. Like other recommendation reports the choice is between organizing by criteria and by alternatives. Yet the authors of an EIS must be particularly sensitive to problems of equal comparison and they must order the EIS so that the audience can make the necessary comparison easily.

WRITING THE EIS

An EIS is a group-writing project, with the additional complication that the value judgments of professionals from many fields must somehow be coordinated. In order to manage this formidable task, the CEQ has established special guidelines for authors (*CEQ Regulations,* Section 1500.4, p. 3). Eight of these guidelines are reproduced below; as you can see, their general intention is to help the authors focus on the critical issues and avoid needless repetition or lengthy coverage of facts that are not in dispute.

Agencies shall reduce excessive paperwork by:

1. Reducing the length of environmental impact statements by means such as setting appropriate page limits
2. Preparing analytic rather than encyclopedic environmental impact statements
3. Writing environmental impact statements in plain language
4. Following a clear format for environmental impact statements
5. Emphasizing the portions of the environmental impact statement that are useful to decision makers and the public and reducing emphasis on background materials
6. Using the scoping process, not only to identify significant environmental issues deserving of study, but also to deemphasize insignificant issues, narrowing the scope of the EIS process accordingly
7. Incorporating by reference
8. Requiring comments to be as specific as possible

Bibliography of Sources and Regulations

Council on Environmental Quality. *Regulations for Implementing the Procedural Provisions of the National Environmental Policy Act, November 29, 1978.* Washington, D.C.: U.S. Government Printing Office.

Environmental Impact Statement Guidelines. U.S. Environmental Protection Agency, Region 10. Revised edition, April 1973. [An older set of guidelines that places more emphasis on the proposed action, less emphasis on the alternatives.]

Seattle, City of. *Request for Proposals: Illabot Creek Small Hydro Project.* Seattle: City Light Department, Environmental Affairs Division, August 30, 1982.

Trzyna, Thaddeus C., and Arthur W. Jokela. *California Environmental Quality Act: Innovations in State and Local Decisionmaking.* Washington, D.C.: U.S. Government Printing Office, October 1974 (EPA-600/5-74-023).

Trzyna, Thaddeus C. *Environmental Impact Requirements in the States: NEPA's Offspring.* Washington, D.C.: U.S. Government Printing Office, April 1974 (EPA-600/5074-006).

SUMMARY

1. The environmental impact statement is an "action-forcing" document that assesses the environmental consequences of a proposed project. The National Environmental Policy Act, mandating the EIS process, was enacted in 1969.
2. Federal EISs are filed with the Council on Environmental Quality; state and local environmental impact legislation varies.
3. In an EIS, the term *environment* includes the physical, biological, social, human, historical, architectural, aesthetic, and even spiritual environments.
4. The EIS process includes five stages: an assessment to determine need for an EIS, a draft statement, public response and comment, a final EIS that responds to the comments, and a record of decision.
5. Current regulations emphasize discussion of alternative actions, including the "no action" alternative.
6. The EIS writer must carefully choose the most effective and fairest method of organization for comparing alternatives.
7. The EIS writer must focus on significant issues, use plain language, avoid needless length and complexity, incorporate additional information by reference, and remain thorough and impartial in the presentation of judgments.

EXERCISES

1. Find an EIS in your college library and write a brief report evaluating the adequacy of the comparison of alternatives.
2. Research the environmental legislation in force in your area. Are there state, local, or regional laws?
3. In a report of five pages, evaluate the environmental consequences of a construction project on your campus or in your community. Briefly consider all of the environmental factors mentioned in this chapter.

4. Read the comments and responses to comments at the end of any published EIS and write a report on the adequacy of the commenting process.

Exhibit 16-1 is an excerpt from an EIS discussing the selection of a site for ocean dumping of acid waste. We include the introduction and the first section of one chapter to give you an idea of the scope and detail the chapter contains (the entire chapter continues for 20 additional pages).

CHAPTER · 17

Proposal Budget Writing

INTRODUCTION

Budget writing is a straightforward process, even though specific line items (or budget amounts) can sometimes be difficult to estimate. This chapter presents one standard method for organizing a budget and two important rules of thumb to follow while you are preparing your total bid. The type of budget you submit will depend on what is customary in your field. Here we will discuss only bids that estimate actual costs (including fair earnings). Other types of bids include cost-plus, incentive, and percentage of savings bids. In a cost-plus proposal, the bidder asks for all costs plus a guaranteed rate of profit. In other words, cost overruns are built-in. An incentive proposal asks for more money for early delivery, just as some contracts specify penalties for late deliveries. A percentage of savings bid asks for a percentage of what the client saves as a result of having the bidder do the work. For example, a computer programmer might be able to design a bookkeeping system that can save a company $20,000 a year in salaries, equipment, and materials. Rather than ask for an hourly consulting fee or a set price for the job, the bidder might ask, say, for twenty percent of the first year's savings to the company.

FIGURE 17-1 *Sample project budget*

1. Personnel Services		
Salary	8000	
Overtime	0	
Benefits	0	
Total		$8000
2. Supplies		
Office supplies	100	
Other	0	
Total		$ 100
3. Other Services and Charges		
Subcontract services	$800	(typing and graphics)
Communications	50	
Data processing	0	
Travel	50	
Printing	50	
Total		$ 950
4. Capital Expenditures	0	
Total		0
PROJECT BUDGET TOTAL		$9050

BEING IN RANGE AND COVERING EMERGENCIES

Beginning consultants, and even veterans making their first bid in a new field, sometimes do not know what to charge. You might be able to do a job for $2000, yet have your bid rejected because the customer expected to spend $5000 to have the job done well. The solution to this problem is to ask your sources about the current rate for this type of work. You might also be able to find out what the client is thinking of spending; companies are often surprisingly frank about their spending expectations. Still, be tactful and cautious when you inquire.

Unexpected expenses are a common problem. Almost never will you submit a bid that is accurate to the last penny. Something will need to be done twice, or take longer, or cost more than you planned. As you write your budget, build in a little leeway, and then go back to be sure that you haven't put yourself out of serious competition by adding too much to your initial bid. A few dollars added to each budget category can add up to a large sum. The sponsoring agency will often make this process easy by providing a printed form that includes many budget categories you might not have considered. In the bid for a technical writing contract printed in Figure 17-1, the consultant added small amounts to the travel, printing, and supply budget lines in order to build in a little reserve. When the client required more artwork, these funds came in handy.

BUDGET CATEGORIES

There are three basic budget categories: personnel costs, nonpersonnel costs, and indirect costs. In addition, some government and foundation programs offer funds on the condition that the bidders donate some amount of their own funds. These donations are usually called "matching funds," but they need not be actual cash. They can reflect the value of services. For example, one crisis clinic telephone hot line receives funds to pay for phone lines, utilities, and space on a matching basis. The organization that runs the hot line has no source of money with which to match the cash from the sponsoring foundation, but the crisis clinic is allowed to count volunteer time as a match. With five volunteers available 24 hours a day, seven days each week, this volunteer match has a high dollar value, even if each volunteered hour is assigned a low cash value.

Personnel Costs

Personnel costs include salaries, fringe benefits, consultant's fees, and subcontracts. The services of volunteers can be counted as a donation in this category.

Nonpersonnel Costs

Nonpersonnel costs include the cost of the space you use; the rental, lease, or purchase of equipment; office supplies; travel (including daily housing and food allowances); the cost of communications services such as telephone, postage, and publications; and insurance coverage. The most sensitive items here are space and the purchase of equipment. An agency that hires your services for six months does not want to pay the full cost of the new word processor or computer that you will keep and use for years. Other genuine costs are even more difficult to recover than the cost of equipment. Basic utilities and janitorial services can be prorated and added to the services budget, yet there comes a point at which it is hard to assign a particular cost to one project, especially when a firm may be carrying out several contracts at one time and working on the proposal stages of several more jobs. Your accounting department, for example, will keep the books, issue checks, and handle equipment orders for every project manager, but how do you calculate the cost of these services to each project budget? This problem raises the issue of what are called indirect costs.

EXHIBIT 17-1

Budget from the Proposal for the California Farmlands Project

BUDGET
FISCAL YEAR 1982-1983

Project Title: California Farmlands Project   Amount: $100,000

Contact Person: (Name, mailing address, and telephone)
EXPENDITURE DETAIL (attach additional pages as necessary)

SALARIES AND WAGES

CLASSIFICATION	SALARY RANGE	AMOUNT REQUESTED
Project Director (½); Principal Investigator (⅓)	$40,000-45,000	$35,000
Coordinator/Research Assoc. (1½)	$12,000-14,000	19,000
Clerical Support (½)	$10,000-12,000	6,000
TOTAL SALARIES AND WAGES		$60,000

CAPITAL EXPENDITURES
Acquisition . . . . $______
Design . . . . $______
Construction . . . . $______
TOTAL CAPITAL EXPENDITURES $0

OPERATING EXPENSES AND EQUIPMENT
Materials and Supplies $1000
Printing $6000
Communications $3000
Travel $11,000
Facilities Operation $0
Data Processing $500
Consultant and Professional Services . . . . $18,500 (special section appended)
Equipment (identify) $0
Other $0
TOTAL OPERATION EXPENSES AND EQUIPMENT $40,000

TOTAL EXPENDITURES $100,000

SOURCE OF FUNDS: Environmental Fund [subject of this proposal]
$100,000
Council for the Humanities, for conference support
$12,500

Indirect Costs

Indirect costs are those costs incurred for the general support and management of a project. In a university, to cite a setting where indirect costs can be a sensitive issue, these costs include the operation of the accounting, payroll, and purchasing offices; support of the grant-writing and administration offices; maintenance of common research facilities; some utilities; building and equipment maintenance services; library facilities; and depreciation. In short, indirect costs are those things that everyone takes for granted but that are not part of the research or development activities of any single project. Some funding agencies provide formulas according to which they will reimburse some of these costs. University and research institute grant-management offices have responded to the generally low level of these reimbursements by claiming a straight percentage of every grant and contract. Before you charge for any indirect costs, inquire about the policy of the agency to which you are applying for funds; many agencies refuse to pay any indirect costs.

SAMPLE BUDGETS

Figure 17-1 on page 404 shows a typical budget breakdown. Alternative budget systems are illustrated by the Design Associates proposal in Chapter 11 (Exhibit 11-4) and by the California Farmlands Project budget in Exhibit 17-1. Note that the Design Associates proposal breaks down the budget according to chronological stages of work rather than by the type or category of expense. Notice, too, that the California Farmlands budget on the previous page asks for nothing for the indirect cost and capital equipment categories: the two most sensitive budget areas. The California Farmlands project also lists a second source of funds, the Council for the Humanities, in order to show broad support for the project. While this amount is not a match, the persuasive strategy is similar.

SUMMARY

1. Costs should be separated into personnel costs, nonpersonnel costs, and indirect costs.
2. Personnel costs include salaries, fringe benefits, and subcontracts.
3. Nonpersonnel items include supplies, travel, transportation, and other material costs directly incurred by a project.

4. Indirect costs are the expenses of maintaining general institutional support services. These costs include accounting, library services, depreciation, and so on. Many clients and sponsors will not pay indirect costs.
5. Equipment, space, and indirect cost charges are highly sensitive items in proposal budgets.
6. Your bid should be in the range the client expects to pay.
7. Add a reasonable amount to your budget to cover unexpected expenses.

EXERCISES

1. Consider a small project you would like to see undertaken at your school. This could be something like landscaping an area on campus, adding lounge furniture to a study area, or improving access from a parking lot to a main thoroughfare. Now consider what expenses are involved in terms of personnel costs, nonpersonnel costs, and indirect costs. Write a budget for the project.
2. Write a budget for either of the proposal cases in Chapter 11. Try to divide your costs into personnel, nonpersonnel, and indirect.
3. Write a budget for the project described in Short Reports Case 1, The Student Lounge (p. 206). What personnel, nonpersonnel, and indirect costs are involved?

I N D E X